# Above Ground Storage Tank Oil Spills

# Above Ground Storage Tank Oil Spills

## Applications and Case Studies

Edited by

**MERVIN FINGAS**

*Spill Science, Edmonton, AB, Canada*

ELSEVIER

Gulf Professional Publishing

An imprint of Elsevier

Gulf Professional Publishing is an imprint of Elsevier
50 Hampshire Street, 5th Floor, Cambridge, MA 02139, United States
The Boulevard, Langford Lane, Kidlington, Oxford, OX5 1GB, United Kingdom



ISBN: 978-0-323-85728-4

For Information on all Gulf Professional Publishing publications
visit our website at https://www.elsevier.com/books-and-journals

*Publisher:* Charlotte Cockle
*Senior Acquisition Editor:* Katie Hammon
*Editorial Project Manager:* Michelle Fisher
*Production Project Manager:* Nirmala Arumugam
*Cover Designer:* Christian J. Bilbow

Typeset by MPS Limited, Chennai, India

# Contents

# Part 2 Case histories

# Part 3 Legislation

# Part 4 Risk analysis

# List of contributors

**Mariano P. Ameijeiras**
Institute for Advanced Studies in Engineering and Technology, CONICET/UNC, Córdoba, Argentina; Faculty of Exact, Physical and Natural Sciences, National University of Córdoba, Córdoba, Argentina

**Merv Fingas**
Spill Science, Edmonton, Alberta, Canada

**Luis A. Godoy**
Institute for Advanced Studies in Engineering and Technology, CONICET/UNC, Córdoba, Argentina; Mechanical and Aerospace Engineering Department, West Virginia University, Morgantown, WV, United States

**Rossana C. Jaca**
Faculty of Engineering, National University of Comahue, Neuquén, Argentina

**John Joeckel**
Seaconsult Inc., VA, United States

**Sabarethinam Kameshwar**
Department of Civil and Environmental Engineering, Louisiana State University, Baton Rouge, Louisiana

**Zhan Liu**
School of Mechanics and Civil Engineering, China University of Mining and Technology, Xuzhou, China

**Adriana Mesa-Gómez**
Civil and Environmental Engineering Department, Universidad de los Andes, Bogotá, Colombia; CERTEC, Universitat Politècnica de Catalunya, Barcelona, Catalonia, Spain

**Mostafa Omidi Bidgoli**
Department of Mechanical Engineering, Islamic Azad University, Badroud, Iran

**Michal Petrů**
Institute for Nanomaterials, Advanced Technologies, and Innovation, Technical University of Liberec, Liberec, Czech Republic

**Jean-Paul Pinelli**
Florida Institute of Technology, Melbourne, FL, United States

**Seyed Saeid Rahimian Koloor**
Institute for Nanomaterials, Advanced Technologies, and Innovation, Technical University of Liberec, Liberec, Czech Republic

**Oscar J. Ramirez**
Risk and process safety management, Bureau Veritas, Bogotá, Colombia

**Mehran S. Razzaghi**
Department of Civil Engineering, Qazvin Branch, Islamic Azad University (QIAU), Qazvin, Iran

**Kazem Reza Kashyzadeh**
Department of Transport, Academy of Engineering, Peoples' Friendship University of Russia (RUDN University), Moscow, Russian Federation

**Ernesto Salzano**
Dipartimento di Ingegneria Civile, Chimica, Ambientale e dei Materiali, Università di Bologna, Bologna, Italy

# Preface

The exploration, production, and consumption of oil and petroleum products are increasing worldwide and the threat of oil pollution increases accordingly. The movement of petroleum from the oil fields to the consumer involves as many as 10–15 transfers between many different modes of transportation including tankers, pipelines, railcars, and tank trucks. Oil is stored at transfer points, terminals, and refineries along the route. Accidents can happen at any of these steps or storage facilities.

Obviously, an important part of protecting the environment is ensuring that there are as few spills as possible. Both government and industry are working to reduce the risk of oil spills, with the introduction of strict new legislation and stringent operating codes. Industry has invoked many operating and maintenance procedures to reduce accidents that could lead to spills. In fact, the rate of spillage has decreased in the past 20 years. Intensive training programs have been developed to reduce the potential for human error. Despite these efforts, spill experts estimate that 30%–50% of oil spills are either directly or indirectly caused by human error, with 20%–40% of all spills caused by equipment failure or malfunction.

Oil spills are a frequent occurrence, particularly because of the extensive use of oil and petroleum products in our daily lives. About 450,000 tons of oil and petroleum products are used in Canada every day. The United States uses about 10 times this amount and, worldwide, about 20 million tons are used per day. In the United States, about half of the approximately 4 million tons of oil and petroleum products used per day is imported, primarily from Canada, Saudi Arabia, and Africa. About 40% of the daily demand in the United States is for automotive gasoline, and about 15% is for diesel fuel used in transportation. About 40% of the energy used in the United States comes from petroleum, 25% from natural gas, and 20% from coal. In both Canada and the United States, much of the refined oil goes into powering transportation. Spill statistics are collected by a number of agencies in any country. In the United States, the US Coast Guard maintains a database of spills in navigable waters, while state agencies keep statistics on spills on land, which are sometimes gathered into national statistics. The US Bureau of Safety and Environmental Enforcement maintains records of spills from offshore exploration and production activities. In Canada, spill statistics are kept by many organizations, and Environment and Climate Change Canada keeps overall records.

It can sometimes be misleading to compare oil spill statistics, however, because different methods are used to collect the data. In general, statistics on oil spills are difficult to obtain, and any dataset should be viewed with caution. The spill volume or amount is the most difficult to determine or estimate. For example, in the case of a vessel accident, the volume in a given compartment may be known before the accident, but the remaining oil may have been transferred to other ships immediately after the accident. Some spill accident databases do not include the amounts burned, if or when that occurs, whereas others include all the oil lost by whatever means. Sometimes the actual character or physical properties of the oil lost are not known, which leads to different estimates of the amount lost. Spill statistics compiled in the past are less reliable than more recent data because few agencies or individuals collected spill statistics before about 1975. More recently, techniques for collecting statistics are continually improving. Reporting procedures vary in different jurisdictions and organizations, such as government or private companies. Minimum spill amounts that must be reported vary according to different regulations, depending on the spill source and location.

The number of spills reported also depends on the minimum size or volume of the spill. In both Canada and the United States, most oil spills reported are more than 4000 L (about 1000 gallons). In Canada, there are about 12 such oil spills every day, of which only about one is spilled into navigable waters. These 12 spills amount to about 40 tons of oil or petroleum product. In the United States, there are about 15 spills per day into navigable waters and an estimated 85 spills on land or into freshwater. Despite the large number of spills, only a small percentage of oil used in the world is actually spilled. In Canada and the United States, most spills take place on land, and pipeline spills account for the highest volume. In terms of the actual number of spills, most happen at petroleum production facilities, wells, production collection facilities, and battery sites (storage tanks). On water, the greatest volume of oil spilled comes from marine or refinery terminals, although the largest number is from the same source as in the United States—vessels other than tankers, bulk carriers, or freighters.

An important question concerns the amount of spillage that is likely to occur in the future. There are several indications of the trend. There are government databases showing trends on oil production.

There are significant trends that should be noted:

1. In Canadian production, the amount of bitumen being marketed is rapidly increasing. In addition to being sold as a diluted product called "Dilbit," bitumen is also being upgraded into synthetic crude. Spills of Dilbit have caused concern because some of the unique spill properties that it has shown, namely that once weathered for a period of time, it may sink in fresh water.
2. In North Dakota, the Bakken oil field is currently producing oil and the expansion of this field is quite rapid. This oil is also of concern when spilled as it is very flammable and has caused considerable damage in spills such as the Lac Megantic spill in Quebec.
3. There is a pipeline shortage to transport oil and oil is increasingly being transported by rail, which also has increased the risk of spills from this source. Spills of oil from trains are believed to pose a higher risk than from pipelines.
4. Pipelines themselves are being built at a very rapid pace and many have been modified to carry products to the North American south rather than carrying other products north. The spills from pipelines have been decreasing in size. It should be noted that the number of pipeline spills might increase as a result of the increasing number of pipelines.
5. The volume and number of spills from tanker vessels has been constantly decreasing over the past 20 years. Tanker spills contribute very little to the spillage in many countries.
6. The spills from storage tanks remain either as second or third as the most frequent source of spillage in North America.
7. The most frequently spilled products are crude oils, followed by light refined products such as diesel and No. 2 fuel oils.

## Oils spilled on land

Oils spilled on land do not spread quickly, unlike on water, and the effects remain localized. Most types of oil will penetrate the soil and contaminate anything living there. A full coating of fresh crude oil or diesel fuel will kill most plants and small trees on contact. Because of the usually limited area of impact, however, the effects of oil on land environments are not as great a concern as for marine environments. Most soils consist of small fragments or grains that form openings or pores when compacted together. If these pores are sufficiently large and interconnected, the soil is said to be permeable and oil or water can pass through it. Sand and gravel are the most permeable types of soil. Materials such as clay, silt, or shale are termed impervious as they have extremely small, poorly interconnected pores and allow only limited passage of fluids. Soils also vary in terms of long-term retentivity. Loam tends to retain the most water or oil due to its high organic content. As most soils are a heterogeneous mixture of these different types of sediments, the degree of spreading and penetration of oil can vary considerably in a given location.

The types of soil are often arranged in layers, with loam on top and less permeable materials such as clay or bedrock underneath. If rock is fractured and contains fissures, oil can readily pass through it. The oil's ability to permeate soils and its adhesion properties also vary significantly. Viscous oils, such as heavy fuel oil, often form a tarry mass when spilled and move slowly, particularly when the ambient temperature is low. Non-viscous products, such as gasoline, move in a manner similar to water, in both summer and winter. For such light products, most spreading occurs immediately after the spill. Crude oils have intermediate adhesion properties. In an area with typical agricultural loam, spilled crude oil usually saturates the upper 10–20 cm of soil and rarely penetrates more than 60 cm. Generally, the oil only penetrates to this depth if it has formed pools in dry depressions. If the depressions contain water, the oil may not penetrate at all.

The fate of oil once on soil varies, and generally clean-up is slow and difficult. After clean-up, there is usually residual contamination which may linger for years. On agricultural land, highly weathered residual oil may not be a health problem and often crops can be grown in the soil the following year after clean-up and tilling/nutrient addition.

## Oils spilled on water

Oils spill on water will rapidly spread to form thin slicks. This makes oil spills on water difficult to deal with and expands the area of contamination.

Containment of an oil spill refers to the process of confining the oil, either to prevent it from spreading to a particular area, to divert it to another area where it can be recovered or treated, or to concentrate the oil so it can be recovered or burned. Containment booms are the basic and most frequently used piece of equipment for containing an oil spill on water. Booms are generally the first equipment mobilized at a spill and are often used as long as the oil persists on the water surface. A boom is a floating mechanical barrier designed to stop or divert the movement of oil on water. Booms resemble a vertical curtain with portions extending above and below the water line. They are constructed in sections, usually 15 or 30 m long, with connectors installed on each end so that sections of the boom can be attached to each other, towed, or anchored.

Booms are used to enclose floating oil and prevent it from spreading, to protect biologically sensitive areas, to divert oil to areas where it can be recovered or treated, and to concentrate oil and maintain an adequate thickness so that skimmers can be used or other clean-up techniques, such as in situ burning, can be applied. Booms are used primarily to contain oil, although they are also used to deflect oil. When used for containment, they are often arranged in a U configuration. The U-shape is created by the current pushing against the center of the boom. The critical requirement is that the current in the apex of the U does not exceed 0.35 m/s or 0.7 knots, which is referred to as the critical velocity. This is the speed of the current flowing perpendicular to the boom, above which oil will be lost by entrainment into the water and under the boom. If used in areas where the currents are likely to exceed 0.35 m/s, such as in rivers and estuaries, booms must be deployed at an angle to the current. The oil can then be deflected out of the strong currents to areas where it can be collected or to less-sensitive areas.

If strong currents prevent the best positioning of the boom in relation to the current, several can be deployed in a cascading pattern to progressively move oil toward one side of the watercourse. This technique is effective in wide rivers or where strong currents may cause a single boom to fail. When booms are used for deflection, the forces of the current on the boom are usually so powerful that stronger booms are required that must be anchored along their entire length. A boom's performance and its ability to contain oil are affected by water currents, waves, and winds. Either alone or in combination, these forces often lead to boom failure and loss of oil. The most critical factor is the current speed relative to the boom. Failures will occur when this exceeds 0.35 m/s (0.7 knots).

Recovery is the next step after containment in an oil spill clean-up operation. Even in the case of land spills, the oil most often flows to a water body from where it is recovered. An important objective of containment is to concentrate oil into thick layers to facilitate recovery. In fact, the containment and recovery phases of an oil spill cleanup operation are often carried out at the same time. As soon as booms are deployed at the site of spill, equipment and personnel are mobilized to take advantage of the increased oil thickness, favorable weather, and less weathered oil. After oil spreads or becomes highly weathered, recovery becomes less viable and is sometimes impossible.

Skimmers are mechanical devices designed to remove floating oil from a water surface. They vary greatly in size, application, and capacity, as well as in recovery efficiency. They are classified according to the area where they are used, for example, inshore, offshore, in shallow water, or in rivers, and by the viscosity of the oil they are intended to recover. Most function best when the oil slick is relatively thick. The oil must, therefore, be collected in booms or against a shoreline or floating ice before skimmers can be used effectively. The skimmer is placed wherever the oil is most concentrated in order to recover as much oil as possible. Weather conditions at a spill site have a major effect on the efficiency of skimmers. Depending on the skimmer type, most will not work effectively in waves greater than 1 m in height or in currents greater than 0.5 m/s. Most do not operate effectively in water with ice or debris such as branches, seaweed, and floating waste. Some have screens around the intake to prevent debris or ice from entering, conveyors or similar devices to remove or deflect debris, and cutters to deal with seaweed. Very viscous oils, tar balls, or oiled debris can clog the intake or entrance and make it impossible to pump oil from the skimmer's recovery system. Skimmers are also classified according to their basic operating principles: oleophilic surface skimmers, weir skimmers, suction skimmers or vacuum devices, elevating skimmers, and submersion skimmers. Over time, all skimming systems become less effective because of the oil's spread into thinner slicks and weathering into a more viscous liquid.

Sorbents are materials that recover oil through either absorption or adsorption. They play an important role in oil spill clean-up and are used in the following ways: to clean up the final traces of oil spills on water or land; as a backup to other containment means, such as sorbent booms; as a primary recovery means for very small spills; and as a passive means of clean-up. An example of such a passive clean-up is when sorbent booms are anchored off lightly oiled shorelines to absorb any remaining oil released from the shore and prevent further contamination or re-oiling of the shoreline. Sorbents can be natural or synthetic materials. Sorbents are available in a loose form, which includes granules, powder, chunks, and cubes, often contained in bags, nets, or socks. Sorbents are also available in the form of pads, rolls, blankets, and pillows. The use of synthetic sorbents in oil spill recovery has increased in the last few years. These sorbents are often used to wipe other oil spill recovery equipment, such as skimmers and booms, after a clean-up operation. Sheets of sorbent are often used for this purpose. Sorbent booms are deployed on the water when the oil slick is relatively thin, i.e., for the final polishing of an oil spill, to remove small traces of oil or sheen, or as a backup to other booms. Sorbent booms can be placed off a shoreline to recover oil that is mobilized by wave or tidal action; this strategy is often used for marshes where other response options are likely to cause additional harm. They are not efficient enough to be used as a primary countermeasure technique for any significant amount of oil.

Oil spills, irrespective of how they occur or where they occur, are a serious threat to the environment. Oil spills are difficult to clean up and require extensive techniques and efforts to restore the environment.

**Mervin Fingas**

## Reference

Fingas, M. F. (2012). 245 pp *The basics of oil spill cleanup*. Boca Raton, FL: Taylor and Francis.

# Acknowledgment

A special thanks to the authors, many of whom put in their own time to complete their chapters. This is especially true because many of the authors were working on many other tasks during the preparation of this book. This "double-duty" was greatly appreciated. The author's names appear throughout the text. I would also like to thank the many people who provided support and encouragement throughout this project, including my colleagues and the staff of De Gruyter. All materials in this book have been peer-reviewed by at least two persons. The following peer reviewers are acknowledged (in alphabetical order): Carl Brown, Emanuele Brunesi, Joaquim Casal, Adriana Gomez, Binyuan Hong, Wong King Jye, Sabarethinam Kameshwar, Adam Larussic, Yuanliang Liu, Roberto Nascimbene, Amos Necci, Tony Regalbuto, Ken Russel, Ghorbani Siamak, Graham Thomas, Ann Hayward Walker, Fushou Xie, and Huizhu Yang.

# Introduction

Spills from above-ground storage tanks constitute a serious problem around the world. In Canada and the United States, oil and petroleum spills from storage tanks are the second or third largest source of oil or oil product spills when considering spill volume. It is thought that the situation is similar around the world.

This book will cover several aspects of spills from storage tanks. This includes hazard analysis, design, and most frequent causes. The objective is to provide readers with resources to consider preventive measures as well as understand the situation with respect to spills from above-ground storage tanks.

There are four sections to the book: Preventative Design and Issues; Case Histories; Legislation; and Risk Analysis.

The first section, Preventative Design and Issues, encompasses six chapters, from Chapter 1–6.

Chapter 1 focuses on the common defects in storage tanks, namely corrosion pitting, and cracks. These defects are discussed at length and several aspects of analysis, testing, simulation, and control are given. Standards related to these actions are summarized.

Chapter 2 discusses the effect of extreme winds on tanks. NATECH or Natural Technological Accidents are discussed. This includes hurricanes, tornadoes, and downbursts. These extreme winds can lead to catastrophic failures of storage tanks. Among the most common damage and failure modes caused by a natural hazard are shell buckling, the sliding or floating of the tank, damage to tank foundation, overturning, impact by debris, detachment of pipes, and damage to the bottom plate by buckling due to uplifting. Each of these failure modes is discussed, analyzed, and modeled.

Chapter 3 examines the seismic performance of storage tanks. This is important because some catastrophic storage tank failures occurred from earthquakes. Modes of seismic damage include sloshing of the interior liquids, bottom uplift, welding failure, anchor rupture, bottom buckling, and shell buckling.

Chapter 4 examines the performance of storage tanks in hurricanes. Hurricanes often involve strong winds, storm surges, wave loads, extreme precipitation, wind-borne debris, and storm surge was driven debris. Hurricanes, in addition to strong winds, can cause storm surges and damage such as roof failures, floatation, sliding, wave-induced failures, and precipitation-induced failures.

Chapter 5 is a different view on the engineering design of storage tanks. This approach uses a thin shell engineering design concept to calculate the structure of storage tanks. Calculations, standards, and test methods are given.

Chapter 6 examines buckling, which is a mode of failure that often results in the catastrophic release of the tank contents. The most common causes of buckling are fire and explosion. The authors show that blast loads due to explosions and thermal loads due to fire have the potential of causing structural and/or functional damage in oil storage tanks, and even their failure and collapse.

The second section, Case Histories, encompasses Chapter 7 that describes the case history of the Ashland oil spill, one of the worse single-tank oil spills ever. The 4-million-gallon spill of diesel fuel was released into an inland river system in the United States. The tank failure was caused by a small brittle fracture in one area of the tank.

The third section is Legislation, which encompasses Chapter 8 that describes typical legislation and standards that would apply to storage tanks and the areas in which they are situated

The fourth section, Risk Analysis, encompasses Chapter 9 which is a review of spill statistics in Canada and the United States on the spill statistics. This shows that storage tanks are a significant source of spillage in both countries.

PART1

# Preventative design and issues

CHAPTER ONE

# Assessment of oil storage tanks performance containing cracks and cavities

**Kazem Reza Kashyzadeh[1], Mostafa Omidi Bidgoli[2], Seyed Saeid Rahimian Koloor[3] and Michal Petrů[3]**

[1]Department of Transport, Academy of Engineering, Peoples' Friendship University of Russia (RUDN University), Moscow, Russian Federation
[2]Department of Mechanical Engineering, Islamic Azad University, Badroud, Iran
[3]Institute for Nanomaterials, Advanced Technologies, and Innovation, Technical University of Liberec, Liberec, Czech Republic

## 1.1 Introduction

Nowadays, crude oil is the most important source of energy production in the world, so all countries in the world now use it as a vital tool in their growth and development. On the other hand, one of the vital issues in the oil, gas, and petrochemical industry is the storage of production materials for consumption. In this regard, crude oil, gas, petroleum products, as well as refineries, and petrochemicals feedstock are not used immediately after production and require short-term and longer-term storage, which requires a large number of reservoirs (Bachu & Shaw, 2005; Dhillon, 2016; Myers, 1997).

Storage reservoirs are containers that are used to store fluids and different names, for example, aboveground storage tanks and cylindrical pressure vessels are given to them depending on their use. These giant structures are so large that their grandeur and importance can only be realized up close. Sometimes the diameter of these reservoirs reaches 120 m and their height reaches 20 m that this property provides a storage capacity of about 135,000 barrels of crude oil (Kennedy, 1993; Myers, 1997; Wright et al., 1918).

In general, it can be said that in all large industries, the purpose of periodic repairs is to prevent a crisis for machinery and equipment. The oil industry as one of the parent industries is no exception to this and

*Above Ground Storage Tank Oil Spills.*
DOI: https://doi.org/10.1016/B978-0-323-85728-4.00006-1

since the amount of production of this vital material depends on the capacity of existing equipment, so the maintenance of equipment for oil companies is important and inevitable. Due to the important role of these reservoirs in the oil industry, they must be free of any defects for a certain period, to maintain and store crude oil, and evaluation of reservoir performance is effective to achieve this goal (Babusiaux, 2004; Dhillon, 2016; Lei, 2018; Revie, 2011).

This study is provided in different parts, in which the first section introduces various types of oil storage tanks and their components. The second part looks at the common defects in oil storage tanks and their causes. The third part collects various topics, including design, construction, technical inspection, and repair standards. The next section presents methods of dealing with defect damage to prevent decommissioning of storage tanks. And the last section analyzes the tank behavior with different defects. Eventually, the conclusions and recommendations of the authors are presented.

## 1.2 Various types of oil storage tanks and their components

There are many methods to classify reservoirs. However, as there is no general rule for this purpose, the general classification is performed using standards and rules based on pressure. In this regard, the tanks employed in the industry are described as follows (Myers, 1997; Omidi Bidgoli et al., 2020; Wright et al., 1918):

- Process tanks
- Low-pressure storage tanks (LP-tank)
- Non-pressure storage tanks (NP-tank)
- High-pressure storage tanks
- Compressed air tanks.

Here, a question arises where are the oil reserves in this reservoir classification? Since the pressure applied to these reservoirs is only hydrostatic pressure because of oil fluids, these reservoirs are considered the types of LP-tank or NP-tank. According to the API 650 standard, the pressure of these reservoirs is less than 2.5 psi, which is an insignificant amount (Myers, 1997).

### 1.2.1 Main components of an oil storage reservoir

Many components must be assembled to manufacture a tank or reservoir. Nevertheless, in a general review, it can be said that the most important components of an oil storage reservoir are as follows (Cleveland, 2004; Semadeni, 2004; Standard, 1988):

Tank floor: The floor of the storage tanks includes two parts of the bottom plates and the annular plates so that the thickness of the annular plate is more than the bottom plate. After that, these plates are welded together. Therefore, their welding quality is very important and affects the service life as well as the cost of repairing and maintaining the tanks. During the manufacturing process, manufacturers and contractors use all their care to be able to prepare and uniformly execute the tank floor. Moreover, the tank floor is covered with a layer of fiberglass to prevent and reduce the corrosion phenomenon.

Tank wall: The wall or body of these tanks encompasses several curved plates. These plates are stacked from bottom to top and the thickness of the plates decreases as the height increases. Accordingly, the wall thickness in the tanks is variable (nonconstant).

Roof: The roof of the oil storage reservoir is also a separate part that can be manufactured as either fixed or floating. In the case of fixed type, with the change of oil height, the roof does not move at all. In contrast, in the case of the floating type, with the change of oil height, the tank roof is moved downward or upward, which is at least 1 m from the tank floor when moving downward.

In summary, the internal structure of an oil storage reservoir is shown in Fig. 1.1.

## 1.3 Common defects in the oil storage tank and their causes

Similar to all equipment and machines in various industries, the oil storage tanks are affected by a group of defects and threats. However, the type and the number of these defects depend on the overall condition of the structure, environmental and working conditions. In this matter, those defects to which the oil storage tanks are exposed include wind and earthquake forces, pitting corrosion, and cracks. It should be mentioned that

Figure 1.1 The internal structure of an oil storage reservoir.

the most important of them is the corrosion and crack phenomenon. Moreover, in the area with climatic conditions (e.g., prone to corrosion) such as the countries around the Persian Gulf, the most common problem is the pitting corrosion, while the crack can rarely be observed in the welding area (Omidi Bidgoli et al., 2020; Salinas et al., 2016; Samimi, 2012). In general, the defects above mentioned happen in one of the following steps:

- Lack of accuracy in the stages of manufacture, implementation, and assembly of the storage tank.
- Defects that happen during the material storage (e.g., oil, fuel, gasoline, etc.) and tank operation.

Regarding the importance of pitting corrosion defects as well as cracks that threaten the joint and welding zone, these phenomena, and their various dimensions are discussed in the following.

### 1.3.1 Corrosion

Nowadays, corrosion is one of the problems that has always caused a lot of damage to various industries, particularly oil, gas, and petrochemical industries, based on which a large part of the national capital is always lost due to the corrosion. Furthermore, most of the irreparable accidents in these industries are due to defects caused by corrosion. In 1995, Shell Oil Company announced that corrosion was the main cause of $ 400 million in losses. In the same year, the British Oil Company announced that the corrosion phenomenon was the main reason for the loss of 6% of the total value of its net asset. The mentioned instances above and thousands of other eyewitnesses imply the importance of coping with and preventing corrosion (Dhillon, 2016; Kermani & Harrop, 1996; Shigeno & Okamoto, 1960).

First of all, it is necessary to have sufficient knowledge and understanding of the nature of the corrosion. At first glance at the physics of this phenomenon, it seems that the metal piece has been bitten by an external agent. However, this phenomenon can be defined in another way: the loss or decay of a substance because of an environmental reaction is named corrosion. Corrosion is a simple galvanic cell including an anode, a cathode, an anode-cathode metal bond, and an electrolyte envelope, in which the anode is corroded while the cathode metal is preserved. Fig. 1.2 illustrates a schematic of this definition (Kermani & Harrop, 1996; Papavinasam, 2013; Shigeno & Okamoto, 1960).

#### *1.3.1.1 Classification of corrosion*

Corrosion can be categorized in dissimilar ways (Omidi Bidgoli et al., 2020; Revie, 2011; Shigeno & Okamoto, 1960). A simple and general classification for this phenomenon is as follows:

1. Wet corrosion
2. Dry corrosion

However, a specialized classification based on the appearance and shape of corroded metal by indicating the details and mechanisms of different types of corrosion is as follows:

- Uniform corrosion

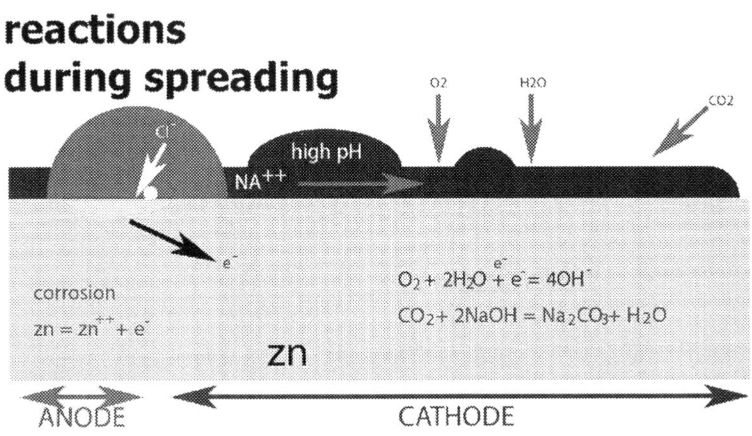


**Figure 1.2** The schematic of a galvanic cell (Shigeno & Okamoto, 1960).

- Galvanic or bimetallic corrosion (dissimilar metal corrosion)
- Crevice corrosion
- Pitting corrosion
- Intergranular corrosion
- Selective leaching
- Erosion corrosion
- Stress corrosion

Among the types of aforementioned corrosions, pitting corrosion is further observed in the petrochemical industry, which is more important for the study. Therefore, a detailed description of this type of corrosion brings in the following.

#### *1.3.1.2 Pitting corrosion*

This type of defect is a severe and localized corrosion that causes the puncture of the metal. These holes may contain different diameters, but their diameters are small in most cases. The cavities are sometimes separated, but sometimes so close together that they create roughness on the surface of the metal. Typically, if the diameter of the crater is approximately equal to or less than its depth, the resulting shape has named the cavity. It is worthwhile to mention that pitting is one of the most destructive and worst types of corrosion, which due to the perforation of equipment or metal parts, would lead to their uselessness (Jirarungsatian & Prateepasen, 2010; Kermani & Harrop, 1996).

While the weight loss from this type of corrosion is negligible, the cavities are often difficult to observe with the eye, so in some cases, they cannot even be observed with the naked eye (their size scale is very small). Indeed, pitting can be considered as an intermediate state between uniform corrosion and complete corrosion resistance. This condition is schematically exhibited in Fig. 1.3.

#### *1.3.1.3 Corrosion in oil storage tanks*

As mentioned before, one of the most important disadvantages of oil tanks is corrosion, especially pitting corrosion. Nonetheless, it is necessary to

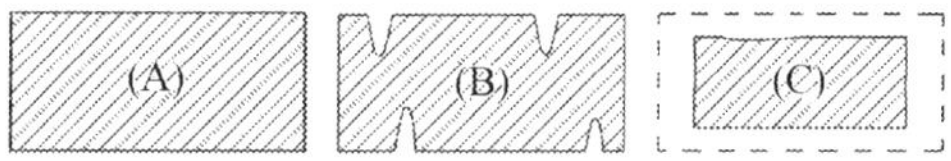


**Figure 1.3** The schematic representation of cavitation as an intermediate state: (A) No corrosion, (B) Pitting corrosion, and (C) Uniform corrosion.

investigate how and in what way, corrosion occurs in the oil storage tanks? In a preliminary classification, the corrosion in the oil storage tanks can be divided into two classes:

1. Internal corrosion
2. External corrosion

Mostly, corrosion occurs at the bottom of the tank, which sometimes completely perforates the floor. Based on this classification, those factors that caused corrosion can also be divided into two categories including internal and external factors (Kermani & Harrop, 1996; Omidi Bidgoli et al., 2020; Shigeno & Okamoto, 1960).

Internal factors: These are factors that cause corrosion and pitting from inside and inward the reservoir structure. It should be mentioned that the most basic internal factor is saline water that settles at the bottom of the reservoir along with the crude oil.

External factors: These factors threaten the reservoir from the outside. The soil under the reservoir, the atmospheric, and environmental conditions of the region are the most important external factors that cause corrosion in these reservoirs. Overall, the interaction of soil with metal has complex mechanisms, in which many parameters are effective to cause soil corrosion. In this way, the most important factors influencing soil corrosion are as follows:

1. Soil type: Factors such as soil particle size, organic compounds, mineral compounds, and structure, determine the soil type.
2. Soil moisture: In this regard, there are three sources of water in the soil:
   - **A.** Water from rain or snow
   - **B.** Water due to capillary properties of soil particles
   - **C.** Groundwater
3. Soil resistance and amount of soluble ions
4. Soil PH

Fig. 1.4 illustrates the images of cavities created in the bottom of a reservoir in the petrochemical industry as pitting corrosion.

## 1.3.2 Cracking

Cracking or the presence of cracks, are among the defects that are rarely observed in oil storage tanks. These small cracks that are in the micro and nanoscale, are mostly observed in the welding areas as shown in Fig. 1.5. In this regard, many studies have been conducted in this field, which

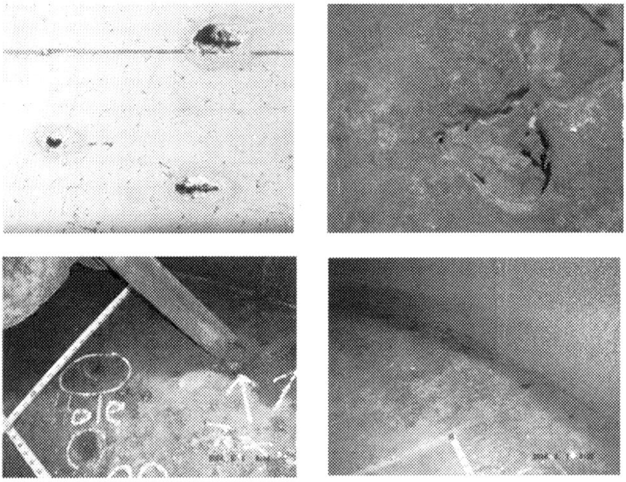

**Figure 1.4** Industrial observations of pitting corrosion in the bottom of the oil storage tank.

**Figure 1.5** Cracks observed in the welding area of an industrial part.

indicates that paying enough attention to how to connect, assembling plates, components in the construction of these reservoirs, and following the standard instructions for preheating and postheating operations in welding can significantly decrease the possibility of cracking (Kim et al., 2009; Pence, 1988).

## 1.4 Design, construction, technical inspection, and repair standards

A standard denotes a regulation or instruction that provides some necessary guidance for design, construction, inspection, and maintenance.

Today, there are various standards in the industry under different goals. In the meantime, oil storage tanks as important equipment in the oil, gas, and petrochemical industry are no exception. In this regard, to obtain better performance and higher efficiency, all stages of design, implementation, inspection, and maintenance of these structures must comply with valid international, domestic, or intra-company standards. In the following, some of the most important standards will be introduced in this field.

There are several mechanical approaches related to oil storage tanks, and each of them is debatable and valuable. Nevertheless, the principles of the available standards are based on one foundation. In other words, in all of them, some common concepts such as material selection, design, construction, and technical inspection methods have been considered for the audience. In this regard, one of the most important and best of these standards belongs to the American Petroleum Institute, known as API and DIN. For oil tanks, this standard is included in two separate parts:

API 650: This part of the standard (Standard, 2013) describes the material requirements, design, construction, installation, and testing for closed- and open-top welded steel storage tanks in various sizes, capacities, and internal pressure, within the atmospheric pressure (internal pressure less than the weight of roof plates). Here, if the additional requirements are satisfied, more internal pressure will be allowed as well. It should be noted that this standard is only for those tanks where the floor can bear the weight uniformly.

API 653: In this part, some issues such as reconstruction, periodic repairs, inspection, and the replacement of various parts of oil storage tanks are fully stated, which can be useful for operators and supervisors.

DIN: Behälter KG offers cylindrical tanks for underground and aboveground storage of water-polluting, flammable, and nonflammable liquids. In this regard, production takes place following DIN 6608, DIN 6616, and DIN 12285. Apart from single-walled horizontal tanks for nonhazardous products, it also covers double-walled cylindrical tanks for hazardous products. The double-shell is continuously monitored through the vacuum or through a liquid leakage indicator to ensure that the tank can safely be operated and is free of defects and leakages. Storage tanks are made of S235JR (steel) as standard, other materials are also possible at extra cost (e.g., stainless steel). Standard tanks all have a DN 600 dome on the top of the tank. Also, additional accessories such as saddle feet, additional domes, or flanges can be offered at an additional cost. This is especially true for services such as sandblasting or painting storage tanks.

Eventually, each tank is delivered with the corresponding test certificate. Storage tanks according to DIN for underground storage have the designations 6608/1 (single-walled) and 6608/2 (double-walled). Also, storage tanks according to DIN for above-ground storage have the designations 6616/1 (single-walled) and 6616/2 (double-walled).

## 1.5 Methods of dealing with defect damage to prevent decommissioning of storage tanks

According to the principles and vision of failure engineering, when the size or number of defects exceeds the allowable limit and enters the critical region, it will cause the failure of the tank. In other words, the reliability coefficient of the structure becomes less than one when the allowable stress is greater than the yield stress of the steel used in the construction of the tanks, and consequently, the structure is endangered and destroyed. In this regard, diagnosing defects, their current size, advance rate towards critical values and finally avoiding creating them or decreasing the growth rate of damage is useful and effective to improve and extend the service life of oil storage tanks.

### 1.5.1 Diagnosis of defects

Identification of the defective parts of the structure is an important issue. In this regard, Non-Destructive Testing (NDT) has significant assistance in the diagnosis of defects in oil storage tanks. NDTs are tests that do not damage the structure in identifying regions with defects and diagnosis of defects and also the structure can continue to serve as before (Kasai & Sekine et al., 2003; Mažeika et al., 2006).

### 1.5.2 Non-destructive methods of identifying locations and corrosion rates in tanks

Since corrosion is one of the most essential problems of different industries, especially key industries such as petrochemical, oil, and gas; it is very important to solve this basic problem and identify the locations and corrosion rate in providing prevention methods. Also, determining the inspection method and its time duration is one of the most important factors in obtaining the desired result. Therefore, the best method can be selected and applied by considering the purpose of the inspection and studying the advantages and limitations of each method (Anvo, 2019; Mažeika et al., 2006).

There are various methods for inspecting equipment before commissioning and during service time as an efficient tool. The defects created during the manufacturing process, which can be considered a source of defects during service, identifies in the precommissioning inspection. Also, the inspection during the repair periods will prevent additional costs, unwanted stops, and possible damages and it will increase production and profit by increasing the system efficiency. In the following, some well-known inspection methods, their advantages, and limitations are introduced (Kermani & Harrop, 1996; Revie, 2011; Shigeno & Okamoto, 1960).

#### 1.5.2.1 Eddy current test

As shown in Fig. 1.6, a magnetic field probe with an alternating impedance current is initially created in the peripheral cross-section of the pipeline. This alternating field induces an eddy current in the pipeline and the coil is affected by the resultant magnetic field caused by the eddy current and is observed in the monitor as a signal with two parameters of phase and amplitude. Since the defect volume is proportional to the signal amplitude and the defect depth is proportional to the signal phase, the defect size can be measured by calibrating the screen with certain sizes of artificial defects mentioned in the relevant standards or by calibrating the device using reference samples (Kasai & Sekine et al., 2003; Revie, 2011).

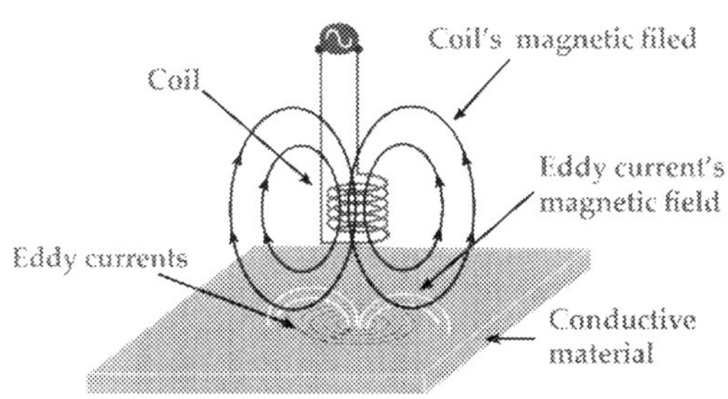


**Figure 1.6** The basis of the eddy current test method. This method is used in the inspection of diamagnetic materials and has very diverse applications, including in the aerospace and military industries. One of the best applications of this method is pipeline and heat exchanger testing, which is one of the most accurate and complete tests of its kind.

The advantages and limitations of this method can be summarized as follows (Revie, 2011):

A: Benefits

A1—Very high speed (about 1 m/s);

A2—Complete diagnosis of corrosion rate;

A3—Complete separation of internal and external defects;

A4—The possibility of simultaneous testing with several various frequencies;

A5—The possibility of simultaneous testing to diagnose the local corrosion and abrasions;

A6—Visual and statistical reporting and automatic documentation;

A7—The ability to review signals and re—evaluate.

B: Limitations

B1—Applicable only on diamagnetic materials;

B2—Lack of peripheral position recognition of the defect in the pipeline;

B3—Lack of peripheral defects recognition (for example, peripheral cracks)

#### 1.5.2.2 Acoustic emissions method

This method is based on the propagation of sound waves from defects under stress (Fig. 1.7). In this way, the specimen test is subjected to forces in various directions and magnitudes. If the active location is in the specimen, its shape and size change under the applied stresses, which is accompanied by the propagation of sound waves. Therefore, the specimen test can receive these waves by placing sensors in different positions and it

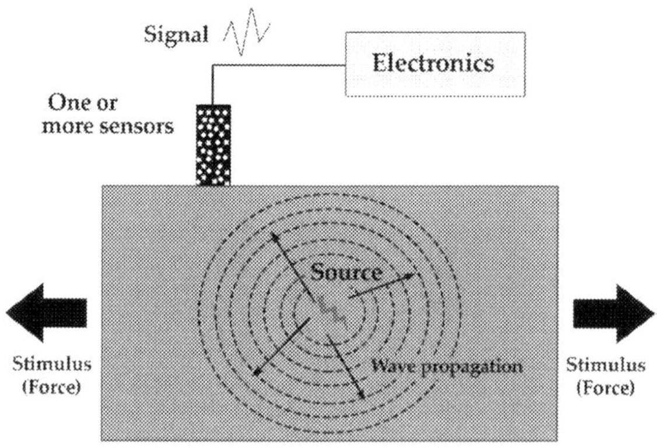


Figure 1.7 The basis of the acoustic emissions method.

converts these mechanical waves into electrical waves. In detecting corrosion positions in tanks and pipelines, the desired forces are applied by changing the working pressure and waiting for their reaction. Since the corrosion is the location of stress concentration, the mentioned positions are discovered and reported by receiving the signals and evaluating them (Kasai & Sekine et al., 2003; Papavinasam, 2013; Revie, 2011).

This method has a very wide range of applications including huge tanks, ships, pipelines, and even small and complex parts. Today, this test is fully automated by using advanced hardware and software because the technology development, is very economical in terms of test and cost, and has a very significant difference compared to other methods. In summary, the method of implementing the inspection in the tanks is as follows.

At first, the desired shape of the tank is created in the relevant software and the maximum distance between the sensors is calculated according to the measurements that determine the sound damping of the tank material. In practice, the sensors are installed in the mentioned places using the appropriate components. To test the accuracy and correct operation of the set, the received waves can be observed on the monitor by hitting very gently on the body of the tank. Now, it receives and stores the resulting waves by altering the amount of pressure according to a diagram that depends on the capability of the system. Then, all data is examined and the places that were the source of the waves are determined as defective positions based on the times of receiving the wave from those positions and their intensity. If necessary, the corrosion rate or other defects can be detected by an ultrasonic method depending on its location. Also, this method has its unique advantages and limitations (Kasai & Sekine et al., 2003; Revie, 2011):

A: Benefits

A1—Ability to inspect complex structures and inaccessible positions during equipment service

A2—Ability to test simultaneous multiple subjects

A3—Identify the locations of the stress concentration and evaluate their importance

A4—Determining the location of defects with certain accuracy and in proportion to the accuracy of the sensors and their installation location

A5—High speed of the test (for example, a large tank can be inspected in 24 h)

B: Limitations

B1—Failure to detect inactive defects

B2—Need to change the system mode from fixed mode to service mode

#### 1.5.2.3 Digital radiography

This method, as a replacement for the conventional radiography with reduced many of the existing costs, can be used from weak sources and less time for radiography using ionizing radiation-sensitive plates that are much more sensitive than radiographic film. And the most important advantage of these two parameters (weak sources and less time for radiography) is a significant decrease in the risk of radiation. Therefore, this method is used as an appropriate candidate for inspection by the radiographic method. This method can be inspected in all cases that can be radiographed using digital radiosensitive plates. A calibration block, usually a stepped block with digital radiography that is examined to inspection for corrosion by specified thicknesses of the same material, is placed on the side of the desired position. The density of each thickness is calibrated with the relevant software by using the corresponding block image. Therefore, the mentioned image is transformed into a color image in that each color is a characteristic of a certain thickness according to the radiographic method. This method can be used in all cases that can be radiographed and are mostly used in cases where the shape of the corrosion is important and the study of corrosion development in terms of time is considered (Kasai et al., 2003; Revie, 2011). Fig. 1.8 shows the radiography test.

**Figure 1.8** Process of a radiography test.

The advantages and limitations of this method include the following (Revie, 2011):

A: Benefits

A1—Thicknesses measurement with an accuracy of 0.1 mm

A2—Applicable to all materials

A3—Applicable for systems operating up to 300°C

A4—Ability to study corrosion development in terms of time

B: Limitations

B1—Applicable in limited positions

B2—Positions must be accessible in two directions

### 1.5.3 Methods for dealing with crack defects in oil storage tanks

In this chapter, while examining the nature of defects in oil storage tanks, some of their detection methods have been studied. But there is a question what solutions should be applied to deal with and prevent the occurrence of these defects to minimize their percentage?

Usually, the crack defects in storage tanks are induced by manufacturing and execution processes, which include welding operations and sheet connections. Since the main defect in storage tanks is related to corrosion, in answer to this question, it should be said that there are various solutions to control and deal with corrosion, some of which are (Omidi Bidgoli et al., 2020; Papavinasam, 2013; Revie, 2011):

- Cathodic and anodic protection
- Use of inhibitors
- Coatings
- Changing the corrosive environment
- Use of retarders, etc.

It can be said about the prevention of corrosion in oil storage tanks that most the crude oil storage tanks (ground) have large dimensions and their construction and maintenance cost a lot of money. The main cause of the destruction of these reservoirs is saltwater along with crude oil, which is a highly corrosive electrolyte. It causes leakage and waste of oil and contaminants, resulting in heavy costs of work stoppage, and rebuilding these reservoirs is inevitable. For example, the oil storage tanks in the south of Iran (Khark Island) are very large metal structures and have cost millions of dollars to build. However, their operation has finally stopped due to severe corrosion and oil leaks from them over the past years and then has led to the construction of new tanks.

At the same time, while wasting the country's capital, the same amount has been damaged to the environment and a large amount of crude oil has been wasted. Therefore, it is necessary to conduct comprehensive research to protect their corrosion due to the importance of these reservoirs, both economically and environmentally. To achieve this goal, the phenomenon of corrosion and its mechanism in reservoirs are investigated. The corrosion starts from the inner floor of the tank according to experts and experiences of the past years. The corrosion occurs less on the outer surface of the tank floor because of the low groundwater level and humidity and on the other hand, proper protection of the outer floor of the tank. The onset of corrosion in the inner bottom of the tank is created by local cells in the bottom of the tank. The saline water in the bottom of the tank is the electrolyte of these localized cells and causes small holes in the bottom of the tank, which is the main cause of saline water leakage from these holes to the outer surface of the tank. The corrosion begins at this surface with the penetration of this saline water to the outer floor of the tank and gradually the damage grows. The induced holes get bigger and bigger as so far as the diameter of the hole up to 10 cm has been observed in this industry.

Initially, this active electrolyte does not have any electrical connection to the cathodic protection system of the outer surface due to the low groundwater level and compaction of the soil under the reservoirs until the leakage rate and volume of saline water discharged from the reservoir are so high that it creates sufficient connection. In fact, this connection rarely occurs because the bottom of the tank is first prepared after compaction layer by layer of soil, then covered with a layer of asphalt, and canals are considered to drain any leaks. This asphalt prevents the corrosive soil from contacting the bottom of the tank and on the other hand, it prevents the formation of local cells by equalizing the electrical resistance under the tank. In other words, this asphalt is considered a good electrical insulator and like a strong but porous network, it passes any leakage and is transferred to the compacted lower surface and then to the canals and drainage pool.

### 1.5.4 Creating a suitable cover for the inner surface of the tanks

Since ancient times, using coatings with protective layers for various purposes to create a barrier and separation between objects and the environment has been common. However, using some methods and the materials have had satisfactory results for many years, but later become obsolete so that the use of newer materials and more appropriate methods have replaced them. The oil industry experts have worked hard to create reliable coatings on the inner of the tanks and have concluded that the production of new paints and coatings has also been very effective for the inner of the tanks. But most corrosion experts believe that the best coatings guarantee only 70% protection, and come close to 100% protection if combined with Cathodic protection (no coating is free of tiny and microscopic holes) (Papavinasam, 2013; Revie, 2011). In this regard, in the construction of new tanks in Khark Island, experts have realized the mentioned above and have used the most expensive and most reliable coatings on the inner surfaces of the tanks. Thus, the entire inner surface is covered with a thick layer of fiberglass (about one centimeter) and also the Cathodic protection of the inner surface of the tank is neglected due to the reliability of this coating. This is a big risk for these tanks.

### 1.5.5 Cathodic protection inside tanks

Methods used to prevent or control corrosion include coatings, chemical inhibitors, design modification, etc. Cathodic protection is one of the most effective methods used. Cathodic protection is the reduction or cessation of corrosion by the application of a one-way external current or the connection of an anode to the target metal so that the metal becomes a cathode. Cathodic protection is widely used in various industries so today it is used for protection in pipelines and underground installations, underground cables, canal valves, ships, water and oil tanks, and all marine installations.

In Cathodic protection, the cathode is preserved by using a metal that acts as an anode against the metal to be protected (Fig. 1.9). This method is known as sacrificial protection in which the anode is sacrificed. Its main problem is the replacement of anodes every once in a while. To solve this problem, the method of impressed current cathodic protection is used in which the anode is no longer sacrificed in this method, but the required electron of the cathodic protection system is injected using a current rectifier. Fig. 1.10 shows an industrial case in which a cathodic protection system is implemented at the bottom of an oil storage tank.

It should be said about the cathodic protection system of oil storage tanks that:

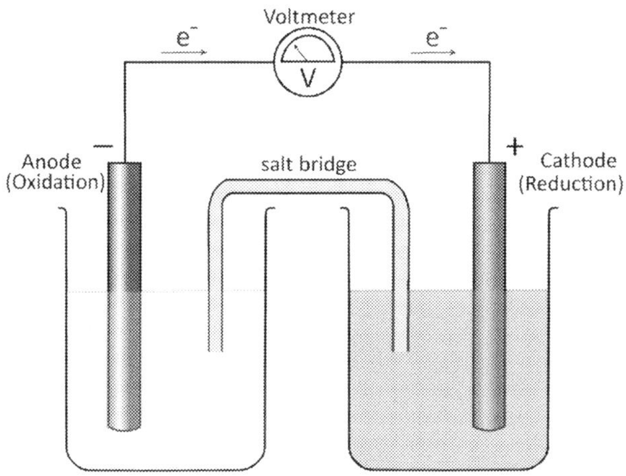


**Figure 1.9** General process of the cathodic protection system.

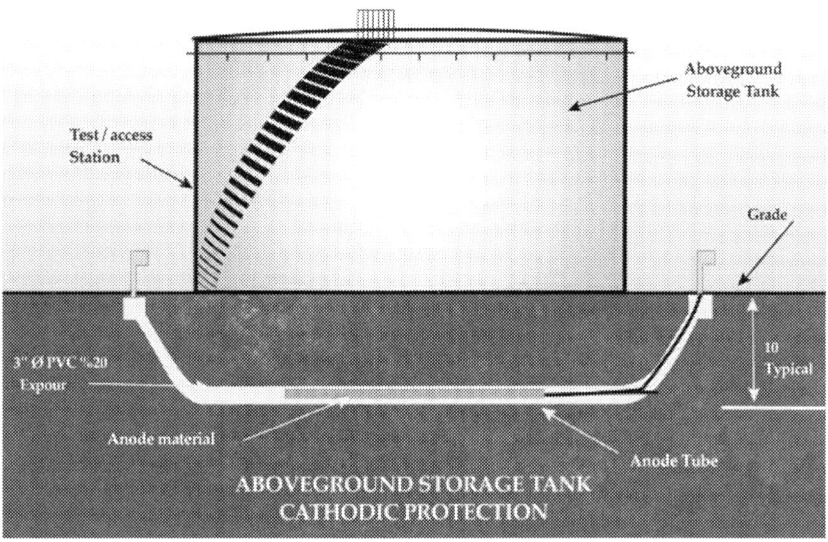


**Figure 1.10** The implemented cathodic protection system at the bottom of an industrial oil storage tank.

First, the diameter of these huge tanks reaches 100 m. The difference between the shrinkage and expansion of iron and fiberglass is significant in this long length. This shell is easily broken and torn by the weight of about 1,300,000 barrels of crude oil in these tanks and the penetration of saline water to the bottom of the tank is practical.

Second: The saline water collected in the bottom of the tank is under pressure with a height of about 20 m of crude oil on it. This saline water under pressure will easily pass through the smallest holes in the cover and will reach the metal floor of the tank. On the other hand, in reality, it is not possible to create a space with a diameter of 100 m and a height of 20 m from integrated fiberglass without defects induced by discontinuity error. Therefore, cathodic protection inside the tank is necessary as a complementary operation of the protected surface with paint inside the tanks.

## 1.6 Analysis of tank behavior with defects

Nowadays, reservoirs are broadly utilized in different industries, especially the oil, gas, gasoline, fuel, and petrochemical companies. Therefore, their accurate performance and stability are important for storing various materials such as crude oil, and defect detection in the structure using various tools and periodic inspections is vital.

In the present research, different defects including cracking and pitting corrosion in one of the tanks with a capacity of 1,350,000 barrels were studied and their critical sizes were determined. Finite Element Simulation, Taguchi Approach (TA), Multiple Regression Technique (MRT), and Response Surface Method (RSM) for the defects of cracking and pitting corrosion have been applied for the first time. The achievements of this study will be useful for improving inspections and preventing future events.

### 1.6.1 Finite element simulations

In this study, the dimension of the oil storage tank is stated in Table 1.1, while Fig. 1.11 demonstrates the image of the tank.

The oil storage tank was modeled in a finite element software, while the element type of S4R as a shell body was used. The sensitivity analysis of response to mesh number was performed to minimize the estimation error of mechanical parameters including stress and deformation (Omidi Bidgoli et al., 2020; Rahimian Koloor et al., 2018; Rahimian Koloor et al., 2020). Eventually, Fig. 1.12 shows the final FEM of the oil storage

**Table 1.1** The dimension of the oil storage tank (Omidi Bidgoli et al., 2020).

| Parameter | Value | Unit |
|---|---|---|
| Diameter | 119.5 | Meters |
| Height | 19 | Meters |
| Type of roof | Floating | |

**Figure 1.11** An image of the oil storage tank studied in this research.

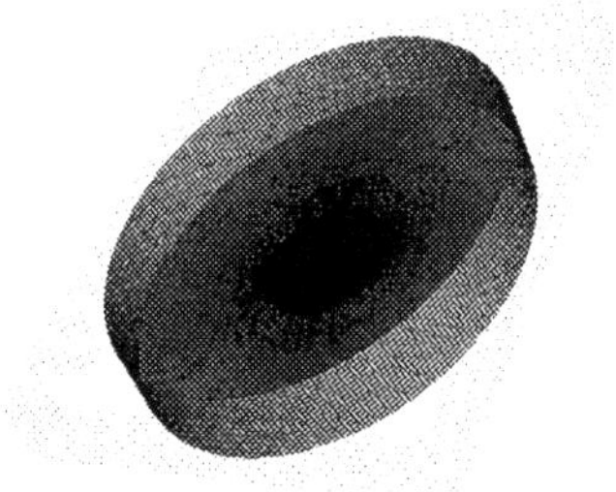

**Figure 1.12** Final FEM of the oil storage tank with the number of 30,000 elements (Omidi Bidgoli et al., 2020).

tank with the number of 30,000 elements. Next, the boundary conditions were considered as follows (Manual, 2006):

Around of tank floor:

$$u_1 = u_2 = u_3 = u_{R1} = u_{R2} = u_{R3} = 0 \quad (1.1)$$

The external surface of the tank floor:

$$u_3 = 0 \quad (1.2)$$

The tank with no defects was analyzed by considering the hydrostatic pressure ($p = \rho gh$) to specify the critical area. Next, the defect behavior was studied and the critical stress is extracted as the result of FE

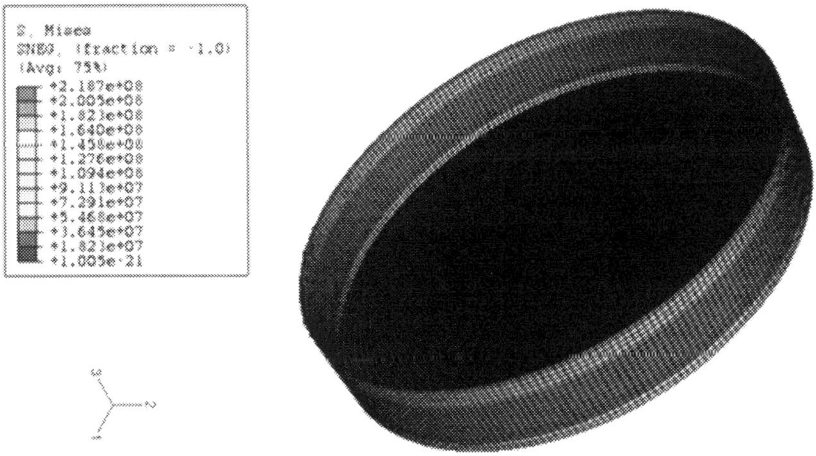


**Figure 1.13** The von Mises stress contour of the oil storage tank under the hydrostatic pressure (Omidi Bidgoli et al., 2020).

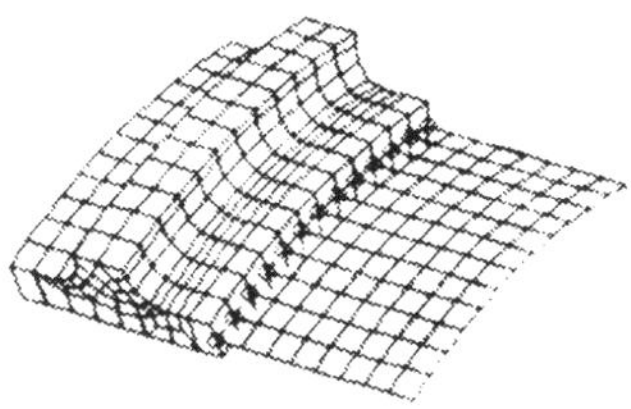

**Figure 1.14** Schematic of the shell to solid coupling elements (Omidi Bidgoli et al., 2020).

simulation. In this regard, the contour of equivalent von Mises stress is demonstrated in Fig. 1.13. As shown in this figure, the maximum critical stress is about 218.7 MPa.

#### *1.6.1.1 Finite element model of crack and pitting corrosion*

There are different methods to model both crack and pitting corrosion defects using the finite element model. One of these strategies is the shell to solid coupling elements in which multipoint constraints are used as depicted in Fig. 1.14. In other words, finite element software provides continuity and integrity between the shell and solid elements by assigning special constraints (Komzsik, 2001).

In the first step, to examine the validity of this technique, a complete model was provided using shell elements. After that, a new model including a combination of shell and solid elements was analyzed under the same boundary conditions and loading. To this end, a middle part of the shell was modeled as the volume. Fig. 1.15 presents the von Mises stress contours in both models.

From Fig. 1.15, the deformation is the same in both models. Nevertheless, there is a difference between the von Mises stresses created in the two components (about 1.3%). Accordingly, this approach can be utilized with the right accuracy for modeling and examining defects.

Geometric parameters of various defects including length, diameter, and depth were analyzed to obtain critical dimensions of oil storage tanks' defects. For this purpose, the yield stress of the target area (416.66 MPa) was considered as a criterion. Surface defects are modeled using C3D8R

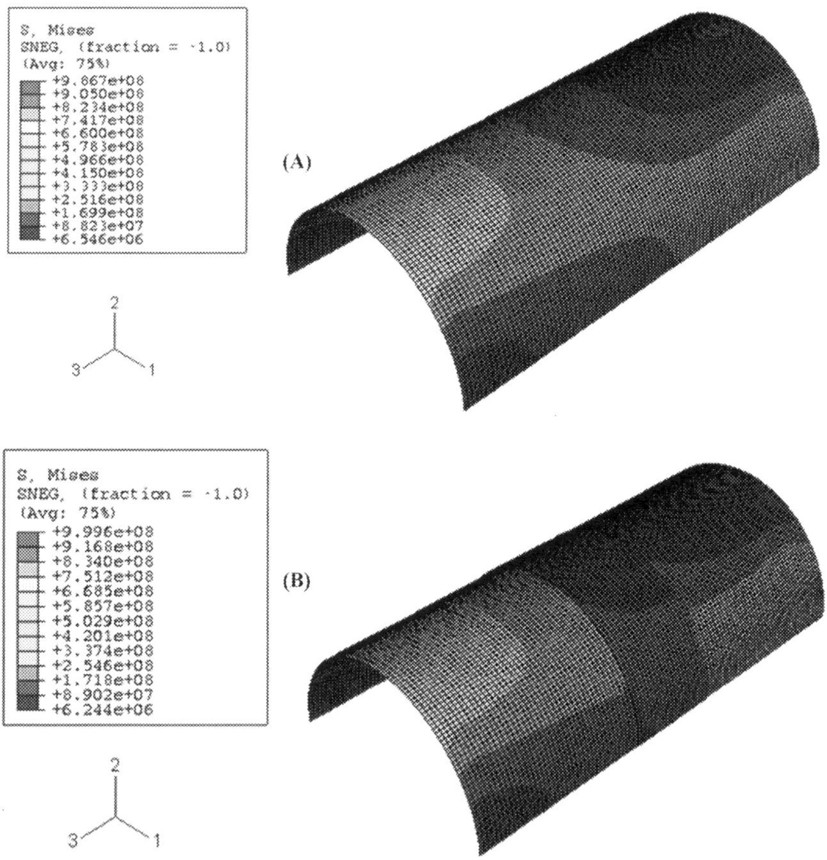


**Figure 1.15** The von Mises stress for both FE models: (A) Completely shell element and (B) combination of shell and solid elements (Omidi Bidgoli et al., 2020).

elements in the critical region. The stress-depth curves for crack and pitting corrosion in various situations are shown in Figs. 1.16 and 1.17, respectively.

The critical sizes of the crack and the pitting corrosion defects based on the finite element simulation results are obtained and reported in Tables 1.2 and 1.3, respectively.

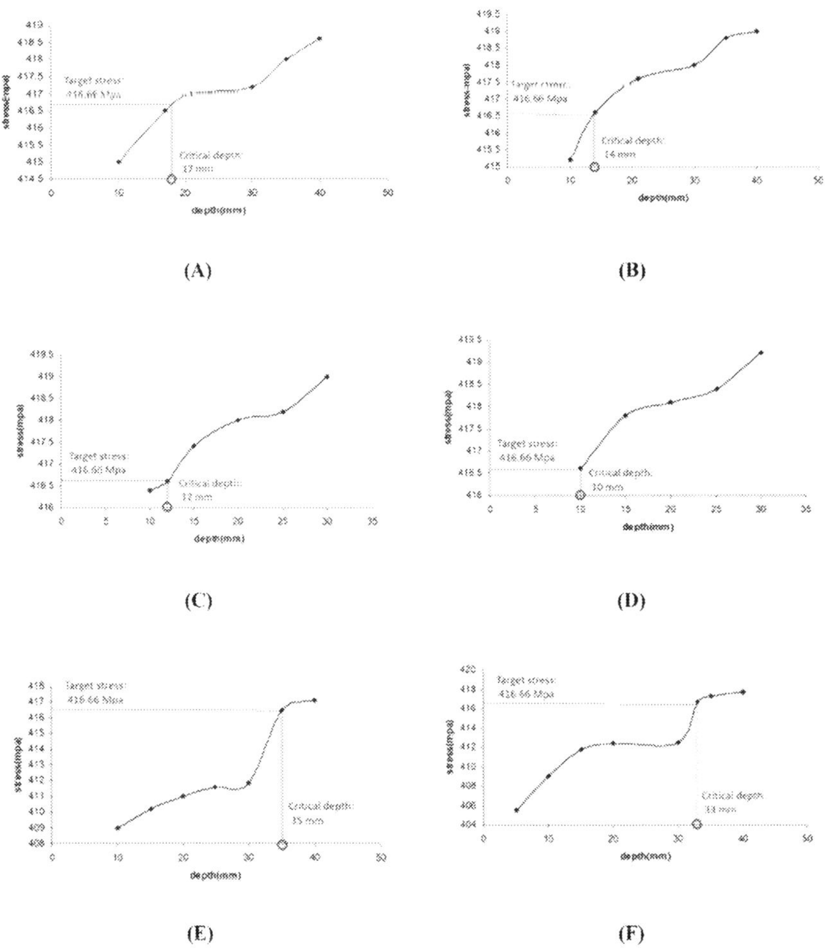


**Figure 1.16** The stress-depth curves for crack in various length and angle: (A) length = 300 mm and angle = 0; (B) length = 400 mm and angle = 0 degree; (C) length = 150 mm and angle = 45 degrees; (D) length = 200 mm and angle = 45 degrees; (E) length = 300 mm and angle 90 degrees; (F) length = 400 mm and angle = 90 degrees (Omidi Bidgoli et al., 2020).

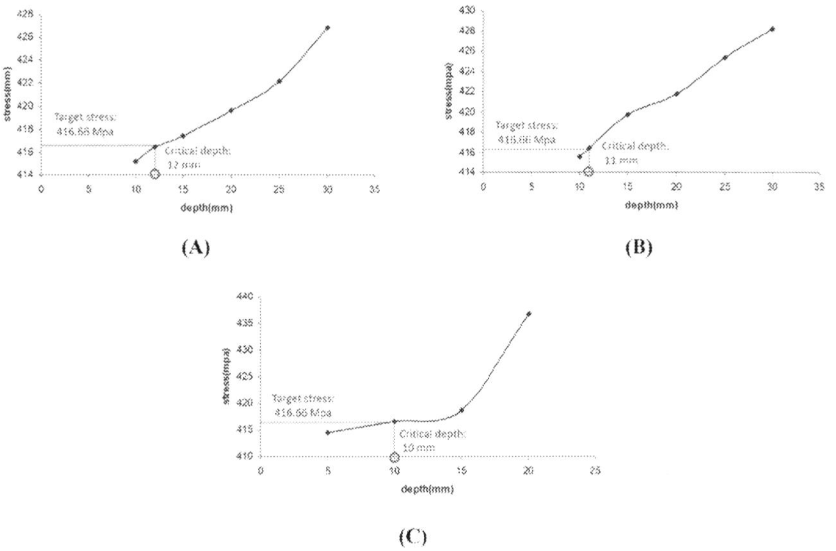


**Figure 1.17** The stress-depth curves for pitting corrosion in various diameter: (A) diameter = 50 mm; (B) diameter = 100 mm; (C) diameter = 150 mm (Omidi Bidgoli et al., 2020).

**Table 1.2** The critical sizes for the crack defect based on the results of FE simulation (Omidi Bidgoli et al., 2020).

| Crack type | Critical dimensions | | |
|---|---|---|---|
| | **Length (mm)** | **Depth (mm)** | **Angle** |
| Environmental crack | 300 | 17 | 0 |
| Environmental crack | 400 | 14 | 0 |
| Crack | 150 | 12 | 45 |
| Crack | 200 | 10 | 45 |
| Longitudinal crack | 300 | 35 | 90 |
| Longitudinal crack | 400 | 33 | 90 |

**Table 1.3** The critical sizes for the pitting corrosion defect based on the results of FE simulation (Omidi Bidgoli et al., 2020).

| Defect type | Critical sizes | |
|---|---|---|
| | **Diameter (mm)** | **Depth (mm)** |
| Corrosion cavity | 50 | 12 |
| Corrosion cavity | 100 | 11 |
| Corrosion cavity | 150 | 10 |

### 1.6.2 Taguchi approach

Different techniques of Design of Experiments (DOE) are used to decrease the number of experiments, cost, and time for studying the response variations in terms of different variables. In this regard, the Taguchi approach presents the least number of experiments to assess the impacts of input parameters (Chauhan et al., 2017; Ghorbani & Reaz Kashyzadeh, 2020; Maleki et al., 2019; Pandey et al., 2017; Reza Kashyzadeh et al., 2020; Zhao et al., 2017). In this technique, a set of tables is set up as an orthogonal array. These arrays make it possible to evaluate the main impacts of different parameters by performing the minimum number of runs. Therefore, this is the most important advantage of TA over different DOE techniques. In general, there are two strategies for analyzing results (standard and signal-to-noise ratio). The standard model depends on computing the effects of factors and on doing an analysis of variance. The subsequent model looks at the scattering close to a certain value. As the value of this ratio (signal/noise) increases, the scattering decreases, and in this case, the effect of that parameter will be more important. Also, the sensitivity analysis of the Taguchi method can additionally predict responses for different modes. The previously published papers confirmed that the Taguchi prediction algorithm is very suitable for industrial components and has an acceptable accuracy (Farrahi et al., 2020; Ghorbani & Reaz Kashyzadeh, 2020; Kashyzadeh & Ghorbani, 2020; Reza Kashyzadeh et al., 2020; Reza Kashyzadeh et al., 2021). Hence, the TA was used to determine the influence of different parameters of various defects (cracks and pitting corrosion) on the von Mises stresses created in the critical region of an oil storage tank. Defect geometry is one of the most important parameters affecting the behavior of structures under different loads and working conditions. Therefore, this part of the study aimed to calculate the critical value of the geometric parameters of different defects and also the effect of geometric parameters on the behavior of the structure. In the 2D problem, the crack length and crack angle are important for the crack defect, and the diameter is also important for the defect of pitting corrosion. But in the 3D problem, the depth of the defect plays an important role. Therefore, in this study, three (length, angle, and depth) and two (diameter and depth) geometric parameters were considered as input variables of TA for the crack and pitting corrosion defects, respectively. This study was performed based on a real oil storage tank, accordingly, the values considered for each of the input variables of the Taguchi analysis were selected based on events occurring in reality. The variable parameters considered in the TA for the pitting corrosion defect and crack, along with their different levels, are reported in Tables 1.4 and 1.5, respectively.

The Taguchi orthogonal matrices with the characteristics L8 (4^1 & 2^1) and L9 (3^2) were used to form TA for the crack and pitting corrosion defects, as shown in Tables 1.6 and 1.7, respectively (Omidi Bidgoli et al., 2020).

The most and least effective parameters based on TA were determined. Given the need to reduce the equivalent von Mises stress at the critical region of the oil storage tank, a smaller term is best used for the data analysis, in accordance with the following formula (Farrahi et al., 2020):

$$\frac{S}{N} = -10Log\left[\frac{1}{n}(y_1^2 + y_2^2 + \ldots + y_n^2)\right] \tag{1.3}$$

**Table 1.4** Characteristics of the variable parameters considered in the Taguchi algorithm for the pitting corrosion defect (Omidi Bidgoli et al., 2020).

| Levels | Variables | |
|---|---|---|
| | **Parameter I: depth** | **Parameter II: length** |
| A | 50 | 10 |
| B | 100 | 15 |
| C | 150 | 20 |

**Table 1.5** Characteristics of the variable parameters considered in the Taguchi algorithm for the crack defect.

| Levels | Variables | |
|---|---|---|
| | **Parameter I: length** | **Parameter II: depth** |
| A | 150 | No |
| B | 200 | No |
| C | 300 | No |
| D | 400 | No |
| A1 | No | 17 |
| B1 | No | 14 |
| A2 | No | 12 |
| B2 | No | 10 |
| A3 | No | 35 |
| B3 | No | 33 |

**Table 1.6** The layout of the orthogonal matrix L8 (stress is in MPa) (Omidi Bidgoli et al., 2020).

| Angle | Run No. | Inputs | | Output |
|---|---|---|---|---|
| | | Parameter I | Parameter II | Stress |
| Zero | 1 | A | A1 | 415.00 |
| | 2 | A | B1 | 415.20 |
| | 3 | B | A1 | 416.96 |
| | 4 | B | B1 | 417.60 |
| | 5 | C | A1 | 417.20 |
| | 6 | C | B1 | 418.00 |
| | 7 | D | A1 | 418.50 |
| | 8 | D | B1 | 419.00 |
| 45 | 1 | A | A2 | 416.40 |
| | 2 | A | B2 | 416.58 |
| | 3 | B | A2 | 417.40 |
| | 4 | B | B2 | 417.80 |
| | 5 | C | A2 | 418.20 |
| | 6 | C | B2 | 418.40 |
| | 7 | D | A2 | 419.00 |
| | 8 | D | B2 | 419.20 |
| 90 | 1 | A | A3 | 409.00 |
| | 2 | A | B3 | 409.11 |
| | 3 | B | A3 | 410.97 |
| | 4 | B | B3 | 412.35 |
| | 5 | C | A3 | 411.78 |
| | 6 | C | B3 | 412.42 |
| | 7 | D | A3 | 417.10 |
| | 8 | D | B3 | 417.73 |

**Table 1.7** The layout of the orthogonal matrix L9 (stress is in MPa) (Omidi Bidgoli et al., 2020).

| Run No. | Inputs | | Output |
|---|---|---|---|
| | Parameter I | Parameter II | Stress |
| 1 | A | A | 415.13 |
| 2 | A | B | 417.41 |
| 3 | A | C | 419.63 |
| 4 | B | A | 415.61 |
| 5 | B | B | 419.70 |
| 6 | B | C | 421.83 |
| 7 | C | A | 416.44 |
| 8 | C | B | 418.55 |
| 9 | C | C | 436.9 |

where $y_1$, $y_2$, ..., and $y_n$ represent the measured bent angles in the bending process, and each bending condition is repeated n times. Subsequently, the main effects of the S/N ratios and mean ratios at every parameter level were analyzed and plotted in Figs. 1.18 and 1.19 for crack and pitting corrosion defects, respectively (Omidi Bidgoli et al., 2020).

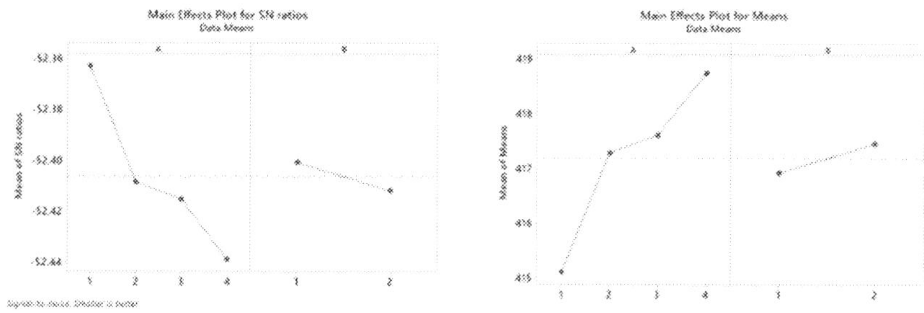


(A)

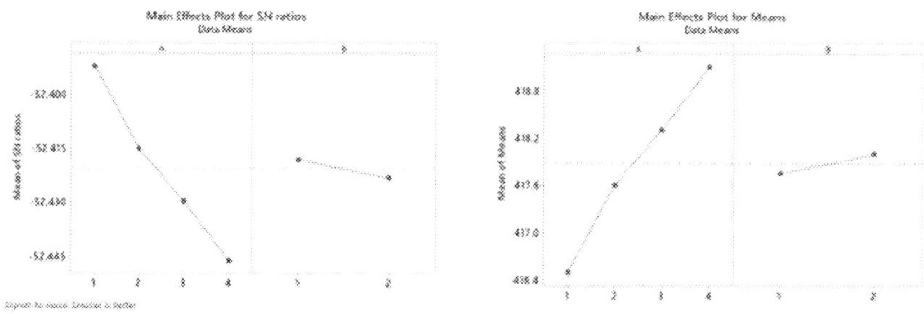


(B)

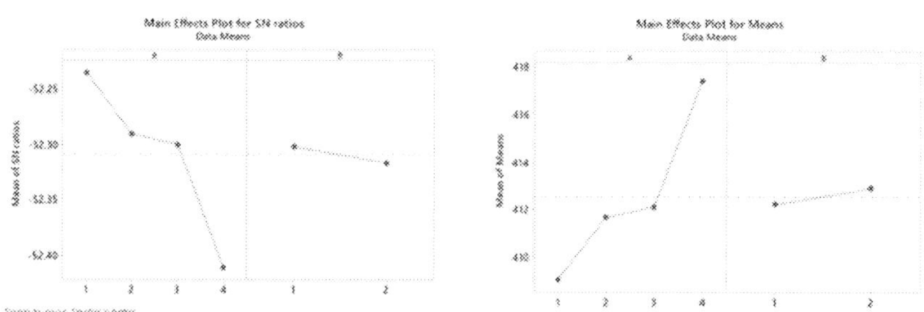


(C)

**Figure 1.18** Effects of the S/N and mean ratios of all parameters related to the crack defect with different angles: (A) zero, (B) 45, and (C) 90 (Omidi Bidgoli et al., 2020).

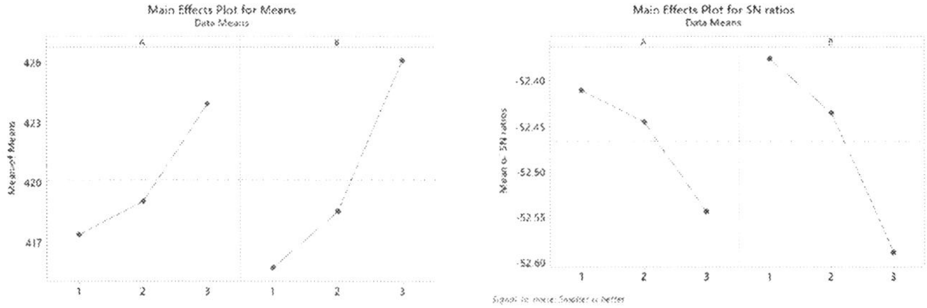


**Figure 1.19** Effects of the S/N and mean ratios of all parameters related to the pitting corrosion defect (Omidi Bidgoli et al., 2020).

Interpretation of the diagrams of TA results states that the run number is sufficient to perform advisement regarding the Taguchi analysis because the diagram pattern of the S/N ratios is reversed in comparison with the diagram pattern of the mean ratios (Ghorbani & Reaz Kashyzadeh, 2020; Reza Kashyzadeh et al., 2020; Zhao et al., 2017). It means that the maximum situation in the diagram of the S/N ratio is the minimum value in the other diagram. From Fig. 1.18, the stress created in the structure increases by increasing both the crack length and depth parameters and it leads to a decrease in the life service of the structure. This result was also predictable, which indicates the correctness of the analysis. The results also show that raising the crack angle from 0° to 90° increases stress. In other words, it turns out that longitudinal cracking is the most destructive mode of cracking in a reservoir. Moreover, failure Mode-I is predominant in this structure. Also, the results show that the stress in the reservoir rises with increasing the diameter and depth of the pitting corrosion defect (Fig. 1.19). However, it is not clear which of the geometric parameters is more effective.

Next, the influence of various parameters of the crack on the maximum von Mises stress is illustrated in Fig. 1.20 (*left*). It shows that the impact of the crack depth increases by raising the crack angle. Also, the impact of the crack length decreases by raising the crack angle. Eventually, the Taguchi sensitivity analysis was performed, and the result is shown in Fig. 1.20, which presents the most effective parameter for both defects. Moreover, the least effective parameter is also determined. Also, the impact of different parameters of the pitting corrosion defect is displayed in Fig. 1.20 (*right*).

Next, Taguchi's prediction algorithm was used for different cases to be able to determine the von Mises stress at the crack tip according to the

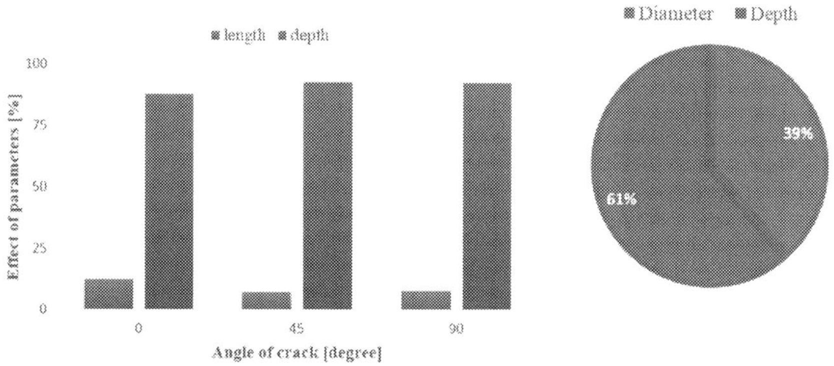


**Figure 1.20** (*Left*) Impact of different parameters of the crack with various angles and (*Right*) Impact of different parameters of the pitting corrosion defect (Omidi Bidgoli et al., 2020).

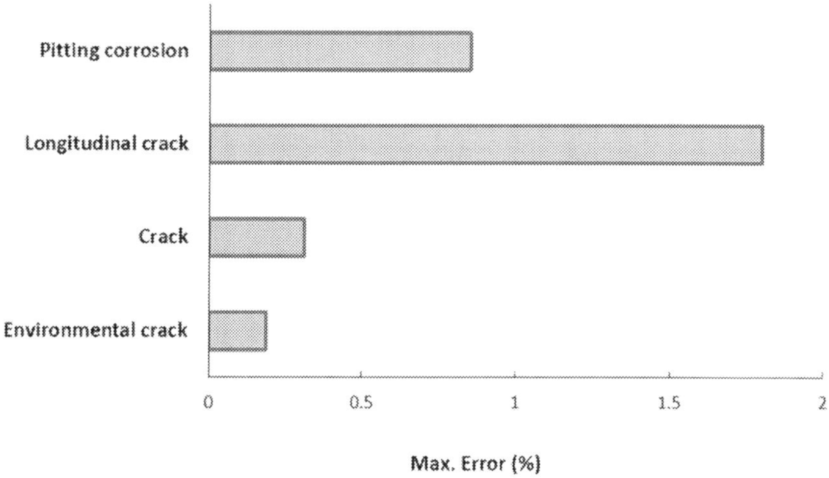


**Figure 1.21** The comparative diagram between the results of finite element simulation and Taguchi's prediction algorithm.

geometric characteristics of the crack. Also, this issue was repeated to study the impact of geometric parameters of the pitting corrosion defect on the maximum value of von Mises stress. The accuracy of this prediction in comparison with the simulation results is shown in Fig. 1.21.

### 1.6.3 Multiple regression techniques

In general, using this technique allows for the study of the linear relationship between a set of independent variables with a dependent variable in a way that the existing relationships among the independent variables are also considered. Its results can be used to predict responses for different states and parameter changes in the shortest possible time compared to performing tests, simulations, or other methods. In addition, the optimal process and interaction between independent parameters can be achieved by relying on MRT results and applying various techniques of optimal parameter estimation such as the least-squares method, random gradient method, and maximum likelihood estimation. For this purpose, the results of finite element simulations were used (Figs. 1.16 and 1.17).

The linear relationship between the von Mises stress created in the crack tip and geometric parameters of the crack including length and depth for three different angles was presented by applying MRT.

For environmental crack ($\theta = 0^\circ$):

$$\sigma_{von-Mises} = 414.47 - 0.00482 \times L + 0.183 \times d - 0.00256 \times d^2 + 0.00017 \times L \times d \tag{1.4}$$

For crack ($\theta = 45^\circ$):

$$\sigma_{von-Mises} = 404.8 + 0.0516 \times L + 0.305 \times d - 0.00193 \times d^2 - 0.000575 \times L \times d \tag{1.5}$$

For longitudinal crack ($\theta = 90^\circ$):

$$\sigma_{von-Mises} = 412.68 + 0.00972 \times L + 0.184 \times d + 0.0011 \times d^2 - 0.000308 \times L \times d \tag{1.6}$$

where L and d are the length and depth of the crack, respectively. To measure the accuracy of the equations presented by MRT, the obtained results were compared with the simulation results (Table 1.8). The maximum error in predicting the von Mises stress at the crack tip is equal to 0.16%, which indicates very good accuracy in using this method.

Next, this technique was used to investigate the pitting corrosion defects. In this regard, the maximum von Mises stress due to the defect of the pitting corrosion is equal to:

$$\sigma_{von-Mises} = 412.5 - 0.158 \times d + 0.0778 \times D + 0.01797 \times d^2 - 0.000386 \times D^2 - 0.00004 \times d \times D \tag{1.7}$$

where *d* and *D* represent the depth and diameter of the pitting defect.

Using this equation results in a maximum error of 0.54% compared to the finite element simulation method. But the time to reach the answer

**Table 1.8** Details of comparison of MRT and FE results.

| Type of crack | Crack angle | Error (%) between MRT & FE | | |
|---|---|---|---|---|
| | | Min | Max | Mean |
| Environmental crack | 0 | 0.003433 | 0.131834 | 0.052259 |
| Crack | 45 | 0.004236 | 0.083147 | 0.037542 |
| Longitudinal crack | 90 | 0.000737 | 0.161671 | 0.063185 |

in this method is a few seconds, which is a very prominent feature in the industry to predict the behavior of the structure with this accuracy.

### 1.6.4 Response surface method

Different strategies of DOE are used to estimate the influence of various factors on the response when there is a wide variety of samples and we do not want to examine all of them. In other words, by performing some tests with predefined settings, the parameters' effects on the response can be achieved with high accuracy. In this regard, TA can only examine the impact of each factor on the response separately (Chauhan et al., 2017; Farrahi et al., 2020; Ghorbani & Reaz Kashyzadeh, 2020; Reza Kashyzadeh et al., 2020; Zhao et al., 2017). But, RSM can check the impacts of changes in two parameters simultaneously on the response. In fact, one of the most important advantages of utilizing RSM compared to different strategies is examining the influence of factors on the response by considering the interaction between them. Hereof, the RSM output is 3D diagrams including changes of two independent parameters and the response. Therefore, this method was used to more accurately investigate the impact of various geometric factors of the defects (crack and pitting corrosion) on the maximum von Mises stress. To this end, three (length, depth, and angle) and two (diameter and depth) factors were considered as input variables for the geometric characteristics of both defects including crack and pitting corrosion, respectively. A schematic of the used algorithm including the input variables and response is illustrated in Fig. 1.22. In this study, the linear effects of each factor and the interaction between them were considered as follows:

L, d, A, Ld, LA, and dA

D, d, and Dd

The results of the RSM analysis are illustrated in Figs. 1.23 and 1.24 for crack and pitting corrosion defects, respectively. In these analyzes, the constant parameters were assumed to be the mean

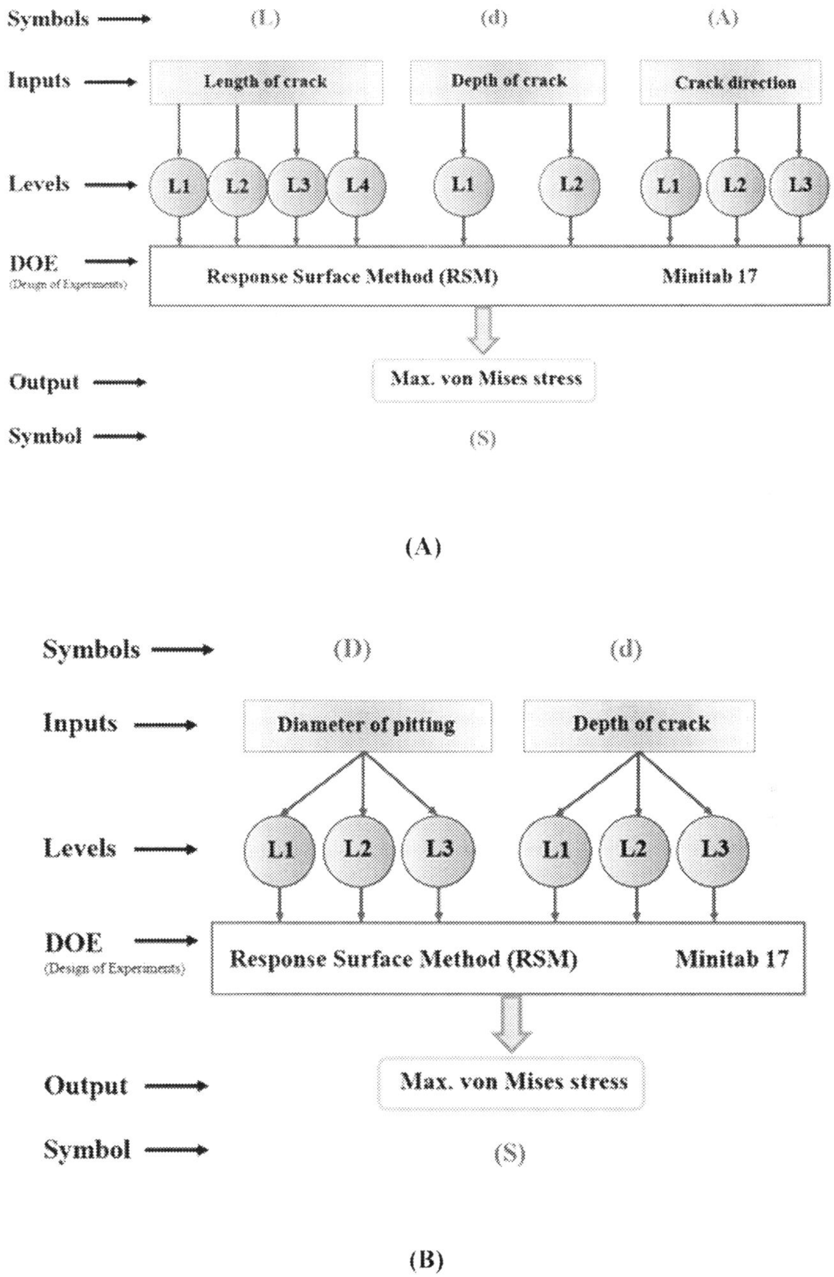


**Figure 1.22** The schematic of the RSM-based design of experiments used in this study includes (A) study of the effect of the crack defect in the oil storage tank and (B) study of the effect of pitting corrosion defect in the oil storage tank.

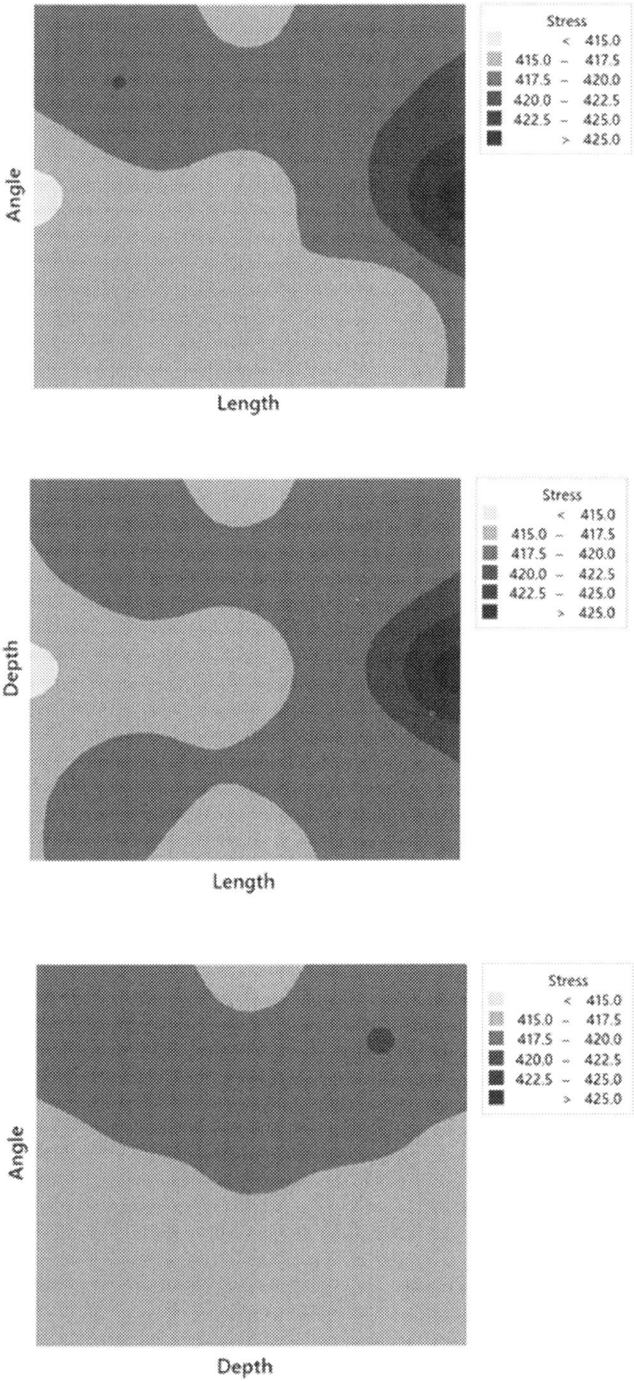


**Figure 1.23** RSM results for maximum von Mises stress related to the crack defect in the oil storage tank.

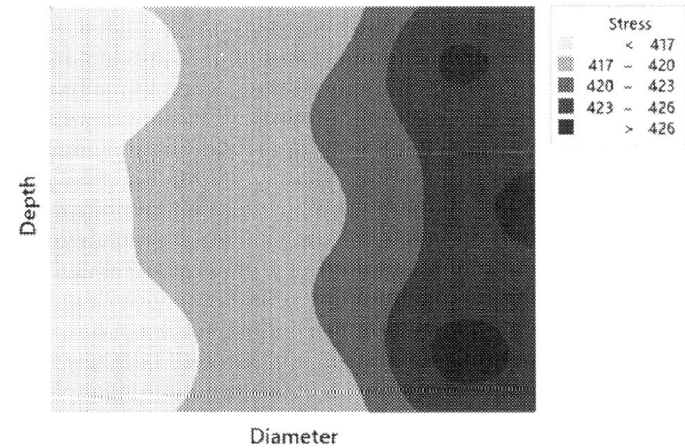


**Figure 1.24** RSM results for maximum von Mises stress related to the pitting corrosion defect in the oil storage tank.

level (Farrahi et al., 2020; Ghorbani & Reaz Kashyzadeh, 2020; Reza Kashyzadeh et al., 2020).

The most important results extracted from the contours of RSM are:

1. Since increasing stress due to a defect is more dangerous for the structure, we are looking for a way to reduce it. In other words, the situation is better when the von Mises stress is less. Therefore, the pale blue range in the contours provided by RSM is appropriate and valuable.
2. The RSM results show that to have the minimum stress due to the crack defect, the crack length should have the least value. But it cannot be said that with increasing the length of the crack, the von Mises stress value increases. Because the dependence of the response on the two parameters of depth and crack angle must also be considered. In other words, in the second part of Fig. 1.23, it is shown that in some ranges, the crack depth prevailed over its length.
3. We cannot talk about the crack angle exactly because it does not have a specific trend. The first part of Fig. 1.23 shows that when the crack angle is 45 degrees and the crack has a minimum length, the value of von Mises stress is much less than when the angle is 0 or 90. But when the length of the crack changes, this relationship is not true. For example, when the crack length is the maximum value, the Max. von Mises stress created in the tank corresponding to angles 0 and 90 is the same. But if the crack angle is 45 degrees, we will have the most value of stress. Also, there is a similar relationship between the two parameters of crack length and depth.

4. The RSM results show that increasing the diameter of the pitting corrosion leads to an increase in the Max. von Mises stress created in the structure. However, in small and medium diameters, changes in depth parameters do not have a significant effect on the value of stress. However, in the area with the largest diameter of the pitting corrosion, some areas show different behavior.

## 1.7 Conclusions

In this chapter, the authors have tried to assess the oil storage tank with different types of defects. Firstly, brief descriptions of the general construction of the oil storage tanks, and various standards based on the application, design, repair, and maintenance were provided. Next, different defects in the various types of storage tanks in the petrochemical industry were introduced and solutions to deal with them were also stated. Eventually, different techniques including finite element simulation and DOEs (TA, MRT, and RSM) were used to assess the oil storage tank in three different modes including defective structure, structure with crack defect, and structure with the defect of pitting corrosion. Also, the effects of geometric parameters of both defects on the behavior of the structure were investigated and the critical dimensions were calculated for them. In summary, the results of this comprehensive study are as follows:

1. The most critical area for crack and pitting corrosion is the second region of the tank body, based on the finite element simulation. However, the maximum hydrostatic pressure is related to the first region of the tank with the largest wall thickness.
2. The effect of the crack depth on the increase in stress is superior to the effect of the crack length.
3. For defects of the pitting corrosion, the impact of the depth on the stress raising is superior to the impact of the diameter. On the other hand, the progression towards the depth has a greater effect on the equivalent von Mises stress than the progression towards its length does.
4. The effect of the crack depth on the maximum von Mises stress at the critical region of the oil storage tank is about 91% and the effect of the crack length is 9%.

5. The effects of the pit depth and diameter on the critical stress for the corrosion defect are about 61% and 39%, respectively.
6. The regression method predicts more accurately than the Taguchi method if the reference is the simulation value. In other words, if the criterion for this comparison is the simulation value. Also, it spends less time predicting.
7. The accuracy of the proposed MRT formula for predicting von Mises stress at the crack tip and due to the pitting corrosion is equal to 0.13% and 0.45%, respectively which are practically negligible in industry.
8. The crack length should have the least value to have the minimum stress. But this is not always true because the dependence of the response on the two parameters of depth and crack angle must also be considered. To understand more about this point, see the second part of Fig. 1.23, it is shown that in some ranges, the crack depth prevailed over its length.
9. Increasing the diameter of the pitting corrosion leads to an increase in the maximum von Mises stress. Moreover, in small and medium diameters, changes in depth parameters do not have a significant effect on the value of stress. Also, in the area with the largest diameter of the pitting corrosion, some areas show different behavior.

## References


Anvo, N. (2019). Internal in-service inspection of petrochemical storage tank floors to detect underside corrosion with non-destructive testing robot, London South Bank University.

Babusiaux, D. (2004). *Oil and gas exploration and production: Reserves, costs, contracts*. Editions Technip.

Bachu, S., & Shaw, J. C. (2005). *$CO_2$ storage in oil and gas reservoirs in western Canada: Effect of aquifers, potential for $CO_2$-flood enhanced oil recovery and practical capacity, Greenhouse gas control technologies* (7, pp. 361–369). Elsevier.

Chauhan, R., Singh, T., Kumar, N., Patnaik, A., & Thakur, N. (2017). Experimental investigation and optimization of impinging jet solar thermal collector by Taguchi method. *Applied Thermal Engineering*, *116*, 100–109.

Cleveland, C. J. (2004). *Encyclopedia of energy*. Elsevier.

Dhillon, B. S. (2016). *Safety and reliability in the oil and gas industry: A practical approach*. CRC Press.

Farrahi, G. H., Kashyzadeh, K. R., Minaei, M., Sharifpour, A., & Riazi, S. (2020). Analysis of resistance spot welding process parameters effect on the weld quality of three-steel sheets used in automotive industry: experimental and finite element simulation. *International Journal of Engineering*, *33*(1), 148–157.

Ghorbani, S., & Reaz Kashyzadeh, K. (2020). Taguchi approach and response surface analysis for design of a high-performance single-walled carbon nanotube bundle interconnects in a full adder. *International Journal of Engineering*, *33*(8), 1598–1607.

Jirarungsatian, C., & Prateepasen, A. (2010). Pitting and uniform corrosion source recognition using acoustic emission parameters. *Corrosion Science, 52*(1), 187–197.

Kasai, N., Sekine, K., & Maruyama, H. (2003). Non-destructive evaluation method for far-side corrosion type flaws in oil storage tank bottom floors using the magnetic flux leakage technique. *Journal of the Japan Petroleum Institute, 46*(2), 126–132.

Kashyzadeh, K., & Ghorbani, S. (2020). Numerical study of free vibration behaviour of filled tool holder using epoxy-granite. *Journal of Physics: Conference Series*, IOP Publishing.

Kennedy, J. L. (1993). *Oil and gas pipeline fundamentals*. Pennwell Books.

Kermani, M., & Harrop, D. (1996). The impact of corrosion on oil and gas industry. *SPE Production & Facilities, 11*(03), 186–190.

Kim, J.-S., An, D.-H., Lee, S.-Y., & Lee, B.-Y. (2009). A failure analysis of fillet joint cracking in an oil storage tank. *Journal of Loss Prevention in the Process Industries, 22*(6), 845–849.

Komzsik, L. (2001). MSC. Nastran.

Lei, Z. (2018). Research on installation defects and accidents in long-distance oil and gas storage and transportation pipeline. *Chemical Engineering Design Communications*, 03.

Maleki, E., Unal, O., & Kashyzadeh, K. R. (2019). Efficiency analysis of shot peening parameters on variations of hardness, grain size and residual stress via Taguchi approach. *Metals and Materials International, 25*(6), 1436–1447.

Manual, A. U. S. (2006). *Version 6.6*. Providence, RI: Hibbitt, Karlsson and Sorenson. Inc.

Mažeika, L., Kažys, R., Raišutis, R., Demčenko, A., & Šliteris, R. (2006). Long-range ultrasonic non-destructive testing of fuel tanks. *DGZfP Proceedings BB*, Citeseer.

Myers, P. E. (1997). *Aboveground storage tanks*. McGraw-Hill Education.

Omidi Bidgoli, M., Reza Kashyzadeh, K., Rahimian Koloor, S. S., & Petru, M. (2020). Estimation of critical dimensions for the crack and pitting corrosion defects in the oil storage tank using finite element method and taguchi approach. *Metals, 10*(10), 1372.

Pandey, N., Murugesan, K., & Thomas, H. R. (2017). Optimization of ground heat exchangers for space heating and cooling applications using Taguchi method and utility concept. *Applied Energy, 190*, 421–438.

Papavinasam, S. (2013). *Corrosion control in the oil and gas industry*. Elsevier.

Pence, A. W. (1988). Failure avoidance in welded fabrication. *National Board Bulletin*, 17–23.

Rahimian Koloor, S. S., Karimzadeh, A., Tamin, M. N., & Abd Shukor, M. H. (2018). Effects of sample and indenter configurations of nanoindentation experiment on the mechanical behavior and properties of ductile materials. *Metals, 8*(6), 421.

Rahimian Koloor, S. S., Karimzadeh, A., Yidris, N., Petrů, M., Ayatollahi, M. R., & Tamin, M. N. (2020). An energy-based concept for yielding of multidirectional FRP composite structures using a mesoscale lamina damage model. *Polymers, 12*(1), 157.

Revie, R. W. (2011). *Uhlig's corrosion handbook*. John Wiley & Sons.

Reza Kashyzadeh, K., Ghorbani, S., & Forouzanmehr, M. (2020). Effects of drying temperature and aggregate shape on the concrete compressive strength: Experiments and data mining techniques. *International Journal of Engineering, 33*(9), 1780–1791.

Reza Kashyzadeh, K., Mousavi Bafrouyi, S. M. S., & Khorsandijou, S. M. (2021). Effects of road roughness, aerodynamics, and weather conditions on automotive wheel force. *International Journal of Engineering, 34*(2), 536–546.

Salinas, R., So, A., Valdez, B., Schorr, M., Bastidas, J., Carrillo, M., & Alvarez, L. (2016). Natural gas industry: Materials and corrosion. *MRS Online Proceedings Library Archive, 1815*.

Samimi, A. (2012). Causes of increased corrosion in oil and gas pipelines in the Middle East. *International Journal of Basic and Applied Science*, 572–577.

Semadeni, M. (2004). *Storage of energy, overview. Encyclopedia of energy* (pp. 719–738). Cleveland, NY: Elsevier.

Shigeno, H., & Okamoto, D. E. K. (1960). *Corrosion of bottom plate of oil storage tank and corrosion control*. Japan: L. T. Nakagawa Corrosion Protecting Company.

Standard, A. (1988). Welded steel tanks for oil storage.

Standard, A. (2013). Welded tanks for oil storage.

Wright, C. A., Bowie, C. P., Burroughs, E. H., Thompson, J. W., Christy, S. B., Smith, S. S., Dykema, W. P., & Geology and Mines, M. B. O. (1918). *Oil-storage tanks and reservoirs: With a brief discussion of losses of oil in storage and methods of prevention*. U.S. Government Printing Office.

Zhao, L., Zhao, Y., Bao, C., Hou, Q., & Yu, A. (2017). Optimisation of a circularly vibrating screen based on DEM simulation and Taguchi orthogonal experimental design. *Powder Technology*, *310*, 307–317.

CHAPTER TWO

# Wind effect on atmospheric tanks

**Adriana Mesa-Gómez[1,2], Jean-Paul Pinelli[3], Oscar J. Ramirez[4] and Ernesto Salzano[5]**

[1]Civil and Environmental Engineering Department, Universidad de los Andes, Bogotá, Colombia
[2]CERTEC, Universitat Politècnica de Catalunya, Barcelona, Catalonia, Spain
[3]Florida Institute of Technology, Melbourne, FL, United States
[4]Risk and process safety management, Bureau Veritas, Bogotá, Colombia
[5]Dipartimento di Ingegneria Civile, Chimica, Ambientale e dei Materiali, Università di Bologna, Bologna, Italy

## 2.1 Introduction

All process industries have the risk of major accidents due to the unwanted release of energy and hazardous materials (hazmat). Those accidents may be due to operational failures, for instance, the incident in a pesticide factory owned by the Union Caribe company in Bhopal-India. In this incident, maintenance errors and working procedures resulted in an undesired release of methyl-isocyanate which spread throughout the city causing serious health problems and even death to thousands of people, and therefore large economic losses to both the company and the nation (Eckerman, 2005). On the other hand, naturals hazards can cause accidents at process facilities as well. A good example of a natural hazard capable of triggering technological accidents was Hurricane Floyd, which caused extensive damage to the oil industry on the east coast of the United States and Canada due to the spillage of thousands of gallons of oil products, a catastrophic environmental impact, and incalculable economic losses (Young et al., 2004).

In recent years, the concern of the general industry about natural hazards has increased worldwide because of the increased risk of damage to industrial installations from extreme natural events, such as floods, earthquakes, forest fires, storms, hurricanes, and landslides. These kinds of incidents cause the unplanned release of hazmat also into the ground and water basins, resulting in accidents with serious impacts on people, property, and the environment (Cruz & Okada, 2008). These accidents are known as

*Above Ground Storage Tank Oil Spills.*
DOI: https://doi.org/10.1016/B978-0-323-85728-4.00005-X

Natech events (technological accidents triggered by a natural event) (Krausmann, Cruz & Salzano, 2017). Different studies have found that earthquakes and floods, followed by landslides, hurricanes and electric shocks, and finally droughts are the natural events that cause the most industrial accident (Young et al., 2004). In addition, hazardous material losses during Natech events can result in fires, explosions, or the dispersion of toxic substances. Historical analysis by Campedel (2008) showed that the industrial equipment type most affected by natural hazards are storage tanks and pipelines, and the substances involved in most Natech events are crude oil, diesel, and gasoline; substances that, at the time of loss of containment (LOC), can cause explosions, fires, and toxic gas dispersions.

In recent decades, the occurrence of natural disasters has been increasing. In 2004, Cruz et al. (2004) reported that different extreme natural phenomena have been increasing over time. The study was conducted throughout the United States between the years 1980 and 1989, which yielded the following results: 228 earthquakes, 26 hurricanes, 16 floods, 15 thunderstorms, 13 blizzards, and 7 storms. Additionally, 1022 flood events occurred worldwide in the decade of the 1990s, while in the last decade, the occurrence of these types of natural events has increased by 74% (CRED, 2019).

In particular, meteorological events have the characteristic of covering very large areas, affecting entire ecosystems in their path, a great variety of industries, and areas of high urban density. Relevant changes in the climatic conditions of the planet used to appear within large periods (centuries or millennia), but in the last century, they have occurred in relatively short time intervals (decades). This climatic variability increases the atmospheric temperature, causing an increase in the frequency and severity of extreme natural events, with consequences on the ecosystems of the planet (Arango et al., 2012; Hardy, 2003).

Therefore, considering the great threat natural hazards present for industrial facilities, especially for equipment that stores large quantities of hazmat, this chapter deals with the effects of strong winds on storage tanks designed under API 620 and API 650 standards.

## 2.2 History of natural events affecting industrial equipment

Due to the increase in the occurrence of Natech events in recent years, different authors have developed studies, analyses, and risk tools associated with these types of events. For example, in 2008 the French Ministry of Sustainable Development carried out a study assessing the distribution of natural phenomena along the European continent which in turn caused serious human, social, and economic losses (French Ministry for Sustainable Development, 2013). The results of this study have shown that storms and floods are the natural phenomena that most affect the majority of European countries (Austria, Belgium, Denmark, France, Germany, Ireland, the Netherlands, and England). Also, similar trends can be assessed in other parts of the world. Hence, as a result of the large number and increase in time of natural events around the world, entities—such as the European Joint Research Center (JRC) at Ispra (Italy)—have begun studies to understand better the impact of natural hazards on chemical industries because of its potential to cause serious accidents (European Commission, 2016). As an example, the JRC developed a RAPID-N tool to map and analyze Natech risks, which contributes to the identification of areas prone to Natech events, analyzing and visualizing their risks, and supports the decision making of authorities before and after the event (Girgin & Krausmann, 2012). There are other sources from where information on Natech events is collected; some of the main European databases are ARIA (Analyze, Recherche et Information sur les Accidents), FACTS (Failure and Accidents Technical Information System), MHIDAS (the Major Hazard Incident Data Service), eMARS (Major Accident Reporting System), and ICHEME (Institution of Chemical Engineers). On the other hand, the most used database in the American continent is the NRC (National Response Center).

Fig. 2.1 presents the distribution of the records in the aforementioned databases.

In terms of released substances, Fig. 2.2 shows that hydrocarbons (oil, diesel, and gasoline) are the substance most released when a Natech occurs. These substances can cause explosions, fires, or toxic dispersions.

Finally, some past Natech events show serious human, environmental, and infrastructure consequences. For instance, flood and contamination caused by Hurricane Katrina in the Gulf of Mexico in 2005 (Krausmann & Salzano, 2017), the fire caused by the impact of lightning on a gasoline storage tank in Oklahoma in 2007 (Krausmann & Salzano, 2017), and damage in spherical storage tanks caused by the Tohoku earthquake in 2011 (Krausmann & Cruz, 2017).

According to the United Nations Office for Disaster Risk Reduction in Colombia (United Nations Office for Disaster Risk Reduction UNISDR, 2015) and (Pinelli et al., 2020), the risk of suffering a

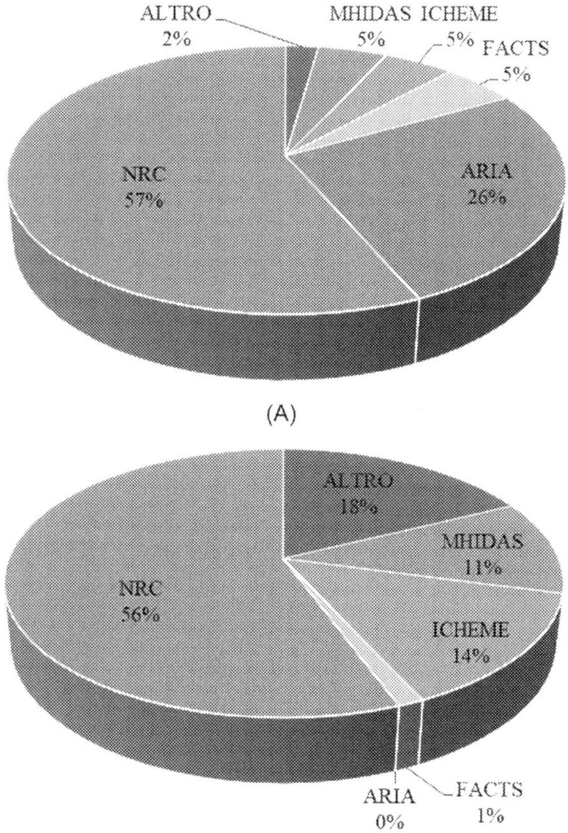


**Figure 2.1** Distribution of the Natech accident events identified in the analysis of the available chemical accident databases: (A) flood events (272 records), (B) seismic events (78 records, 1930–2007) modified from (Campedel, 2008). *Credit: Campedel, M. (2008). Analysis of major industrial accidents triggered by natural events reported in the principal available chemical accident databases. European Commission, Joint Research Centre. <https://ec.europa.eu/jrc/en/publication/eur-scientific-and-technical-research-reports/analysis-major-industrial-accidents-triggered-natural-events-reported-principal-available> Accessed 13.07.21.*

catastrophic loss is at the intersection of natural hazards, vulnerability, and exposure, as shown in Fig. 2.3. The information given in the plot is discussed in detail.

### 2.2.1 Natural hazards

As mentioned by Burton et al. (1978), a natural hazard has an element of human and structural participation. A physical event is an event with no

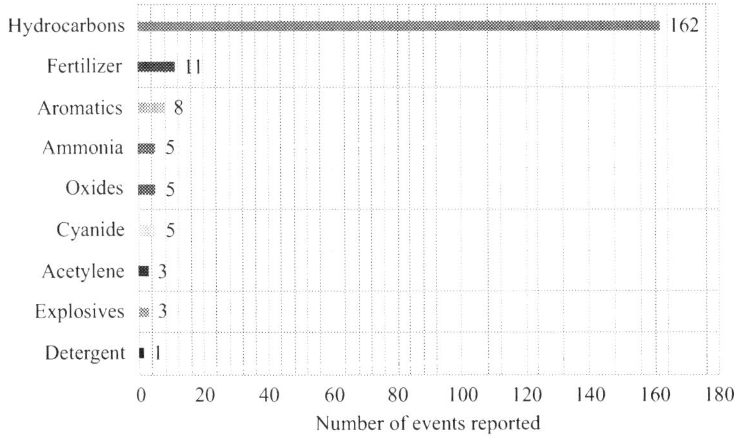


**Figure 2.2** Substances commonly involved in Natech accidents modified from (Campedel, 2008). *Credit: Campedel, M. (2008). Analysis of major industrial accidents triggered by natural events reported in the principal available chemical accident databases. European Commission, Joint Research Centre. <https://ec.europa.eu/jrc/en/publication/eur-scientific-and-technical-research-reports/analysis-major-industrial-accidents-triggered-natural-events-reported-principal-available> Accessed 13.07.21.*

effect either on people or the infrastructure, therefore it is known as a natural event or phenomenon but not as a natural hazard. It becomes a natural hazard when there exist exposure or/and vulnerability (the intersection of Fig. 2.3).

Natural phenomena that occur in large populated or industrial areas are hazardous events, capable of causing a great number of fatalities or incalculable damages to the property, thus resulting in a natural disaster. Therefore, for process safety, in areas where there is no presence of people or industrial facilities, the natural phenomena do not constitute a risk and consequently will not cause a disaster.

Munich (2019) proposed a classification of hazardous natural as geophysical, such as earthquakes, tsunamis, or volcanic activity; meteorological such as tropical cyclones, extratropical storms, convective storms, and hurricanes; hydrological such as floods and mass movements; and climatological such as extreme temperature, drought, and forest fires. In recent years, one of the major concerns of the industry sector worldwide is the increase in the frequency of high-intensity natural events, also due to the climate change effects, coupled with an increase in industrial exposure. As can be seen in Fig. 2.4 (Munich, 2019), the natural phenomena that most frequently affect the planet are those of hydrological and meteorological origin.

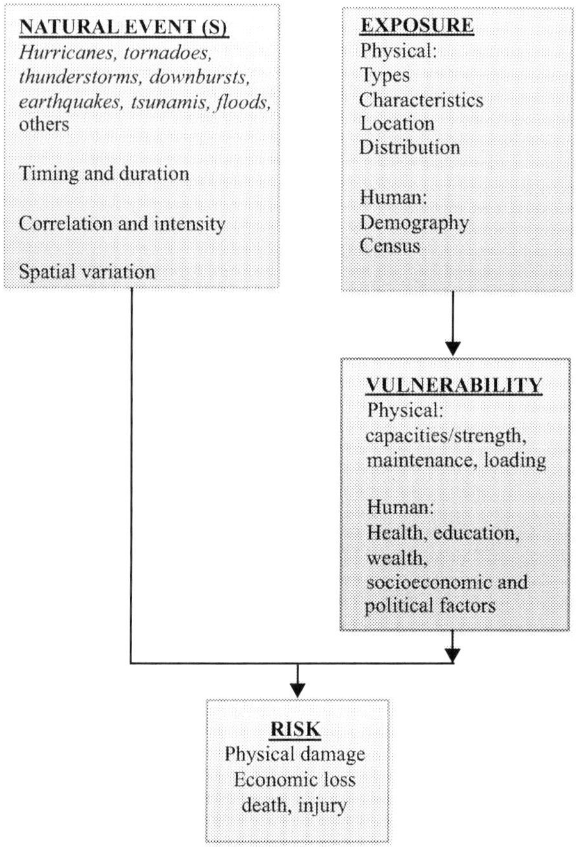


**Figure 2.3** Factors that aggravate or mitigate risk in a Natech event. Modified from (Pinelli et al., 2020). *Credit: Pinelli, J.-P., Esteva, M., Rathje, E.M., Roueche, D., Brandenberg, S., Mosqueda, G., Padgett, J., Haan, F. (2020). Disaster risk management through the designsafe cyberinfrastructure.* International Journal of Disaster Risk Science, 11*(6), 719–734.*

Natural events are characterized by estimating their frequency and severity. These variables allow people to analyze the damage that a natural event could cause. Eq. (2.1) shows that the frequency $f\left(\frac{1}{year}\right)$ of a natural hazard is the inverse of its return period $t_r$, which measures in years the probability of occurrence of a natural event of a certain intensity (Antonioni et al., 2015).

$$f = \frac{1}{t_r} \tag{2.1}$$

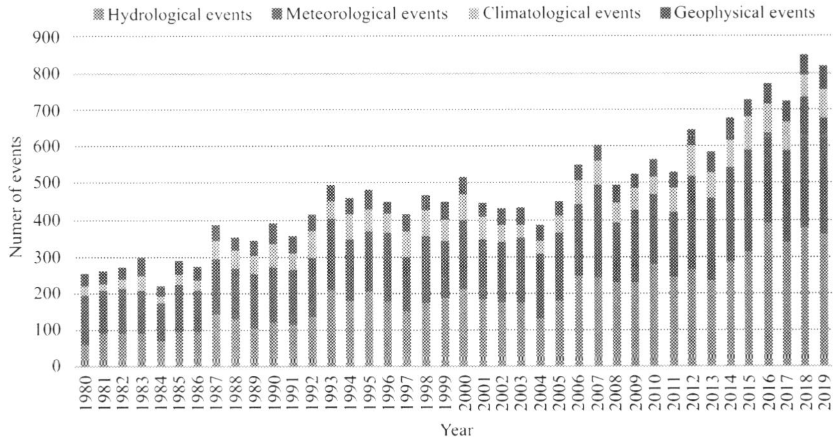


**Figure 2.4** Relevant natural loss events worldwide (1980–2019) (Munich, 2019).

The return period for each natural hazard is usually available from local competent authorities in each country. For instance, the responsible government agency in Colombia for collecting and reporting information related to seismic events is the Colombian Geological System (SGC). The management of scientific information on hydrological and meteorological events is carried out by the Institute of Hydrology, Meteorology and Environmental Studies (IDEAM). Different studies propose regional models for their estimation, including hydrological models for floods (Bryant, 2004; Charlton, 2008), ground-motion models for earthquakes (Malhotra, 2005; Wang, 2008), and meteorological models for wind speeds (Capuano et al., 2017; Lee et al., 2017).

## 2.2.2 Exposure and vulnerability

The exposure measures the number of people or assets located in the area of impact. Theoretically, it might be possible to build infrastructure so strong as to be almost invulnerable. But that is not economically feasible. In practice, a substantial part of the infrastructure will be vulnerable to the hazard.

Vulnerability is associated with physical, socioeconomic, or environmental factors that increase the susceptibility of people or facilities to the consequences of hazards. For the Natech event, the loss of containment of dangerous material will be the one determining factor to estimate the consequences of the event. Other factors associated with these events are population protection (sheltering), meteorological conditions, the amount of hazmat released, the process conditions, and so on.

### 2.2.3 Risk

Risk refers to the relationship between the probability or frequency of the undesirable outcome and a measure of the degree of that undesirable outcome (Porter, 2020). The calculation of risk involves quantities of exposure, hazard, and vulnerability (Porter, 2020).

Taking into account the great threat presented by natural events on industrial facilities, different authors have developed the knowledge for the analysis and management of the risks associated with Natech events (Mesa-Gómez et al., 2020; Mesa-Gómez et al., 2021). Two types of analyses were proposed to understand how Natech events have been studied. The first one is a posteriori analysis, which consists of identifying information that allows characterizing the event from the analysis of historical data or reports of past events (Villalba, 2016). For instance, an analysis of historical data aimed at accidents involving transportation lines, identifying costs associated with a Natech event and volume of material spilled, with the aim of presenting the relevance of Natech events in the risk management (Girgin & Krausmann, 2016). And the second one is a priori analysis, which consists of identifying possible accidental scenarios and analyzing the risks they present (Villalba, 2016). For example, a methodology to assess the risk associated with Natech events due to the impact of lightning on storage tanks (Necci et al., 2016).

Fig. 2.5 presents a timeline with models developed for the estimation of damage to a storage tank by different natural hazards (left side) and the methodologies for risk estimation associated with Natech events (right side). The methodologies presented use damage models for risk assessment.

## 2.3 Storage tanks and strong winds

Many authors agree that storage tanks are the most affected equipment type by extreme winds such as tornadoes, hurricanes, and storms, including downbursts (Burgos et al., 2014). However, for these types of natural events to affect or damage a storage tank, the equipment must be empty or partially full (0% to 10% fill level) because the stored liquid plays an important role in the wind resistance of the tank (Uematsu et al., 2014; Zhao & Lin, 2014). If the tank is empty or partially filled, the net wind forces will damage the tank shell (Uematsu et al., 2014). So, it is convenient to impound sufficient liquid for empty tanks before a

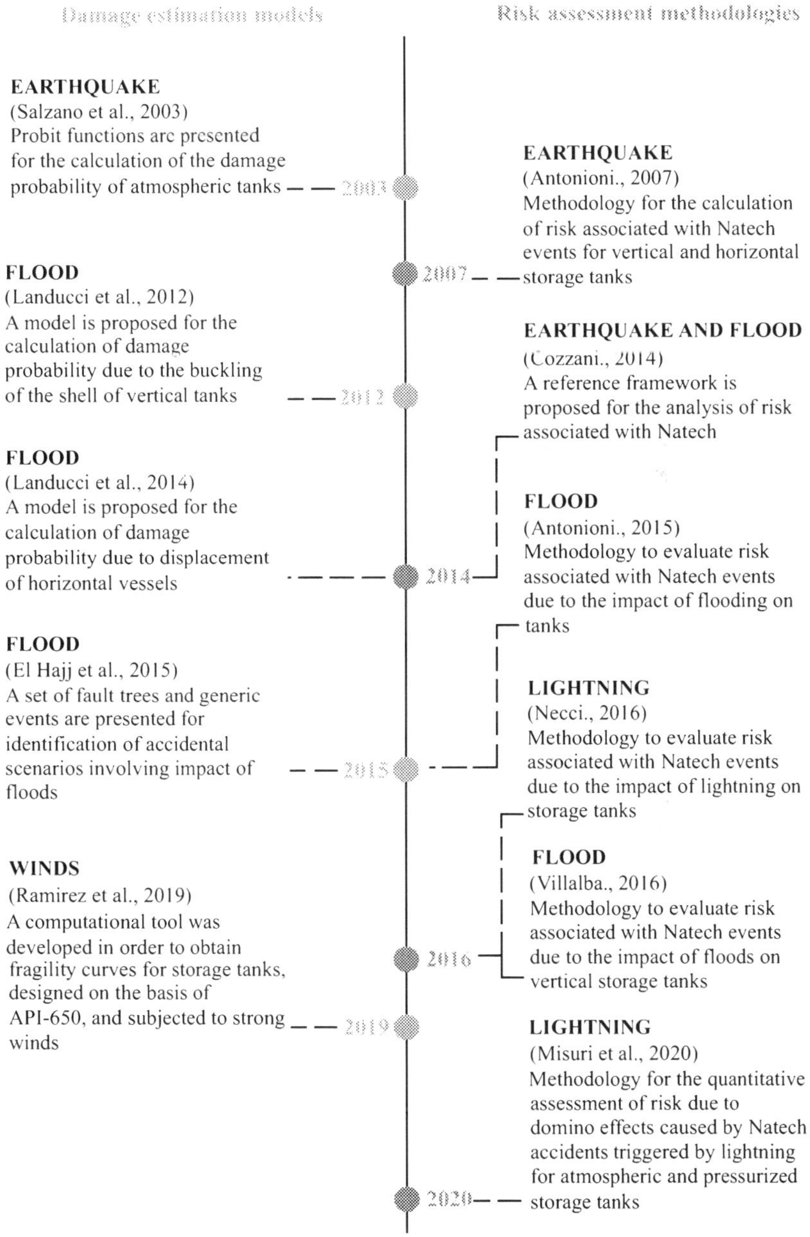


**Figure 2.5** State of the art for the a priori analysis of Natech events (Antonioni et al., 2007, 2015; Cozzani et al., 2014; El Hajj et al., 2015; Landucci et al., 2012, 2014; Misuri et al., 2020; Necci et al., 2016; Ramírez et al., 2020; Salzano et al., 2003; Villalba, 2016).

windstorm to prevent tanks from being damaged due to a strong wind load (Zhao & Lin, 2014). Thus, one relevant final consequence of the Natech event will be the damage or loss of the process equipment (Maraveas et al., 2015) because this condition could lead to a loss of containment of hazmats. It is important to mention that although damage due to strong winds is more frequent for atmospheric storage tanks with low filling degrees, damage can also occur for larger thanks which have a high level of liquid, depending on the tank type (pressurized tanks), geometry (horizontal tanks), size (small tanks), and content (tanks field with gas or other low-density material), but these equipment types are not into the scope of this chapter.

### 2.3.1 Strong winds as hazards

It is important to understand that wind is a random dynamic phenomenon, which varies in time and space (Simiu & Yeo, 2019). Typically, the wind speed is defined by its velocity and its direction. The wind speed profile has a minimum of zero at a ground level and grows with the altitude up to an equilibrium value that defines the boundary layer. The wind speed variation with height typically follows a logarithmic behavior or a power law, although this will not be true for tornadoes and downbursts, which can be also a source of wind-induced Natech accidents. The reference time interval over which the wind speed is averaged defines the magnitude of the velocity. The most common time intervals used to characterize wind loads are 3-s gusts, 1-min sustained winds, or 10 min sustained winds. For the same phenomenon, the longer the averaging period, the smaller the value of the average wind velocity.

Different specifications around the world propose different formulas to compute the wind pressures [e.g., ASCE 7−16 in the United States (American Society of Civil Engineers, 2017)], but in general, they all provide wind maps for specified return periods $t_r$, where the engineer will select a design wind speed depending on location and importance of the facility. The basic wind velocity pressure equations are then adjusted by different factors to take into account issues like topography, terrain exposure, altitude, etc. Nondimensional external pressure coefficients are used to calculate velocity pressures. Internal pressure coefficients, which depend on the openness of the structure also multiply the velocity pressure to yield internal pressures. For low-rise relatively stiff structures like oil tanks, the final result is a distribution of equivalent static design pressures normal to the structure.

**Table 2.1** Extreme winds classification based on intensity or impact vector (Allaby, 2007; Potter & Colman, 2003)

| Wind load | Hurricane category | Hazard classification | Wind speed (km/h) | Storm surge (m) |
|---|---|---|---|---|
| Low load | 1 | Very low | 119–153 | 1.2–1.5 |
| Medium load | 2 | Low | 154.4–177 | 1.8–2.4 |
| High load | 3 | Moderate | 178.5–209 | 2.7–3.6 |
| | 4 | | 210–249 | 3.9–5.4 |
| Very high load | 5 | High | >250 | >5.4 |

Extreme winds characterize storms such as tropical cyclones (hurricanes in the Western Hemisphere; typhoons in the Eastern Pacific; cyclones in the Indian Ocean). These usually generate over warm ocean waters at low latitudes and are particularly dangerous because of their destructive potential. In the Western Hemisphere, hurricanes are classified according to the Saffir/Simpson hurricane scale (Allaby, 2007; Potter & Colman, 2003) (see Table 2.1). Table 2.1 shows that hurricanes have 1 min sustained wind speeds at 10 m, over open waters, equal to or greater than 120 km/h (74 mph).

Because tropical cyclones are rotating storms, the direction of the winds acting upon an industrial facility during a storm will vary depending on the location of the facility concerning the eye. If the facility is in the path of the eye, the wind direction could vary 180 degrees as the eye passes over the facility.

Tornadoes (Simmons et al., 2013) and downbursts (Solari, 2020) can be equally catastrophic. Researchers are actively working on the characterization of the wind loads in these cases. For instance, ASCE 7–16 in the United States includes for the first time some tornado load provisions (Wang & Cao, 2021), and ASCE 7–20 will include tornado risk maps for engineering design.

### 2.3.2 Atmospheric above-ground tanks characterization

This chapter focuses on atmospheric above-ground tanks. As mentioned above, one of the findings by Campedel (2008), is that the equipment mainly affected by natural events is storage tanks. Normally, the consequences of an accident for this type of equipment are quite significant due to the large amount of hazmat stored. The standards API-620 and API-650 (American Petroleum Institute, 2013, 2020) provide the references to characterize and parameterize this type of equipment. These standards

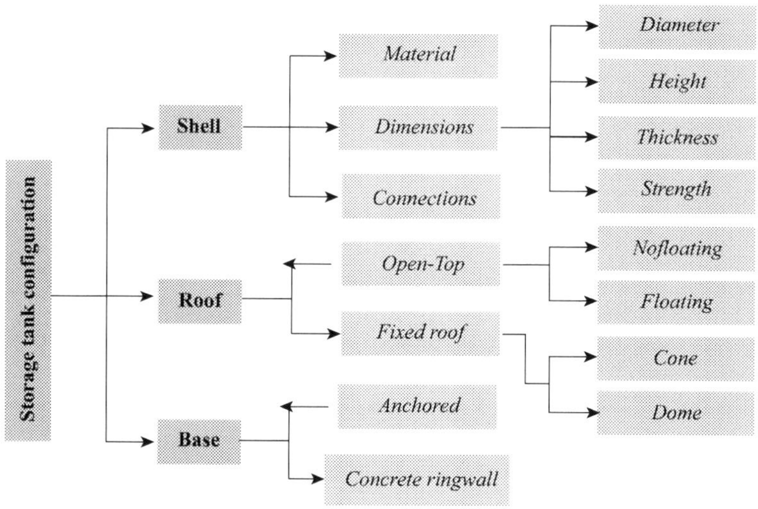


**Figure 2.6** Configuration of a storage tank based on API-620 and API-650 standards.

establish minimum requirements for each of the storage tank components and functionalities for the petrochemical industry. In particular, they set up criteria for the design, construction, inspection, and maintenance of atmospheric storage tanks. Fig. 2.6 presents the main components of a vertical storage tank based on API-650/620.

There is no single criterion to classify storage tanks. Three three main components, the shell of the tank, the type of roof, and the type of base characterize and govern the size of the tank.

#### *2.3.2.1 Storage tank shell*

Atmospheric storage tanks are the most common type of tanks in the chemical and petrochemical industry. These tanks are usually operated at an internal pressure slightly above the atmospheric pressure, no more than 0.5 psig (Myers, 1997a). To parameterize the type of shell that will contain the stored fluid, the following characteristics are needed: shell material, number and type of connections, diameter, height, and thickness of the tank. Each material has its yield strength and tensile strength. To establish the thickness of the tank, API-650 proposes the following expressions in terms of the height and diameter of the tank (American Petroleum Institute, 2020). A course is an individual cylinder that is part of a tank. Two or more courses are joined together to make a tank.

For the first course:

$$t_1 = \left(1.06 - \frac{0.0696D}{h}\sqrt{\frac{hG}{S_d}}\right)\left(\frac{4.9hDG}{S_d}\right) + \mathrm{CA} \tag{2.2}$$

Where:

- $t_1$ is the bottom-course thicknesses (mm),
- $h$ the design liquid level (m),
- $D$ is the nominal tank diameter (m),
- $G$ is the design specific gravity of the liquid to be stored (-),
- CA is the corrosion allowance (mm), and
- $S_d$ is the allowable stress for the design condition (MPa).

For the second course:

$$t_2 = \begin{cases} t_1 \rightarrow if \quad \dfrac{h_1}{(rt_1)^{0.5}} < 1.375 \\ t_{2a} \rightarrow if \quad \dfrac{h_1}{(rt_1)^{0.5}} > 2.625 \\ t_{2a} + (t_1 - t_{2a})\left[2.1 - \dfrac{h_1}{1.25(rt_1)^{0.5}}\right] \rightarrow if \quad 1.375 \geq \dfrac{h_1}{(rt_1)^{0.5}} \leq 2.625 \end{cases} \tag{2.3}$$

Where:

- $t_2$ is the second-course thicknesses excluding any corrosion allowance (mm),
- $h_1$ is the height of the bottom shell course (mm),
- $r$ is the nominal tank radius (mm), and
- $t_{2a}$ is the corroded thickness of the second shell course (mm).

For upper courses:

$$t_i = \frac{4.9D\left(h - \frac{x}{1000}\right)G}{S_d} + CA \tag{2.4}$$

Where $x$ is the distance of the variable design point from the bottom of the course (m).

Eqs. (2.2)–(2.4) allow characterizing the shell of the tank, from geometrical characteristics and the type of tank material.

#### 2.3.2.2 Storage tank roof

There are two types of tank roofs: fixed roof and floating roof. As shown in Fig. 2.7, the tanks can be opened at the top or closed by a fixed roof:

Cone-roof tanks typically have roof rafters and support columns except in very small-diameter tanks (Fig. 2.7B) (Myers, 1997b). According to API-650, the angle of inclination $\theta$ of the roof should be between 9.5 and 37 degrees (slope 2:12 to 9:12).

Eq. (2.5) estimates the nominal thickness for the cone roof.

$$t_{rc} = \frac{D}{4.8\,sin(\theta)}\sqrt{\frac{T}{2.2}} + CA \tag{2.5}$$

Where:

- $t_{rc}$ is the nominal thickness for the cone roof (mm),
- $D$ is the nominal tank diameter (m),
- $\theta$ is the angle of the cone elements to the horizontal (°),
- $T$ is a parameter for load combinations (-), and
- CA is the corrosion allowance (mm).

The dome-roof tanks are similar to tanks with a cone roof, but their shape is similar to an umbrella. These are usually of a size no larger than 20 m in diameter. Unlike tanks with a conical roof, these can be self-supporting structures (Fig. 2.7C). According to API-650, the radius of the dome should be between 0.8$D$ and 1.2$D$.

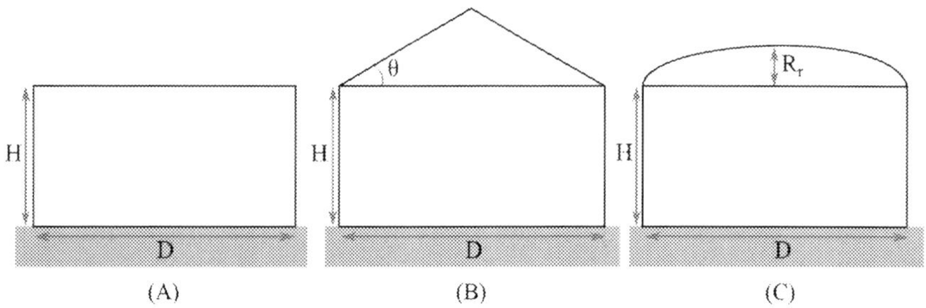


**Figure 2.7** The basic configuration of a storage tank. (A) open-top tank, (B) cone-roof tank, (C) dome-roof tank.

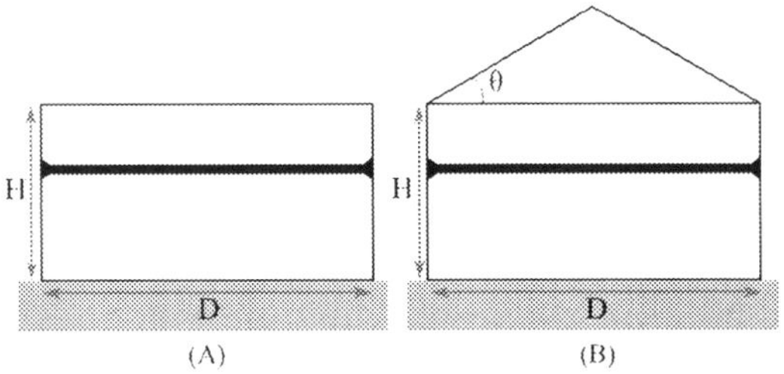


**Figure 2.8** Types of floating roof in a storage tank: (A) external floating roof, (B) internal floating roof.

Eq. (2.6) estimates the nominal thickness of the dome roof.

$$t_{rd} = \frac{R_r}{2.4}\sqrt{\frac{T}{2.2}} + \text{CA} \tag{2.6}$$

Where:

- $t_{rd}$ is the nominal thickness for the dome roof (mm),
- $R_r$ is the roof radius (m),
- $T$ is a parameter for load combinations (-), and
- CA is the corrosion allowance (mm).

All the floating roofs are inside the storage tanks, where they float on the surface of the stored liquid. This cover is a disk-shaped structure that has sufficient buoyancy to ensure that the roof will float. Tanks with a floating roof and without a fixed roof, are known as an external floating roof (Fig. 2.8A), on the other hand, tanks with both a floating roof and a fixed roof, are internal floating roof (Fig. 2.8B).

#### *2.3.2.3 Storage tank base*

Storage tanks have an additional resistance factor in their base. A tank can be anchored or unanchored to the ground, to avoid displacement of the equipment in case of suffering an external lateral load. Additionally, a foundation concrete ring might prevent the tank from sinking into the soil.

Fig. 2.9 presents a detailed schematic of the anchorage of a storage tank. The information required for analyzing the effects of strong winds on storage tanks is the number of tank anchor bolts, their diameter, and the type of material.

### 2.3.3 Definition of possible accidental scenarios

The natural hazard characteristics and the structural configuration of the storage tank govern possible accident scenarios that may occur due to the impact of the hazard on the process equipment.

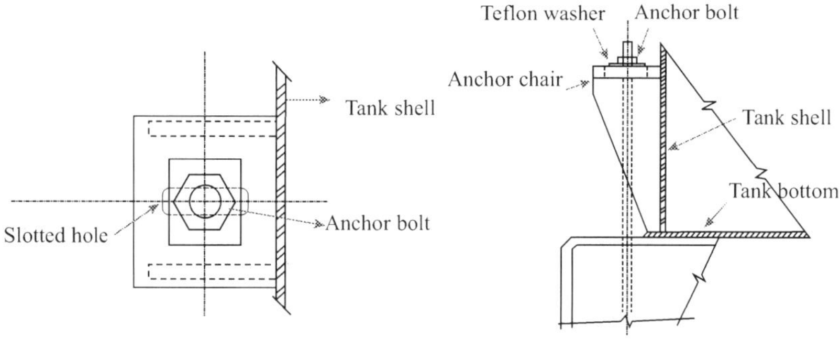


**Figure 2.9** Tank anchor detail.

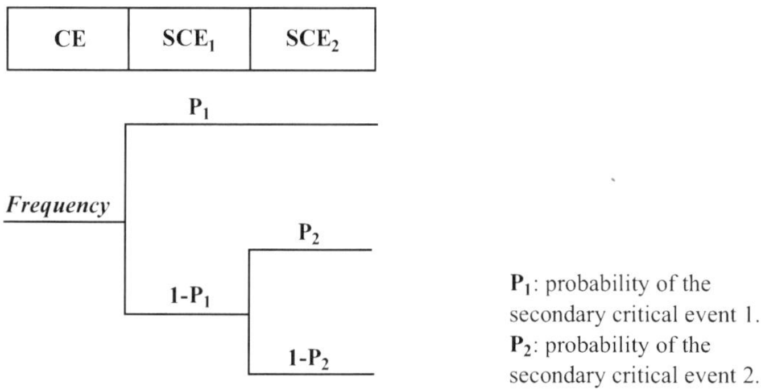


**Figure 2.10** Structure of the event tree. Modified from (Ocampo, 2016). *Credit: Ocampo, F. (2016). Marco para el tratamiento de incertidumbre en el análisis de riesgo cuantitativo en transporte de material peligroso a través de tuberías. In Departamento de Ingeniería Química: Master's. Universidad de los Andes, Bogotá, Colombia.*

The event tree method is used to define the possible final events that extreme winds on a vertical storage tank can trigger. ARAMIS (Salvi et al., 2002) proposes a guide to develop event trees, which are adapted to Natech events caused by natural phenomenon. The objective of the events tree is to identify the possible consequences of a critical event or event (CE), which usually represents the failure of a component or an external failure. Thus, the first step is to define the CE. Then, the sequence of events and security barriers after the CE are identified. These types of events are called secondary critical events (SCE), while the events at the end of each branch are known as final events or major events (FE) (Ocampo, 2016). Final events are defined as the significant effects produced by secondary events capable of affecting people, structures, and the environment. Fig. 2.10 shows a representation of an event tree and its elements.

For the case of storage tanks impacted by strong winds, the selected critical event (CE) is the impact of a wind load on a tank. Regarding the secondary critical events (SCE), their selection depends on the consequences of the impact of a natural hazard on a vertical storage tank. Some authors (Campedel, 2008; Cozzani et al., 2010) present a historical data analysis where they identify different types of structural damage that storage tanks can suffer during a natural phenomenon. These analyzes identified three types of secondary critical events: damage modes, failure modes due to damage, and release modes (LOC).

The storage tank damage modes are shell buckling, displacement or sliding, floatation, overturning, and impact by debris. Not all damage modes apply to all-natural hazards. For strong winds, buckling, overturning, and debris impact are the most common. Table 2.2 presents the types of damage from strong winds and their characterization.

Five different failure modes can result from the damage: the collapse of the structure, total connection failure, partial connection failure, shell rupture, and failure of the tank's roof. These failure modes are common when a strong wind load or debris impacts a storage tank. Fig. 2.11 shows a generic event tree to identify the consequences of a Natech event in storage tanks.

**Table 2.2** Type of damage produced by strong winds

| Type of damage | Solicitation | Storage tanks resistance |
|---|---|---|
| Buckling | Wind pressure ($q_{eq}$) | Resistance pressure ($P_r$) |
| Overturning | Stability factor ($J$) | – |
| Debris impact | Depth penetration ($D_p$) | Thickness ($t$) |
| | Impact force ($F_i$) | Resistance force ($F_r$) |

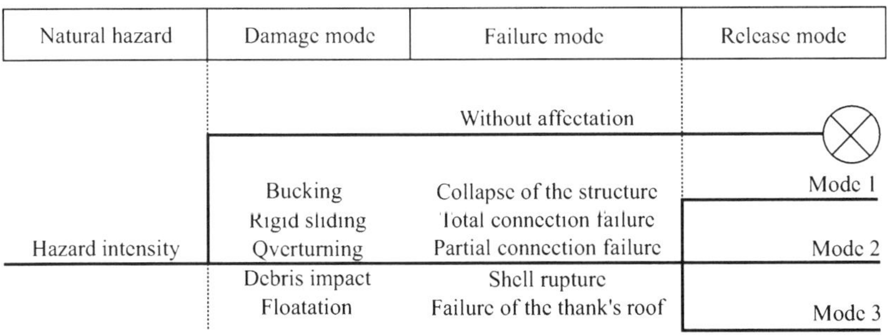


**Figure 2.11** Event tree for the sequence of events due to the impact of a natural event on vertical storage tanks.

Once one of the possible failure modes happens, the loss of containment of hazardous material occurs. Three release modes can result from this secondary critical event, which will be part of the spill volume estimation phase and depend on the typology of failure. Guidelines and international standards have established release modes for different types of process equipment (National Institute of Public Health and the Environment RIVM, 2009; Uijt de Haag & Ale, 1999; van den Bosch & Weterings, 2005). Based on these studies, the release modes for atmospheric storage tanks are:

- Release Mode 1: Instantaneous release of entire contents.
- Release Mode 2: Release of entire contents in 10 min in a continuous and constant stream.
- Release Mode 3: Continuous release from a hole with an effective diameter of 10 mm.

Fig. 2.12, shows the event tree for the identification of consequences or final accidental scenarios (FE) associated with the impact of a wind load, which involve the elements presented above.

## 2.3.4 Structural and natural hazard analysis

Fig. 2.13 presents a simple outline of the scenario to be evaluated when a natural hazard damages a storage tank. An extreme natural event can exert an external load or solicitation (either by pressure, debris impact, or movement) of such magnitude that, when impacting any type of structure, the solicitation could exceed the resistance force to which it was designed, causing some type of structural damage.

Among the most common damages and failure modes caused by a natural hazard on a tank are the shell buckling, the sliding or floating of the tank, damage to the tank foundation, overturning, impact by debris, detachment of pipes, and damage to the bottom plate by buckling due to uplifting.

The next sections present the mathematical models that several authors propose to determine the possibility of different types of damage.

### *2.3.4.1 Storage tanks damaged by strong winds*

The models below estimate extreme winds damage to a storage tank. Fig. 2.14 shows the different types of damage that high wind speeds and debris impact can cause to a storage tank.

Vertical storage tanks with a cylindrical shape are equipment with the capacity to store large quantities of different materials, such as crude, fuel,

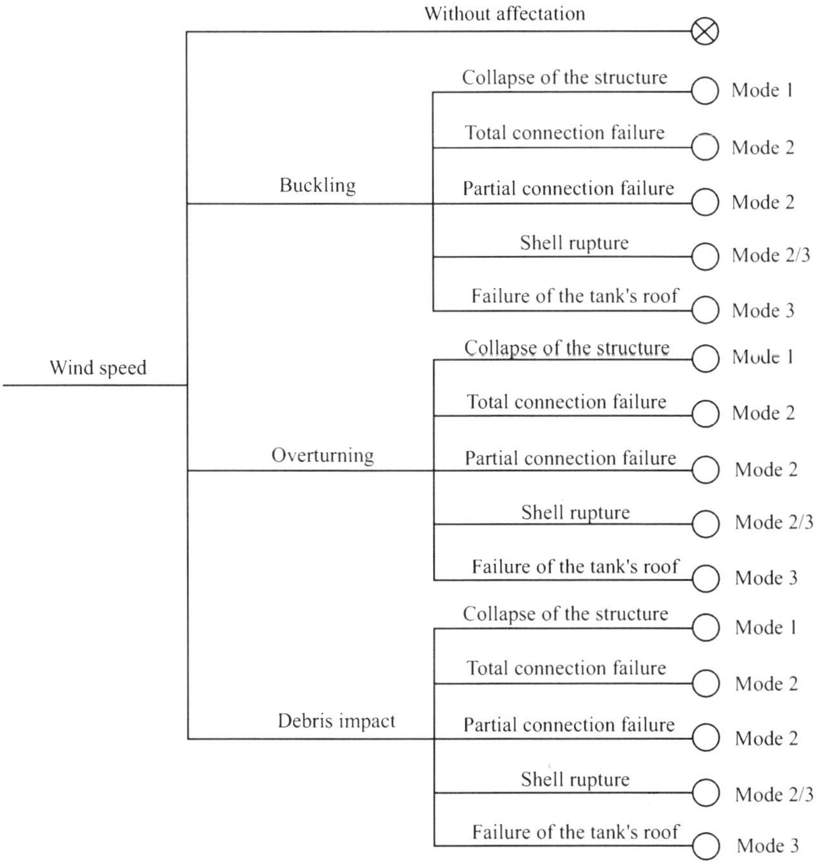


**Figure 2.12** Event tree to identify the sequence of the events of a storage tank impacted by a wind load depending on the wind speed.

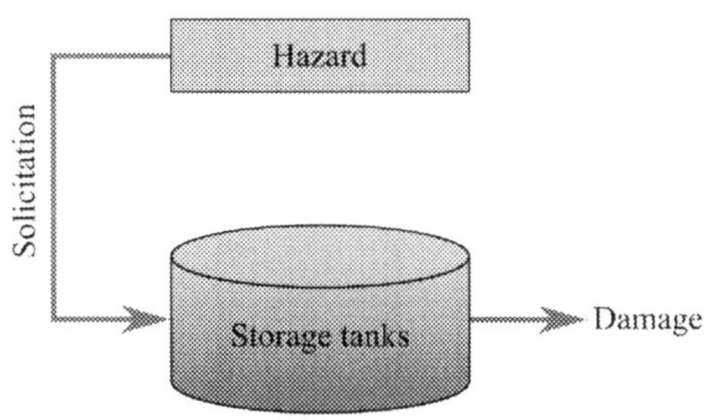


**Figure 2.13** Damage due to natural hazard.

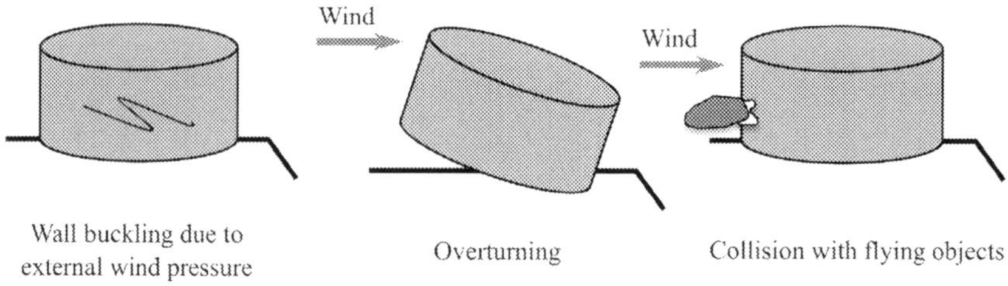


**Figure 2.14** Types of damage to a storage tank exposed to high wind speeds.

and chemicals. They are welded and have a structure with very thin walls with long diameters and heights. The prevalent types of damage are the buckling of the walls due to the external wind pressure and the damage of the tank shell due to the impact of projectiles on the wind drags.

#### 2.3.4.1.1 Shell buckling

Damage due to buckling of the shell usually occurs when the tanks are empty or partially full because the internal liquid pressure balances external pressure due to the wind (Uematsu et al., 2014; Zhao & Lin, 2014). This is why, in most cases, the failure can result in a substantial loss of financial resources but a limited spill of hazmat (Maraveas et al., 2015).

Fig. 2.15 shows the pressure balance over the storage tank, between the resistance pressure of the tank and the wind external pressure acting over the tank. In this case, $q_{eq}$ (Pa) is the equivalent uniform external pressure (for details see Fig. 2.17).

Eq. (2.7) gives the resistance pressures $P_r$, which is the sum of the pressure from the stored fluid $P_f$, as per Eq. (2.8), and the material resistance pressure of the tank $P_{cr}$, as per Eq. (2.9) (Timoshenko & Gere, 2012). The latter depends on the mechanical properties of the tank material.

$$P_r = P_f + P_{cr} \tag{2.7}$$

$$P_f = \rho_f g H \Phi \tag{2.8}$$

$$P_{cr} = \frac{2Et}{D}\left(\frac{1}{(n^2-1)\left(1+\left(\frac{2nH}{\pi D}\right)^2\right)} + \frac{t^2}{3D^2(1-\nu^2)}\left(n^2-1+\frac{2n^2-1-\nu}{1+\left(\frac{2nH}{\pi D}\right)}\right)\right) \tag{2.9}$$

Where:

- $P_f$ is the stored fluid density $\left(\frac{kg}{m^3}\right)$,
- $g$ is the gravity $\left(\frac{m}{s^2}\right)$,

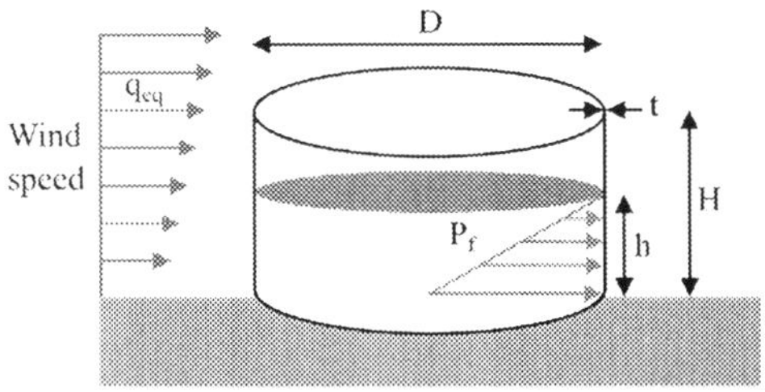


**Figure 2.15** Schematic of the load-resistance forces considered for shell buckling produced by wind.

- $E$ is the modulus of elasticity (Pa),
- $t$ is the shell thickness (mm),
- $\Phi$ is the tank fill level (-),
- $H$ is the height of the tank (m),
- $D$ is the nominal tank diameter (m),
- $n$ is a parameter to minimize critical pressure (-), and
- $\nu$ is the Poisson coefficient (-).

Furthermore, according to international standards, such as the American Petroleum Institute (API-650), American Society of Civil Engineers (ASCE-7), and European Standard (EN 1991-1-4 and EN1993-1-6) the model for the calculation of wind pressure takes into account the type of exposure of the affected structure. The following expressions (American Petroleum Institute, 2020; American Society of Civil Engineers, 2017; European Committee for Standardization, 2005) define the design wind pressure.

Equation (2.10) provides the velocity pressure $q_z$, evaluated at height $z$:

$$q_z = 0.00256 K_z K_{zt} K_d V^2 I G_s \left[\frac{\text{lb}}{\text{ft}^2}\right] \tag{2.10a}$$

$$q_z = 0.613 K_z K_{zt} K_d V^2 I G_s \left[\frac{\text{N}}{\text{m}^2} = \text{Pa}\right] \tag{2.10b}$$

Where:

- $K_z$ is the velocity pressure exposure coefficient (1.04 for open terrain exposure C at a height of 12 m), Exposure C is defined as open terrain with scattered obstructions having heights less than 30 ft. This category includes flat open country and grasslands.
- $K_{zt}$ is the topographic factor (1.0 for all structures except those on isolated hills or escarpments),

- $K_d$ is the wind directionality factor (0.95 for round tanks),
- $V$ is the 3-sec gust wind speed at 10 m for open terrain exposure (exposure C) (mph or m/s),
- $I$ is an importance factor (1.0 for category II structures), and
- $G_s$ is the gust factor (0.85 for exposure C).

Eq. (2.11) provides the wind load or wind design pressure $p$ (in Pa or $\frac{\text{lb}}{\text{ft}^2}$) on the structure surfaces over a storage tank (Uematsu et al., 2014; Zhao & Lin, 2014):

$$p = C_p q_z \tag{2.11}$$

Where $C_p$ is the wind pressure coefficient.

For cylindrical tanks, the wind design pressure ($p$) usually varies both along the circumference and the height. Zhao and Lin (2014) established that the variation in height is not as pronounced compared to the variation in the circumference. Therefore, the assumption is that the variation of the pressure coefficients is constant along the height and only varies with the longitude, (see Fig. 2.16). To estimate the wind pressure coefficients, several authors and design codes have proposed an expression (Eq. 2.12) based on Fourier series decomposition. Table 2.3 shows the representative Fourier coefficients proposed by some authors (Zhao & Lin, 2014):

$$C_p(\theta) = \sum_{i=0}^{m} a_i \cos(i\theta) \tag{2.12}$$

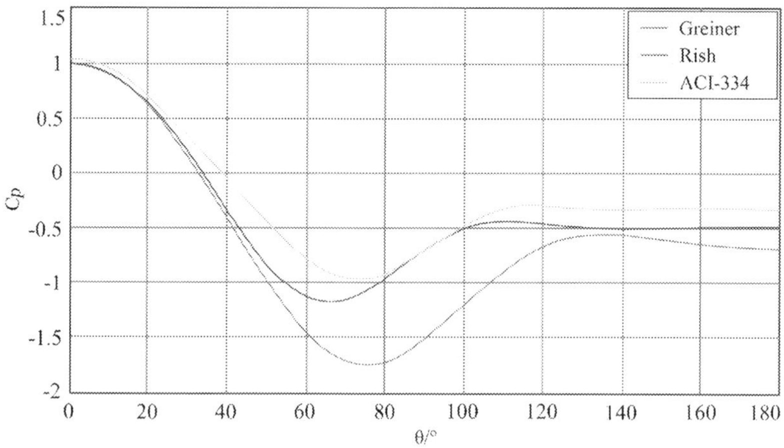


**Figure 2.16** Extremal wind pressure coefficients along the circumference of cylinders. Modified from (Zhao & Lin, 2014). *Credit: Zhao, Y., & Lin, Y. 2014. Buckling of cylindrical open-topped steel tanks under wind load.* Thin-Walled Structures, *79, 83–94.*

**Table 2.3** Fourier coefficients proposed by different authors for tanks with close-tops

| Parameter | Author | | | |
|---|---|---|---|---|
| | Greiner | Rish | ACI-334 | EN 1993–4-1 |
| $\alpha_0$ | −0.65 | −0.387 | −0.2636 | −0.54 + 0.16(D/H) |
| $\alpha_1$ | 0.37 | 0.338 | 0.3419 | 0.28 + 0.04(D/H) |
| $\alpha_2$ | 0.84 | 0.533 | 0.5418 | 1.04 + 0.20(D/H) |
| $\alpha_3$ | 0.54 | 0.471 | 0.3872 | 0.36 + 0.05(D/H) |
| $\alpha_4$ | −0.03 | 0.166 | 0.0525 | −0.14 + 0.05(D/H) |
| $\alpha_5$ | −0.07 | −0.066 | −0.0771 | |
| $\alpha_6$ | | −0.055 | −0.0039 | |
| $\alpha_7$ | | | 0.0341 | |

Where:

- $\theta$ is the longitude measured from windward, and
- $a_i$ is the Fourier coefficient.

The Fourier coefficients in Table 2.3 are only for tanks with closed-tops, and therefore no wind internal pressure is included. For tanks with an open-top, a uniform negative wind pressure coefficient should be included to take into account the internal suction, as per Eq. (2.13).

$$C_p = \begin{cases} -0.8 \rightarrow \dfrac{H}{D} \geq 2 \\ -0.5 \rightarrow \dfrac{H}{D} \leq 1 \end{cases} \tag{2.13}$$

The nonuniform distribution of pressure $p$ resulting from external wind loading on cylindrical tanks may, for shell buckling design, be substituted by an equivalent uniform external pressure $q_{eq}$ (Pa) as shown in Fig. 2.17 (European committee for standardization, 2005), estimated through Eq. (2.14):

$$q_{eq} = k_w p_{max} \tag{2.14}$$

Where

- $k_w$ is a wind direction factor (-), and
- $p_{max}$ is the maximum nonuniform pressure (Pa).

This is the maximum positive pressure from the wind, as can be seen in Fig. 2.17.

$$k_w = 0.46\left(1 + 0.1\sqrt{\frac{(C_\theta r)}{(\omega t)}}\right) \tag{2.15}$$

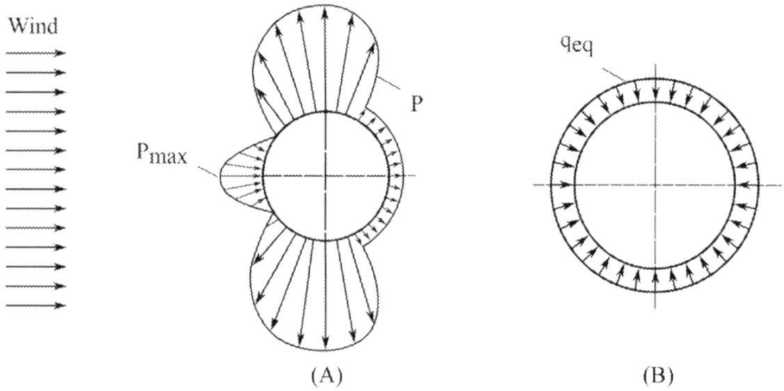


**Figure 2.17** (A) wind pressure distribution around shell circumference, (B) equivalent axisymmetric pressure distribution around shell circumference. Modified from (European committee for standardization, 2005). *Credit: European Commitee for Standardization (2005) Eurocode 1: actions on structures—Part 1–4: general actions—wind actions. European Commitee for Standardization.*

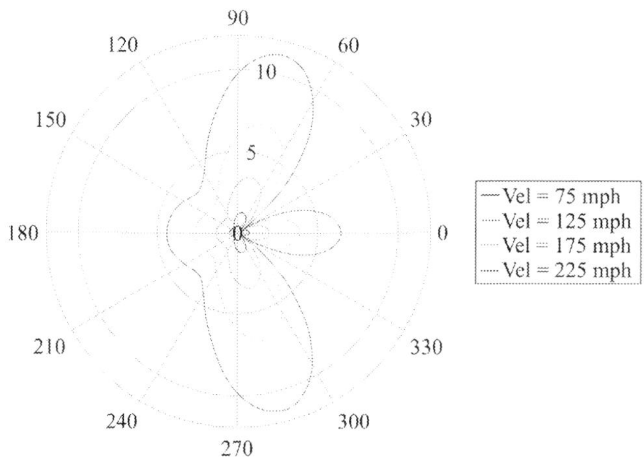


**Figure 2.18** Wind pressure distribution around shell circumference at different velocities.

Where:

- $C_\theta$ is an external buckling factor for medium-length cylinders (-),
- $\omega$ is a relative length parameter for the shell (-),
- $r$ is the nominal tank radius (m), and
- $t$ is the thickness of the tank (m).

Fig. 2.18 shows the nonuniform profile of wind pressure for different wind velocities. The direction of the wind corresponds to the

0 degree angle. Table 2.4 shows the corresponding uniform external equivalent pressures.

$$\text{Damage} = \begin{cases} if\ q_{eq} - P_r > 0 & Buckling \\ if\ q_{eq} - P_r \leq 0 & No\ buckling \end{cases} \tag{2.16}$$

When extreme winds affect the tank, the balance between the wind loads acting on the tank (Eq. 2.14) and the resistance pressure of the tank (Eq. 2.7), determines if the equipment will suffer damage by buckling or deformation of its shell (Eq. 2.16).

#### 2.3.4.1.2 Overturning

International entities have collected information about storage tanks affected by extreme wind. One of the most recent cases is Hurricane Katrina, which produced winds up to 280 km/h, and had the potential to overturn an onshore tank. This type of damage is the least likely to occur when the tank is empty and without anchoring. However, the API-650 standard establishes various stability criteria for a specific wind load.

This chapter analyses overturning for storage tanks without anchorage to the ground. The API-650 standard establishes stability criteria for tanks without anchoring, see Fig. 2.19.

**Table 2.4** Equivalent axisymmetric pressure at different wind velocities

| Wind speed (mph) | $q_{eq}$ (Pa) |
|---|---|
| 75 | 0.7135 |
| 125 | 1.9820 |
| 175 | 3.8846 |
| 225 | 6.4215 |

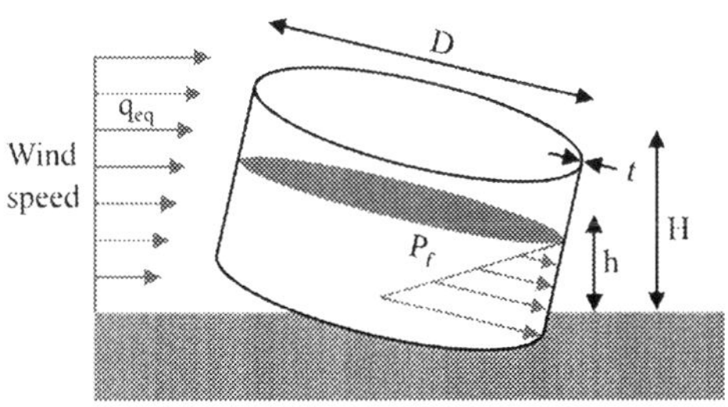


**Figure 2.19** Schematic of load-resistance forces considered the overturning by a wind load.

The stability criteria for overturning due to an external wind load on a non-anchored tank are represented by Eqs. (2.17)–(2.18):

$$0.6M_w + M_{pi} < \frac{M_{DL}}{1.5} + M_{DLR} \quad \rightarrow \quad F_{o1} < F_{r1} \tag{2.17}$$

$$M_w + F_p(M_{pi}) < \frac{(M_{DL} + M_F)}{2} + M_{DLR} \quad \rightarrow \quad F_{o2} < F_{r2} \tag{2.18}$$

Where:

- $F_p$ is a pressure combination factor,
- $M_{pi}$ moment about the shell-to-bottom joint from design internal pressure,
- $M_w$ overturning moment about the shell-to-bottom joint from horizontal plus vertical wind pressure,
- $M_{DL}$ moment about the shell-to-bottom joint from the nominal weight of the shell and roof structural supported by the shell that is not attached to roof plate,
- $M_F$ moment about the shell-to-bottom joint from the liquid weight,
- $M_{DLR}$ moment about the shell-to-bottom joint from the nominal weight of the roof plate plus any attached structural.
- $F_{oi}$ is the overturning forces produced by the wind on the tank, and
- $F_{ri}$ is the resistance force of the tank.

Through Eqs. (2.17)–(2.18), it is possible to establish a relationship between the overturning forces ($F_{oi}$) produced by the wind on the tank at the time of being affected by extreme winds and the resistance force of the tank ($F_{ri}$), to determine if the equipment will suffer damage by overturning, Eq. (2.19).

$$\text{Damage} = \begin{cases} \textit{if } F_{o1} - F_{r1} > 0 \text{ and } F_{o2} - F_{r2} > 0 \text{ Overturn} \\ \textit{if } F_{o1} - F_{r1} \leq 0 \text{ or } F_{o2} - F_{r2} \leq 0 \text{ No overturn} \end{cases} \tag{2.19}$$

#### 2.3.4.1.3 Debris impact

Hurricanes, downbursts, or tornados have great potential for destruction. The waste from the destruction turns into debris or flying projectiles with the potential to impact other structures and cause considerable damage (Pathirana et al., 2017). The area covered by dangerous winds can be significantly wide (depending on the meteorological phenomenon), which results in a domino effect on other structures, with debris from upstream structures impacting downstream structures.

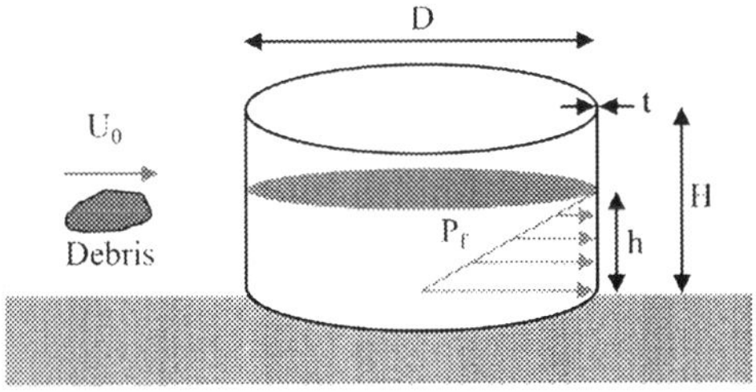


**Figure 2.20** Schematic of load-resistance forces considered for the impact of debris drag by the wind.

Extreme winds can drag objects (debris) which represent a hazard to the integrity of a storage tank. An object dragged by the wind can carry enough kinetic energy to damage the components of a storage tank. A balance of forces calculates the intensity of debris impact on a tank. The model varies according to the debris characteristics and wind conditions.

Fig. 2.20 shows that the wind kinetic energy, which drives the debris, increases with the wind speed. When debris impacts the tank, the damage depends on both the debris kinetic and the tank resistance. Salzano and Basco (2015) propose a different methodology to evaluate the vulnerability of a storage tank based on the severity of the impact, determined by Johnson's number $J'$ and the depth of penetration $h_p$ by the impact. This methodology relates details and information about the process equipment, the impact object, and the impact speed $U_0$. In impact dynamics, Eq. (2.20) for Johnson's number estimates the severity of the impact on a continuum with impulsively loads with an initial velocity pulse.

$$J' = \frac{U_0^2 M}{\sigma_D t r_p^2} \tag{2.20}$$

Where:

- $U_0$ is the speed of the debris $\left(\frac{m}{s}\right)$,
- $M$ debris mass (kg),
- $\sigma_D$ dynamic yield stress (Pa),
- $t$ course shell thickness of the tank (m), and
- $r_p$ debris radius (m).

Table 2.5 shows the range of values of Johnson's number with the corresponding regimes. To evaluate the damage from the impact of an object in a storage tank, Lees (2004) modified Johnson's number as:

On the other hand, Lin et al. (2005) proposed a methodology to perform a risk assessment for urban structures impacted by flying objects dragged by the wind. This chapter applies the methodology proposed by

**Table 2.5** Threshold values for damage for Johnson's damage number J′ (Salzano & Basco, 2015)

| J | Regime | Probability of damage |
|---|---|---|
| $1 \times 10^{-3}$ | Quasi-static elastic | 0 |
| $1 \times 10^{-2}$ | Moderate plastic behavior | 0.1 |
| $1 \times 10^{+1}$ | Extensive plastic deformation | 0.5 |

Lin to a vertical storage tank. The determining factor for estimating whether an object can buckle or penetrate a storage tank will be the aerodynamic force on a static debris object in a wind field $F_{ob}$ ($N$), which Eq. (2.21) calculates from the physical properties of the object and the impact velocity (Lin et al., 2005).

$$F_{ob} = \frac{1}{2}\rho_w U_0^2 A_p C_F \tag{2.21}$$

Where:

- $\rho_w$ is the wind density $\left(\frac{\text{kg}}{\text{m}^3}\right)$,
- $A_p$ is the reference debris area ($\text{m}^2$). The surface area of the object that could impact the tank, and
- $C_F$ is an aerodynamic force coefficient (-).

Eq. (2.21) indicates that as wind speed increases, the aerodynamic force on the debris increases. If the debris is unattached, the debris is picked up when the aerodynamic force is greater than the debris gravity force ($F_{ob} > mg$) (Lin et al., 2005). Eq. (2.22) yields the speed at which debris starts moving. Since $mg = A_p h \rho_p g$.

$$U_0^2 = \frac{2h_{ob}\rho_{ob} g I_S}{\rho_w C_F} \tag{2.22}$$

Where:

- $h_{ob}$ is a debris characteristic dimension ($m$), it represents the typical dimension of compact-like objects, the thickness of plate/sheet objects, or the thickness of slender cylinders,
- $\rho_{ob}$ is the density of the debris material $\left(\frac{\text{kg}}{\text{m}^3}\right)$,
- $\rho_w$ is the wind density $\left(\frac{\text{kg}}{\text{m}^3}\right)$,
- $g$ is the gravity $\left(\frac{\text{m}}{\text{s}^2}\right)$,
- $C_F$ is an aerodynamic force coefficient (-), and
- $I_S$ is a fixed strength integrity parameter, calculated as the ratio between the wind force required to overcome the friction force, divided by debris weight.

Finally, once the debris impact damages the tank, a possible and useful way to validate Johnson's damage number is by calculating the penetration depth $h_p$ (m) of an object from its impact parameters. It should be noted that for industrial accidents, the penetration depth on the tank shell from the impact of a flying projectile is an important factor of reference to evaluate a loss of containment of industrial equipment. If $h_p$ exceeds the thickness $t$ of the affected equipment, the unwanted release of the stored hazmat will occur. Lees' textbook reports a simplified approach to estimating $h_p$ in terms of minimum thickness (Lees, 2004).

$$h_{p,small} = k_s M^a U_0^b \quad M \leq 1\text{kg} \tag{2.23}$$

$$h_{p,large} = k_L \frac{M}{A_p} \log\left(1 + 5 \times 10^{-5} U_0^2\right) \quad M > 1\text{kg} \tag{2.24}$$

Where

- $M$ debris mass (kg),
- $a$ and $b$ are constants that depend on the target material (-),
- $k_s$ and $k_L$ are constants for small and large debris respectively,
- $A_p$ is the reference debris area ($m^2$). The surface area of the object that could impact the tank, and
- $U_0$ is the speed of the debris $\left(\frac{m}{s}\right)$.

Equations (2.23) and (2.24) show that the model for calculating $h_p$ does not take into account the characteristics of the affected process equipment. Table 2.6 presents the parameters for Eqs. (2.23) and (2.24).

Nguyen et al. (2009) proposed a more robust model for the calculation of the penetration depth $h_p$ (m) of a projectile. Fig. 2.21 shows the penetration scheme of a rod projectile, where $t$ is the target thickness, $h_p$ is the penetration depth, $d_p$ is the fragment diameter, $h_{ob}$ is the fragment length, $U_o$ is the fragment velocity, $\beta$ is the fragment inclination, and $e_{cr}$ is the critical thickness plate (Nguyen et al., 2009).

The standard API-620 establishes minimum thicknesses according to the diameter of the tank. This value (Table 2.7) will be assumed as the critical thickness ($e_{cr}$) for the shell of the storage tank.

**Table 2.6** Constant values for fragment penetration reported in Lee's textbook (Lees, 2004)

| Target material | $k_S$ | $k_L$ | a | b |
|---|---|---|---|---|
| Concrete | $1.8 \times 10^{-5}$ | $1 \times 10^{-3}$ | 0.4 | 1.5 |
| Steel | $6.0 \times 10^{-5}$ | $5 \times 10^{-5}$ | 0.3 | 1 |
| Brickwork | $2.3 \times 10^{-5}$ | $2.5 \times 10^{-3}$ | 0.4 | 1.5 |

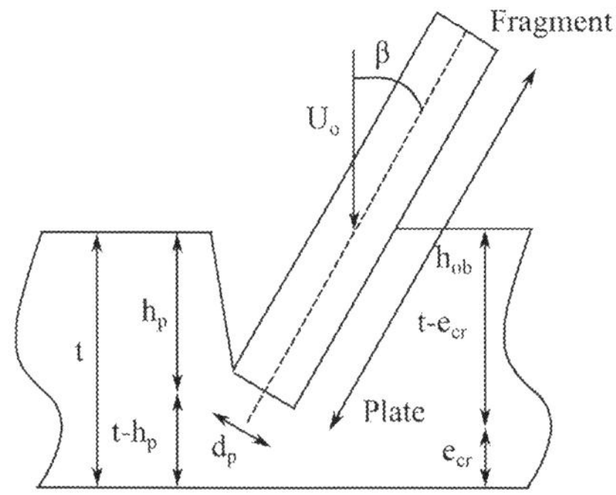


**Figure 2.21** Impact of a projectile (fragment) on a target (a plate). Modified from (Nguyen et al., 2009). *Credit: Nguyen, Q.B., Mebarki, A., Saada, R.A., Mercier, F.M., Reimeringer, M. (2009). Integrated probabilistic framework for domino effect and risk analysis.* Advances in Engineering Software, 40*(9), 892–901.*

**Table 2.7** Minimum plate thickness for different diameters (American Petroleum Institute, 2013)

| Tank diameter (m) | Minimum thicknesses (mm) |
|---|---|
| ≤ 15.2 | 4.8 |
| >7.6–18.3 | 6.4 |
| >18.3–30.5 | 8 |
| >30.5 | 9.6 |

The model takes into account both the characteristics of the impact material and the properties of the target material, Eqs. (2.25) and (2.26) (Nguyen et al., 2009).

Penetration depth for the case $\beta \neq 0$:

$$h_p = \frac{\left(-d_p\cos(\beta) + \sqrt{\left(d_p\cos(\beta)\right)^2 + \frac{4}{\pi}\tan(\beta)\left(\frac{E_c}{f_u\varepsilon_u}\right)^{\frac{2}{3}}}\right)}{2\tan(\beta)} \tag{2.25}$$

Penetration depth for the case $\beta = 0$:

$$h_p = \left(\frac{E_c}{f_u \times \varepsilon_u}\right)^{\frac{2}{3}} * \frac{1}{\pi \times d_p} \tag{2.26}$$

Where:

- kinetic energy is defined as $E_c = \frac{M \times U_o^2}{2}\left(\frac{\text{kg} \times \text{m}^2}{\text{s}^2}\right)$,
- $d_p$ is the debris diameter (m), and

- $f_u$ and $\varepsilon_u$ are the ultimate strength and ultimate strain of the target's constitutive material (Pa), respectively.

The debris and objects dragged by the wind have irregular geometries, therefore, instead of considering real fragments, the model assumes that the projectiles have spherical or rods shapes. In the case of rods or projectiles, it is necessary to calculate the equivalent diameter in the function of their length $l_p$ and area $A_p$. Eq. (2.27) calculates the equivalent diameter:

$$d_p = \frac{\left(\sqrt{(\pi \times h_{ob})^2 + 2 \times \pi \times A_p} - \pi \times h_{ob}\right)}{\pi} \tag{2.27}$$

Furthermore, Eq. (2.28) gives the resistance force $F_r$ (N) of the tank:

$$F_r = P_r A_p \tag{2.28}$$

Where:

- $P_r$ is the resistance pressure of the tank calculated with Eq. (2.7) (Pa), and
- $A_p$ is the reference debris area ($m^2$). The surface area of the object that could impact the tank.

Eq. (2.29) presents the damage criteria for debris impact.

$$\text{Damage} = \begin{cases} if\ F_{ob} - F_r > 0 \text{ Damage} \\ if\ F_{ob} - F_r \leq 0 \text{ No damage} \end{cases} \tag{2.29}$$

#### *2.3.4.2 Definition of limit state equations*

The limit state equations (LSE) compare the resistance of the system ($R$) under study against a solicitation or external load ($S$) (Sánchez-Silva, 2010a). The basic reliability problem in Eq. (2.30) compares the resistance $R$ of a tank against a solicitation $S$, from a natural hazard, which affects the tank. To define the damage condition:

$$R - S \leq 0 \tag{2.30}$$

Eq. (2.31) represents the limit state equation $g(R, S)$ (Sánchez-Silva & Klutke, 2016):

$$g(R, S) = R - S \leq 0 \tag{2.31}$$

In reliability and risk analysis, the calculation of the probability of damage to a system is a fundamental factor in decision making. The

**Table 2.8** Limit State Equation for different types of damage due to wind

| Damage mode | Tanks resistance | Solicitation | LSE |
|---|---|---|---|
| Buckling | $P_r$ | $q_{eq}$ | $g(P_r,q_{eq}) = P_r - q_{eq}$ |
| Debris impact | $t$ | $h_p$ | $g(t,hp) = t - hp$ |
| | $F_r$ | $F_i$ | $g(F_r,F_i) = F_r - F_i$ |

probability of damage to a storage tank will depend on its capacity to resist external load and solicitation. Eq. (2.32) defines the probability of different types of damage to a storage tank:

$$p_d = p(g(R, S) \leq 0) \tag{2.32}$$

For natural hazards, the parameters in the mathematical models have a random variability or uncertainty associated with the natural behavior of the hazards. Equations (2.31) and (2.32) include this uncertainty.

Table 2.8 presents each of the LSEs for each of the types of damage considered.

The probabilistic approach of Eq. (2.32) applies to each of the LSEs defined for the different damage modes. The next section deals with the construction of the fragility curves from the damage probability.

## 2.3.5 Storage tanks fragility analysis

A common definition of fragility is "the quality of being easily broken or damaged". Kennedy et al. (1980) were one of the first to introduce the concept of fragility in the engineering field, based on fragility functions for structures affected by earthquakes. Their fragility functions were defined as a probabilistic relationship between the frequency of damage to a structural component (in nuclear plants) and the peak ground acceleration of an earthquake. From this, a fragility function (Fig. 2.22), can be defined as a mathematical function that expresses the probability that some unwanted event occurs (such as the damage of process equipment or its elements) based on some measure of environmental excitation (solicitation or natural hazard intensity) (Porter, 2020).

To estimate the probability of damage to a storage tank by the impact of strong winds, the theory of fragility curves is used.

### *2.3.5.1 Fragility curves*

In the derivation of the fragility curves of a storage tank impacted by strong winds, Monte Carlo simulations deal with both the aleatory

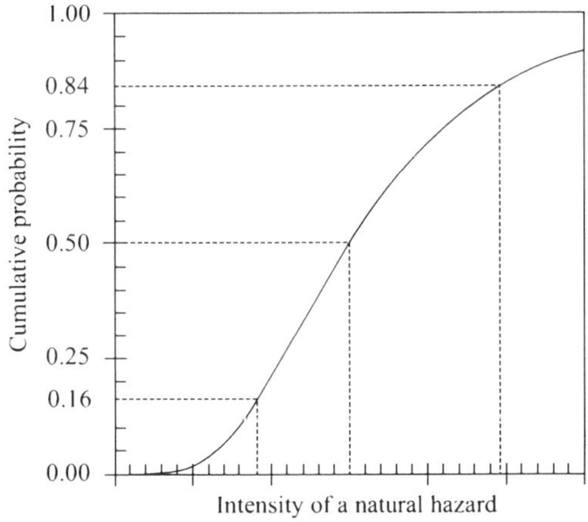


**Figure 2.22** Representation of a fragility function with a lognormal cumulative distribution.

uncertainty, associated with the parameters of the models with natural random behavior and the epistemic uncertainty, due to the lack of knowledge about both the natural phenomenon and the characteristics and behavior of the storage tank (Hofer et al., 2002). The fragility curves are a function of the hazard intensity and represent the probability of exceeding a certain level of damage, given a certain level of hazard intensity. Fig. 2.23 presents a diagram summarizing the general methodology to estimate the damage probability, based on an iterative process that includes the uncertainty of the parameters (Ramírez et al., 2019; Ramírez et al., 2020).

The process presented in Fig. 2.23 requires the probabilistic characterization of the parameters that define the natural hazard, the resistance of the storage tank, its geometry, and the type and height of liquid inside the tank. Table 2.9 summarizes the random properties of some of the parameters for wind hazards. The choice of distributions, mean values, and coefficients of variance comes from a combination of observations, experiments, and engineering judgment (Ramírez et al., 2019; Ramírez et al., 2020).

Similarly, probability distribution functions can characterize the geometry of the tank, the mechanical properties of the tank, the height of liquid inside the tank, and the density of the stored fluid. The distributions associated with the parameters are normal, uniform, lognormal, exponential, Weibull, and gamma (Ramírez et al., 2019).

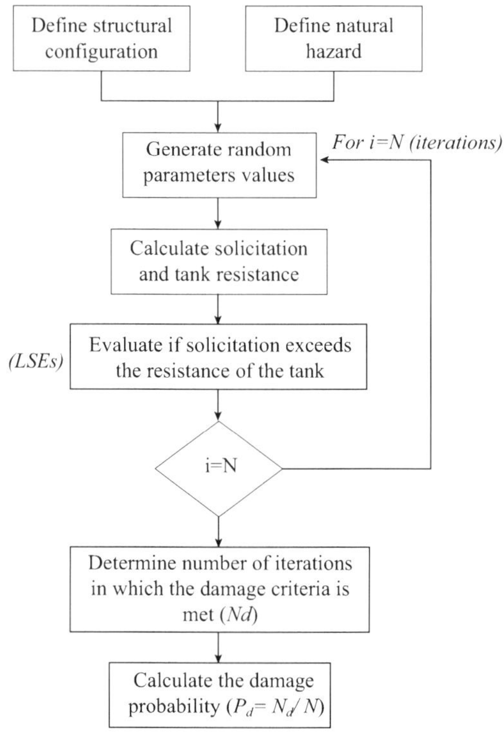


**Figure 2.23** Methodology to calculate the damage probability of a storage tank integrating the uncertainty within a purely probabilistic framework (Ramírez et al., 2019; Ramírez et al., 2020). *Credit: Ramírez, O., Mesa, A., Zuluaga, S., Muñoz, F., Sánchez-Silva, M., Salzano, E. (2019). Fragility curves of storage tanks impacted by strong winds.* Chemical Engineering Transactions, 77, *91–96; Ramírez, O., Zuluaga, S., Muñoz, F., Sánchez-Silva, M., Pinelli, J.-P., Salzano, E. (2020). The effects of extreme winds on atmospheric storage tanks.* Reliability Engineering & System Safety, 195, *106686.*

**Table 2.9** Parameters with random behavior for wind load

| Parameter | Unit | Type of distribution | Mean (μ) | Coefficient of variation (*cov*) |
|---|---|---|---|---|
| $\rho_W$ | kg/m3 | Normal Uniform | Air density of the affected area | 9.6% |
| $\rho_p$ | kg/m3 | Lognormal | Debris density | 10.2% |
| $\rho_f$ | kg/m3 | Exponential Weibull | Density of the stored at 1 atm and 25°C | 9.1% |
| $K_z$ | – | Gamma | 1.26 | 11.9% |
| $K_{zt}$ | – | | 1.0 | 5% |
| $K_d$ | – | | 0.95 | 8.2% |

With the parameters and variables of both the natural hazard and storage tank defined, a basic reliability analysis valuates the probability that the solicitation exceeds the resistance of the equipment. Eq. (2.33) defines the damage probability of a storage tank impacted by a natural hazard.

$$p_d = \frac{1}{N_{sim}} \sum_{i=1}^{N_{sim}} g(i) = \frac{N_d}{N_{sim}} \tag{2.33}$$

$$(i) = \begin{cases} 1 \; if \; LSE \leq 0 \\ 0 \; if \; LSE > 0 \end{cases} \tag{2.34}$$

Where:

- $N_{sim}$ is the number of simulations or iterations of the Monte Carlo algorithm, and
- $N_d$ is the number of iterations in which the damage criteria from Eq. (2.34) are met.

The algorithm calculates the damage probability independently for each of the types of damage Table 2.8 proposes.

### *2.3.5.2 Failure probability*

The probability of failure depends on a loss of containment. If total collapse does not occur, the failure of some component of the tank will define the type of leak for the tank's fluid. On the other hand, the total collapse of a tank will result in a type 1 mode for the fluid discharge, which represents an instantaneous release of the entire content. In the event trees for each natural hazard, the release modes are defined for each of the failure modes.

Damage is different than failure. Damage relates to an impairment (e.g., a dent) that does not necessarily lead to losses of its content. While failure is associated with a crack or opening caused by the damage, by which the contents escape. As a result, there is a dependence between the damage probability and the failure probability, where the tank must be damaged first before it fails.

Several databases such as FACTS, MHIDAS, Maritime mobile Access and Retrieval System (MARS), ICHEME, NRC, and so on collect information related to industrial accidents caused by different natural hazards. Analyses of the historical data from these databases, complemented by Natech events investigations, and engineering models and analyses, lead to an estimate of the probability of failure of storage tank components following the tank's damage.

#### 2.3.5.3 Probit functions to estimate damage probability

This section presents the calculation of the damage probability as a Probit function, which is commonly adopted in the quantitative risk assessment of industrial installation and allows for the simple definition of threshold values for the effect of wind on industrial equipment.

The behavior of the cumulative probability functions has a certain similarity to dose-response curves. Crowl and Louvar (2011a) state that the Probit functions estimate the probability that the exposure to a given dose (e.g., the intensity measure of an event) shall affect a living being or a structure (buildings or houses). Among its most common uses, is the calculation of the probability of personal injury from fire, explosion, or toxic dispersion. In this chapter, Probit functions estimate the damage probability of a tank impacted by a natural hazard. This method relates the damage probability and the Probit points (Eq. 2.35), yielding as a result, a linear function that estimates the Probit points according to the dose (intensity of the event).

$$\hat{p} = 0.5\left(1 + \frac{Y-5}{|Y-5|}\right) erf\left(\frac{|Y-5|}{\sqrt{2}}\right) \tag{2.35}$$

Where:

- $Y$ are the Probit points, and
- *erf* is the Gauss error function.

Due to the similarity between cumulative probability curves and dose-response curves, the dose is the vector of impact or intensity of the natural hazard, and the response is the probability that the tank shall suffer some kind of physical damage due to the impact of the natural hazard. For the application of the Probit model, Crowl and Louvar (2011a) proposes a logarithmic function that relates the Probit points ($Y$) with the intensity ($V_i$) of each of the natural hazards, Eq. (2.36).

$$Y = k_1 + k_2 Ln(V_i) \tag{2.36}$$

Where:

- $k_1$ and $k_2$ are the constants of the model and depend on the geometry of the tank and the type of natural hazard, and
- $V_i$ is the intensity parameter for each natural hazard, wind speed for wind loads.

Fig. 2.24 presents a Probit function (blue line) for a tank impacted by a natural hazard. The Red line represents the probability cumulative function.

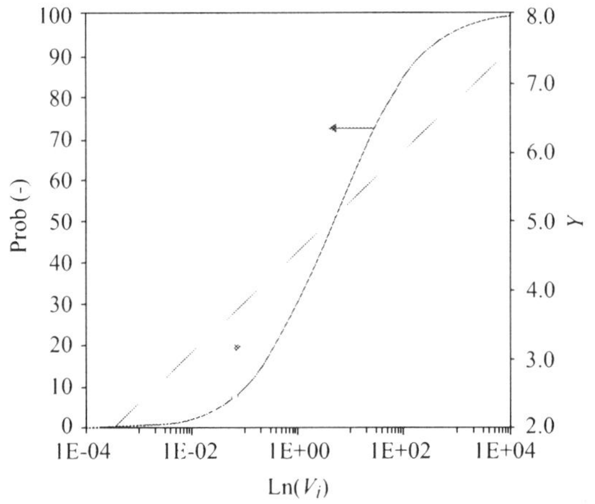


**Figure 2.24** The probability cumulative function and the corresponding Probit function $Y$ vs. hazard intensity $Ln(V_i)$.

As was mentioned, storage tanks are only affected if they are empty or partially empty (less than 10% of the filling level). A general Probit function is proposed to calculate the probability of damage in a storage tank impacted by a wind dose. The function applies to tanks with a level close to 0% and 10% and for substances stored with a density between 750 kg/m$^3$ and 1100 kg/m$^3$.

To estimate the constants of Eq. (2.36), multiple simulations of different tank configurations were performed, and through the method of least squares, the following expressions are proposed to calculate the constants of the model according to the geometrical characteristics of the tank.

Probit constant for an empty tank:

$$k_1 = -143075.25x^2 + 3964.9x - 100.8 \quad (2.37)$$

$$k_2 = 19043.4x^2 - 423.2x - 16.65 \quad (2.38)$$

Probit constants for a tank with a 10% fill level.

$$k_1 = 714608.5x^2 - 10999.8x - 63.996 \quad (2.39)$$

$$k_2 = -72026x^2 + 1149.61x + 13.253 \quad (2.40)$$

$$x = \frac{Ht}{D} \quad (2.41)$$

Where:

- $H$ is the height of the tank (m),
- $D$ is the nominal tank diameter (m), and
- $t$ is the shell thickness (m),

The development of the Probit functions and the fragility curves includes the treatment of uncertainties associated with the parameters of the models to calculate the different damage probabilities. This process improves the results of the risk calculation in Natech events in storage facilities due to the new knowledge of the uncertainty for the analyzed case, and it also helps in the making decisions.

### 2.3.6 Storage tanks vulnerability analysis

Vulnerability is defined as the expected damage of an element or group of elements, under risk resulting from a hazard of a given magnitude or intensity (Hamid et al., 2011; Sánchez-Silva, 2010b). Some important aspects to perform a vulnerability assessment of process equipment are:

1. The evaluation criteria refer to the strength and shape of the equipment so that the process equipment can maintain its integrity and functionality.
2. The type of loss, such as a reduction in the capacity to provide a service, number of injured, lost stored material, repair costs, among others.
3. Scenarios of losses tend to be the most probable or frequent final events or the one that entails maximum social or economic consequences.

Vulnerability is not fragility. In a vulnerability matrix, the columns represent the probability mass functions (pmf) of damage for a given interval of hazard intensity measure. The mean values of this pmf for each hazard intensity interval yield the vulnerability curves. Whereas the fragility curves represent the probability of exceeding a certain level of damage (Pita et al., 2013; Pita et al., 2014; Porter, 2020). For a given class of tanks, and a particular Hazard, there is one vulnerability curve but an infinite number of fragility curves.

The type of damage that this chapter evaluates is the loss of containment of the stored fluid. A loss of containment can generate different scenarios such as explosions, fires, or toxic dispersions. These scenarios can cause serious consequences in the surroundings (people, structures, and environment). Each of these scenarios has different effects, a fire can produce high heat radiation, an explosion can generate an overpressure blast wave and dispersion of hazmat can generate toxicity levels highly harmful to living beings.

The methodologies for risk analysis in Natech events commonly use existing models for the estimation of the loss of containment. This chapter focuses on the loss of containment of hazmats in the liquid state. Considering the above, the volume of spilled material will be the parameter that will allow the assessment of the aforementioned scenarios in the future.

To estimate the amount of material released after the system has been damaged, it is necessary to establish the different modes of material release. Fig. 2.12, the release modes for atmospheric storage tanks are the following, which are the ones usually assumed in quantitative risk analysis:

- Release Mode 1: Instantaneous release of entire contents.
- Release Mode 2: Release of entire contents in 10 min in a continuous and constant stream.
- Release Mode 3: Continuous release from a hole with an effective diameter of 10 mm.

Once the release modes have been identified, they should relate to the types of failure caused by damage to a storage tank. The event tree of Fig. 2.12 presents the different types of failure associated with the different types of damage.

For each release mode, source models or release models describe the material release process. Crowl and Louvar (2011b) and the National Institute of Public Health and the Environment RIVM (2009) propose different methods to calculate the material released.

- For release mode 1, no model exists, since it is an instantaneous total discharge.
- For release mode 2, it is necessary to establish what type of failure causes the LOC, that is, it can be a failure due to total rupture of the connections or rupture of the tank shell. The release by both failures can be modeled as a flow-through orifice, the difference lies in its discharge coefficient. In the total rupture of the connections, the diameter of the pipe or connection is known, while the rupture of the shell of the tank or a partial failure of the connections has an indefinite orifice geometry.
- Release mode 3, can also be modeled as a leak through a hole because the model is flexible with the size of the orifice.

To model the flow of a liquid through a hole in a tank $Q_m$ (kg/s), it is convenient to start from the principle of conservation of mass and a mechanical energy balance. Where $h_L^0$ represents the height at which the release hole is located. Eq. (2.42) (Crowl & Louvar, 2011b) calculates the leakage mass flow rate at any time:

$$Q_m = \rho_f C_o A \sqrt{2\left(\frac{g_c P_g}{\rho_f} + g h_L^0\right)} - \frac{\rho_f g C_o^2 A^2}{A_t} T \qquad (2.42)$$

Where:

- $p_f$ is the stored fluid density $\left(\frac{\text{kg}}{\text{m}^3}\right)$,
- $C_o$ is the release coefficient (-),
- $A$ is the orifice area ($\text{m}^2$),
- $g_c$ is the gravitational constant $\left(\frac{\text{m} \times \text{Kg}_\text{m}}{\text{kg}_\text{f} \times \text{s}^2}\right)$,
- $P_g$ is the gauge pressure (Pa),
- $g$ acceleration due to gravity $\left(\frac{\text{m}}{\text{s}^2}\right)$,
- $h_L^0$ is the liquid height above the leak ($m$),
- $A_t$ is the surface area of the tank ($\text{m}^2$), and
- $T$ is the release time (s).

The following equation computes the liquid level height in the tank:

$$h_L = h_L^0 - \frac{C_o A}{A_t}\sqrt{2\left(\frac{g_c P_g}{\rho_f} + g h_L^0 T\right)} + \frac{g}{2}\left(\frac{C_o A}{A_t} T\right)^2 \tag{2.43}$$

Eq. (2.44) determines the emptying time of the tank up to the level of the discharge orifice $T_e$(s).

$$T_e = \frac{1}{C_o g}\left(\frac{A_t}{A}\right)\left[\sqrt{2\left(\frac{g_c P_g}{\rho_f} + g h_L^0\right)} - \sqrt{\frac{2 g_c P_g}{\rho_f}}\right] \tag{2.44}$$

Finally, the following equation estimates the total volume release $V_L$ ($\text{m}^3$):

$$V_L = \int_0^{T_e} \frac{Q_m}{\rho_f} dt \tag{2.45}$$

Equations (2.42)–(2.45) estimate the volume of unplanned spilled hazardous material. These Equations represent a good approximation to calculate the volume leaking through a hole in a storage tank. The volume of spilled material is important since the estimation of the consequences will depend on this value. The most common consequences of a Natech event are environmental contamination due to liquid dispersion or a toxic cloud, fire of flammable material, or explosions. So, the greater the amount of spilled volume, the greater the consequence associated with the Natech event.

#### 2.3.6.1 Frequency of final accidental scenario

The event tree shows that the occurrence of a series of situations that lead to the loss of containment of hazmat limits a Natech event. The combination of secondary events of the tree branches determines the calculation of the probability of the final accidental scenario ($p_{as}$) in Eq. (2.46).

$$p_{as} = p_d \times p_f \tag{2.46}$$

Where:

- $p_d$ is the damage probability, and
- $p_f$ is the failure probability.

Based on this, the frequency of the final accidental scenario $f_{as}$ (1/year) will be the multiplication of the frequency of the initiating event $f$ by the probability of the accidental scenario $p_{as}$, as shown in Fig. 2.14.

$$f_{as} = f \times p_{as} \tag{2.47}$$

The Center for Chemical Process Safety defined the risk for industrial accidents, as the measure of economic, human, or environmental losses, in terms of the probability of occurrence or frequency of the incident and the magnitude of the losses or injuries. From this, the calculation of the frequency of the final accidental scenario is fundamental given that it is an input parameter for the risk calculation. The frequency must relate to the consequences, which are a function of the volume of hazmat spilled.

## 2.4 Conclusions

The focus of this chapter is to analyze the undesired events that can occur if an atmospheric storage tank is damaged due to the impact of strong winds. A systematic methodology that integrates both qualitative and quantitative information related to the Natech scenarios evaluates the failure and damage probability. This methodology allows analysts to perform a preliminary fragility and vulnerability assessment associated with Natech events regarding strong winds, integrating both the parameterization of a storage tank based on the API-650 and API6-620 standards and the solicitation of extreme winds according to damage models or international standards.

The probability of damage relates to the possible accidental scenario. New event trees present a series of situations that model in sequence the cause of the loss of containment of the liquid stored in the tank. The event trees take into account different types of damage.

The probability of failure relates to the loss of containment due to the impact of the natural hazard. A methodology for generating fragility curves includes the variability or uncertainty of parameters associated with the process equipment and the natural hazard (strong winds). A combination of fragility curves and the application of Probit curves leads to an estimate of the probability of failure. These two aspects allow analysts to include the uncertainty associated with the input parameters to each of the models considered. The treatment of uncertainty is carried out through Monte Carlo simulations for calculating the failure probability in any structural configuration of the tank (new or existing) against the intensity or impact vector of the natural hazard. The model allows analysts to define the number of simulations to be performed until they obtain convergence of results. Probit functions are carried out through the fragility curves. Probit models are defined for storage tanks impacted by extreme winds to estimate the failure probability considering geometric characteristics of the tank and hazards solicitation.

## References

Allaby, M. (2007). *Encyclopedia of weather and climate* (2nd ed.). New York: Facts on File Science Library.

American Petroleum Institute (2013). *API STD 620—Design and construction of large, welded, low-pressure storage tanks*. American Petroleum Institute.

American Petroleum Institute (2020). *API STD 650—Welded tanks for oil storage*. American Petroleum Institute.

American Society of Civil Engineers (2017). *ASCE 7–16—Minimum design loads and associated criteria for buildings and other structures*. American Society of Civil Engineers.

Antonioni, G., Landucci, G., Necci, A., Gheorghiu, D., & Cozzani, V. (2015). Quantitative assessment of risk due to Natech scenarios caused by floods. *Reliability Engineering and System Safety*, *142*, 334–345.

Arango, C., Dorado, J., Guzmán, D., & Ruíz, F. (2012). Variabilidad climática de la precipitación en Colombia asociada al ciclo El Niño, La Niña—Oscilación del Sur (ENSO).

Antonioni, G., Spadoni, G., & Cozzani, V. (2007). A methodology for the quantitative risk assessment of major accidents triggered by seismic events. *Journal of Hazardous Materials*, *147*(1), 48–59.

Bryant, E. (2004). *Natural hazards* (2nd ed.). New York, United States: Cambridge University Press. (Chapter 6).

Burgos, C. A., Jaca, R. C., Lassig, J. L., & Godoy, L. A. (2014). Wind buckling of tanks with conical roof considering shielding by another tank. *Thin-Walled Structures*, *84*, 226–240.

Burton, I., Kates, R., & White, G. (1978). *The environment as hazard* (1st ed.). New York, United States: Oxford University Press. (Chapter 2).

Campedel, M. (2008). Analysis of major industrial accidents triggered by natural events reported in the principal available chemical accident databases. European Commission, Joint Research Centre. <https://ec.europa.eu/jrc/en/publication/eur-scientific-and-technical-research-reports/analysis-major-industrial-accidents-triggered-natural-events-reported-principal-available> Accessed 13.07.21.

Capuano, P., Basco, A., Ruocco, A., Di., Esposito, S., Fusco, G., Garcia-Aristizabal, A., Mercogliano, P., Salzano, E., Solaro, G., Teofilo, G., Scandone, P., & Gasparini, P. (2017). The ARGO project: assessing Natech risks on offshore oil platforms. *Energy Procedia, 125*, 145–152.

Charlton, R. (2008). *Fundamentals of fluvial geomorphology*, 1st ed. New York (Chapter 3–6).

Cozzani, V., Antonioni, G., Landucci, G., Tugnoli, A., Bonvicini, S., & Spadoni, G. (2014). Quantitative assessment of domino and Natech scenarios in complex industrial areas. *Journal of Loss Prevention in the Process Industries, 28*, 10–22.

Cozzani, V., Campedel, M., Renni, E., & Krausmann, E. (2010). Industrial accidents triggered by flood events: analysis of past accidents. *Journal of Hazardous Materials, 175*(1), 501–509.

CRED (2019). EM-DAT: The Emergency events database—Universite catholique de Louvain (UCL)—CRED, D. Guha-Sapir [Data file]. Universite catholique de Louvain (UCL)—CRED, D. Guha-Sapir. <http://emdat.be/emdat_db> Accessed 10.02.20.

Crowl, D. A., & Louvar, J. F. (2011a). *Chemical process safety: fundamentals with applications* (3rd ed.). Boston, MA: Pearson Education. (Chapter 2).

Crowl, D. A., & Louvar, J. F. (2011b). *Chemical process safety: fundamentals with applications* (3rd ed.). Boston, MA: Pearson Education. (Chapter 4).

Cruz, A. M., & Okada, N. (2008). Methodology for preliminary assessment of Natech risk in urban areas. *Natural Hazards, 46*(2), 199–220.

Cruz, A. M., Steinberg, L. J., Vetere-Arellano, A. L., & Pisano, F. (2004). *State of the art in Natech risk management: (NATECH: Natural Hazarad Triggering a Technological Disaster)*. Italy: European Commission, Joint Research Centre.

Eckerman, I. (2005). *The Bhopal saga: causes and consequences of the world's largest industrial disaster*. India: Universities Press.

El Hajj, C., Piatyszek, E., Tardy, A., & Laforest, V. (2015). Development of generic bow-tie diagrams of accidental scenarios triggered by flooding of industrial facilities (Natech). *Journal of Loss Prevention in the Process Industries, 36*, 72–83.

European Commission (2016). Technological accidents triggered by natural disasters. <https://ec.europa.eu/jrc/en/research-topic/technological-accidents-triggered-natural-disasters> Accessed 08.07.21.

European Commitee for Standardization (2005). *Eurocode 1: actions on structures—Part 1–4: general actions—wind actions*. European Commitee for Standardization.

French Ministry for Sustainable Development (2013). The "Natech" risk, or technological accidents triggered by a natural event. <https://www.aria.developpement-durable.gouv.fr/wp-content/uploads/2013/08/FT_impel2013_NaTech_risks.pdf> Accessed 08.07.21.

Girgin, S., & Krausmann, E. (2012). Rapid natech risk assessment and mapping tool for earthquakes: rapid-N. *Chemical Engineering Transactions, 26*, 93–98.

Girgin, S., & Krausmann, E. (2016). Historical analysis of U.S. onshore hazardous liquid pipeline accidents triggered by natural hazards. *Journal of Loss Prevention in the Process Industries, 40*, 578–590.

Hamid, S., Pinelli, J.-P., Chen, S.-C., & Gurley, K. (2011). Catastrophe model-based assessment of hurricane risk and estimates of potential insured losses for the State of Florida. *Natural Hazards Review, 12*, 171–176.

Hardy, J. T. (2003). *Climate change, causes, effects and solutions* (1st ed.). West Sussex, England: Wiley.

Hofer, E., Kloos, M., Krzykacz-Hausmann, B., Peschke, J., & Woltereck, M. (2002). An approximate epistemic uncertainty analysis approach in the presence of epistemic and aleatory uncertainties. *Reliability Engineering and System Safety*, 77(3), 229–238.

Kennedy, R. P., Cornell, C. A., Campbell, R. D., Kaplan, S., & Perla, H. F. (1980). Probabilistic seismic safety study of an existing nuclear power plant. *Nucl. Eng. Des.*, *59*(2), 315–338.

Krausmann, E., & Cruz, A. M. (2017). *Natech risk assessment and management*. Amsterdam: Elsevier. (Chapter 2).

Krausmann, E., Cruz, A. M., & Salzano, E. (2017). *Natech risk assessment and management*. Amsterdam: Elsevier. (Chapter 1).

Krausmann, E., & Salzano, E. (2017). *Natech risk assessment and management*. Amsterdam: Elsevier. (Chapter 3).

Landucci, G., Antonioni, G., Tugnoli, A., & Cozzani, V. (2012). Release of hazardous substances in flood events: damage model for atmospheric storage tanks. *Reliability Engineering and System Safety*, *106*, 200–216.

Landucci, G., Necci, A., Antonioni, G., Tugnoli, A., & Cozzani, V. (2014). Release of hazardous substances in flood events: damage model for horizontal cylindrical vessels. *Reliability Engineering and System Safety*, *132*, 125–145.

Lee, J. K., Lee, K.-H., Kim, S., Il., Yook, D., & Ahn, S. (2017). Weibull parameter calculation and estimation of wind speeds for the return period: a case study in the Kori and Wolsong NPP areas. *Ann. Nucl. Energy*, *108*, 406–412.

Lees, F. (2004). (3rd ed.). *Lees' loss prevention in the process industries: hazard identification, assessment and control*, (2). Burlington, United States: Butterworth-Heinemann.

Lin, N., Letchford, C., & Holmes, J. (2005). Experimental investigation of trajectory of windborne debris with applications to debris impact criteria. In *Proceedings of the 10th Americas Conference on Wind Engineering, ACWE 2005*, Baton Rouge, LA, United States, May 31–June 4, 2005.

Malhotra, P. K. (2005). Return period of design ground motions. *Seismological Research Letters*, *76*(6), 693–699.

Maraveas, C., Balokas, G. A., & Tsavdaridis, K. D. (2015). Numerical evaluation on shell buckling of empty thin-walled steel tanks under wind load according to current American and European design codes. *Thin-Walled Structures*, *95*, 152–160.

Mesa-Gómez, A., Casal, J., & Muñoz, F. (2020). Risk analysis in Natech events: state of the art. *Journal of Loss Prevention in the Process Industries*, *64*, 104071.

Mesa-Gómez, A., Casal, J., Sánchez-Silva, M., & Muñoz, F. (2021). Advances and gaps in Natech quantitative risk analysis. *Processes*, *9*(1).

Misuri, A., Antonioni, G., & Cozzani, V. (2020). Quantitative risk assessment of domino effect in Natech scenarios triggered by lightning. *Journal of Loss Prevention in the Process Industries*, *64*, 104095.

Munich, R. E. (2019). Natural disaster risks: losses are trending upwards. Risks Posed by Natural Disasters. <https://www.munichre.com/en/risks/natural-disasters-losses-are-trending-upwards.html#1995343501> Accessed 05.07.21.

Myers, P. E. (1997a). *Aboveground storage tanks* (1st ed.). New York, United States: McGraw-Hill Education. (Chapter 7).

Myers, P. E. (1997b). *Aboveground storage tanks* (1st ed.). New York, United States: McGraw-Hill Education. (Chapter 8).

National Institute of Public Health and the Environment (RIVM) (2009). Reference manual Bevi risk assessments. <https://www.rivm.nl/documenten/reference-manual-bevi-risk-assessments-version-32> Accessed 06.07.21.

Necci, A., Antonioni, G., Bonvicini, S., & Cozzani, V. (2016). Quantitative assessment of risk due to major accidents triggered by lightning. *Reliability Engineering and System Safety*, *154*, 60–72.

Nguyen, Q. B., Mebarki, A., Saada, R. A., Mercier, F. M., & Reimeringer, M. (2009). Integrated probabilistic framework for domino effect and risk analysis. *Adv. Eng. Softw.*, *40*(9), 892–901.

Ocampo, F. (2016). *Marco para el tratamiento de incertidumbre en el análisis de riesgo cuantitativo en transporte de material peligroso a través de tuberías. Departamento de Ingeniería Química: Master's.* Bogotá, Colombia: Universidad de los Andes.

Pathirana, M., Lam, N., Perera, S., Zhang, L., Ruan, D., & Gad, E. (2017). Damage modelling of aluminium panels impacted by windborne debris. *J. Wind Eng. Ind. Aerodyn., 165*, 1–12.

Pinelli, J.-P., Esteva, M., Rathje, E. M., Roueche, D., Brandenberg, S., Mosqueda, G., Padgett, J., & Haan, F. (2020). Disaster risk management through the designsafe cyberinfrastructure. *Int. J. Disaster Risk Sci., 11*(6), 719–734.

Pita, G. L., Pinelli, J.-P., Gurley, K. R., & Hamid, S. (2013). Hurricane vulnerability modeling: Development and future trends. *J. Wind Eng. Ind. Aerodyn., 114*, 96–105.

Pita, G., Pinelli, J.-P., Gurley, K., & Mitrani-Reiser, J. (2014). State of the art of hurricane vulnerability estimation methods: a review. *Natural Hazards Review, 16*, 4014022.

Porter, K. (2020). *A beginner's guide to fragility, vulnerability, and risk.* Colorado: University of Colorado Boulder.

Potter, T. D., & Colman, B. R. (2003). *Handbook of weather, climate, and water: dynamics, climate, physical meteorology, weather systems, and measurements.* New Jersey, United States: Wiley-Interscience. (Chapter 31).

Ramírez, O., Mesa, A., Zuluaga, S., Muñoz, F., Sánchez-Silva, M., & Salzano, E. (2019). Fragility curves of storage tanks impacted by strong winds. *Chemical Engineering Transactions*, 77, 91–96.

Ramírez, O., Zuluaga, S., Muñoz, F., Sánchez-Silva, M., Pinelli, J.-P., & Salzano, E. (2020). The effects of extreme winds on atmospheric storage tanks. *Reliability Engineering and System Safety, 195*, 106686.

Salvi, O., Kirchsteiger, C., Delvosalle, C., Duijm, N.-J., Casal, J., Goosens, L., Mazzarotta, B., Lebecki, K., Wybo, J.L., Dusserre, G., Londiche, H., & Calzia, J. (2002). ARAMIS: accidental risk assessment methodology for industries in the framework of SEVESO II directive. In *Colloque Urbanisation et Prévention des Accidents Majeurs.* <https://hal-ineris.archives-ouvertes.fr/ineris-00972395> Accessed 05.07.21.

Salzano, E., & Basco, A. (2015). Simplified model for the evaluation of the effects of explosions on industrial target. *Journal of Loss Prevention in the Process Industries, 37*, 119–123.

Salzano, E., Iervolino, I., & Fabbrocino, G. (2003). Seismic risk of atmospheric storage tanks in the framework of quantitative risk analysis. *Journal of Loss Prevention in the Process Industries, 16*(5), 403–409.

Sánchez-Silva, M. (2010a). *Introducción a la confiabilidad y evaluación de riesgos: teoría y aplicaciones en ingeniería* (2nd ed.). Bogotá, Colombia: Ediciones Uniandes. (Chapter 8).

Sánchez-Silva, M. (2010b). *Introducción a la confiabilidad y evaluación de riesgos: teoría y aplicaciones en ingeniería* (2nd ed.). Bogotá, Colombia: Ediciones Uniandes. (Chapter 3).

Sánchez-Silva, M., & Klutke, G.-A. (2016). *Reliability and life-cycle analysis of deteriorating systems* (1st ed.). Berlín, Alemania: Springer International Publishing. (Chapter 2).

Simiu, E., & Yeo, D. (2019). *Wind effects on structures: modern structural design for wind* (4th ed.). New Jersey: Wiley-Blackwell.

Simmons, K. M., Sutter, D., & Pielke, R. (2013). Normalized tornado damage in the United States: 1950–2011. *Environmental Hazards, 12*(2), 132–147.

Solari, G. (2020). Thunderstorm downbursts and wind loading of structures: progress and prospect. *Frontiers in Built Environment, 6*, 63.

Timoshenko, S. P., & Gere, J. M. (2012). *Theory of elastic stability* (2nd ed.). New York, United States: McGraw-Hill international student editions. (Chapter 2).

Uematsu, Y., Koo, C., & Yasunaga, J. (2014). Design wind force coefficients for open-topped oil storage tanks focusing on the wind-induced buckling. *Journal of Wind Engineering & Industrial Aerodynamics, 130*, 16–29.

Uijt de Haag, P. A. M., & Ale, B. J. M. (1999). *Guidelines for quantitative risk assessment. Purple Book. Commitee for the prevention of disasters* (1st ed.). Amsterdam: Publicatiereeks Gevaarlijke Stoffen. (Chapter 3).

United Nations Office for Disaster Risk Reduction (UNISDR) (2015). Disaster risk reduction. Terminology on disaster risk reduction. <https://www.undrr.org/terminology/disaster-risk-reduction> Accessed 05.07.21.

van den Bosch, C. J. H., & Weterings, R. A. P. M. (2005). *Methods for the calculation of physical effect. Yellow Book. Commitee for the prevention of disasters* (3rd ed.). Amsterdam: Publicatiereeks Gevaarlijke Stoffen. (Chapter 7).

Villalba, N. (2016). *Marco de referencia para el análisis del riesgo asociado a eventos Natech provocados por inundaciones. Departamento de Ingeniería Química: Master's.* Bogotá, Colombia: Universidad de los Andes.

Wang, J., & Cao, S. (2021). Characteristics of tornado wind loads and examinations of tornado wind load provisions in ASCE 7–16. *Engineering Structures*, *241*, 112451.

Wang, Z. (2008). Understanding seismic hazard and risk: a gap between engineers and seismologists. In *Proceedings of the 14th World Conference On Earthquake Engineering, 14WCEE*, Beijing, China, october 12-october 17, 2008 11.

Young, S., Balluz, L., & Malilay, J. (2004). Natural and technologic hazardous material releases during and after natural disasters: a review. *The Science of the Total Environment*, *322*(1), 3–20.

Zhao, Y., & Lin, Y. (2014). Buckling of cylindrical open-topped steel tanks under wind load. *Thin-Walled Structures*, *79*, 83–94.

CHAPTER THREE

# Seismic performance of liquid storage tanks

**Mehran S. Razzaghi**
Department of Civil Engineering, Qazvin Branch, Islamic Azad University (QIAU), Qazvin, Iran

## 3.1 Introduction

Ground-supported liquid storage tanks are widely utilized in the water distribution network and process industries for storing water, and hazardous liquids. Hence, their appropriate performance due to extreme loading cases is remarkably important from economic, environmental, and human safety points of view. The performance of such structures to the past earthquakes revealed that strong ground motions are potential threats to steel cylindrical tanks. The early studies on the seismic behavior of cylindrical tanks were performed in the late 1940s and early 1950s. Jacobsen (1949) and Housner (1954) were the first researchers who focused on tank hydrodynamics. Their studies were based on the assumption of the rigid behavior of tank shells. The catastrophic damage to tanks during the Chilean earthquake of 1960 and the Alaska earthquake of 1964 attracted the researchers' attention to the tank shell flexibility (Haroun & Housner, 1981; Jaiswal et al., 2005; Veletsos, 1974). Since then, comprehensive research on the seismic analysis of tanks was initiated (Jaiswal et al., 2007). The seismic design of steel tanks started in the 1970s when the earliest seismic design code specifications for liquid storage tanks were published (Razzaghi & Eshghi, 2015). So far, different design codes such as API 650, NZSEE, Eurocode 8, and AWWA D-100 have presented seismic design provisions for liquid storage tanks. Codification of the seismic provisions was a crucial step for loss prevention in water distribution and process industries. Meanwhile, the actual performances of tanks during the earthquakes, such as the Izmit earthquake of 1999 in Turkey, and the Emilia earthquake of 2012 in Italy (Brunesi et al., 2015) have also helped improve deficiencies in the design codes.

*Above Ground Storage Tank Oil Spills.*
DOI: https://doi.org/10.1016/B978-0-323-85728-4.00001-2

Accordingly, many of the precode tanks and those designed for the earlier code editions may be seismically vulnerable. On the other hand, the importance of the appropriate seismic performance of existing tanks has emphasized the requirement for assessing their seismic safety. Hence, several researchers have focused on evaluating the seismic vulnerability retrofitting of tanks (Kildashti et al., 2018; Phan et al., 2016; Tavakoli Joorabi & Razzaghi, 2019).

This chapter is chiefly concerned with the seismic performance of on-ground steel cylindrical tanks. To this end, at first, the seismic response of tanks and their typical failure modes is presented. Then, an overview of the factors affecting the seismic performance of tanks and a brief review of some of the significant seismic design codes are presented. Finally, the tank seismic safety assessment techniques are provided.

## 3.2 Seismic response

### 3.2.1 Hydrodynamic effects

When a liquid storage tank is subjected to seismic excitation, the upper parts of its content (so-called convective liquid) move in a long-period motion known as sloshing response. The period of the convective mode of the tank content is typically long (usually greater than 3 s or so). For large tanks, it can take values as high as five seconds or more (NZSEE, 2009). The rest of the content, which is impulsive liquid, moves rigidly with the tank shell (Jacobsen, 1949). The proportion of the tank content acting like an impulsive or convective liquid is depending on the ratio of the content height to the tank diameter $\left(h_l/D\right)$. On the other hand, subjecting to an earthquake, the distribution of the pressure exerted on the tank shell (known as hydrodynamic pressure) changes from its static state. For a rigid tank rested on a rigid foundation the hydrodynamic pressure can be estimated by a mechanical mass-spring analogy that was independently proposed by Jacobsen (1949) and Housner (1954). Note that in some analogies the convective mass is divided into a series of masses representing different sloshing modes. However, since the tank response can practically be well estimated by the impulsive mass and the first mode of the sloshing liquid, the sloshing submasses are often neglected. Fig. 3.1A illustrates a schematic of such a mas-spring system.

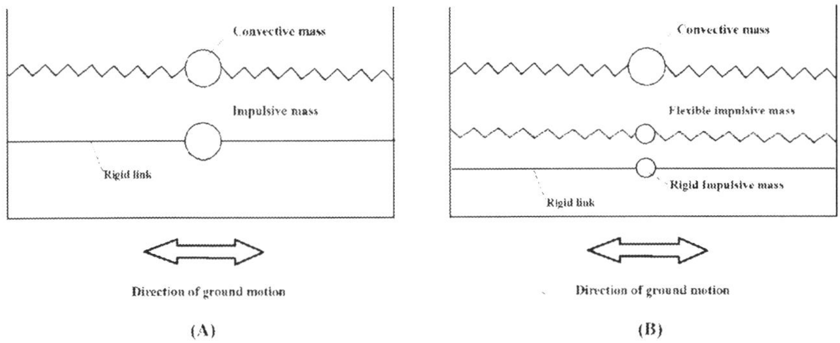


**Figure 3.1** Mass-spring analogy for (A) rigid tank-liquid system (B) flexible tank-liquid system.

For a rigid cylindrical tank the hydrodynamic pressure exerting on the tank wall due to a horizontal seismic excitation can be expressed as follows:

$$P(x,\theta,t) = P_i(z,\theta,t) + P_c(z,\theta,t) \tag{3.1}$$

where $P_i(z,\theta,t)$ and $P_c(z,\theta,t)$ are the impulsive and convective pressure components respectively which are functions of the time, $t$, and the cylindrical coordinates z and $\theta$. It should be noted that the above-mentioned components can be combined in different ways such as SRSS based on the desired analysis technique. The impulsive and convective pressure distribution can then be expressed as (Veletsos & Shivakumar, 1997):

$$P_i(z,\theta,t) = \psi_i(z)\rho R\cos(\theta)\ddot{x}_g(t) \tag{3.2}$$

$$P_c(z,\theta,t) = \rho R\cos(\theta)\sum_{j=1}^{\infty} C_j(z)A_j(t) \tag{3.3}$$

where $R$ is the tank radius, $\rho$ is the density of content, $\ddot{x}_g(t)$ is the ground motion acceleration, $\psi_i(z)$ and $Cj\ (z)$ are dimensionless shape functions describing the distribution of the impulsive and convective pressure along with the cylinder's height respectively, $A_j(t)$ is the pseudo-acceleration response of the j$^{th}$ sloshing mode of vibration and can be calculated as:

$$C_j(z) = \frac{2}{\lambda_n - 1}\frac{\cosh\left[\lambda_n(h_L/R)\left(\frac{z}{h_L}\right)\right]}{\cosh[\lambda_n(h_L/R)]} \tag{3.4}$$

where $\lambda_n$ is the $n$th root of the Bessel's function. Usually, employing the first three roots ($\lambda_1 = 1.841, \lambda_2 = 5.311, \lambda_3 = 8.536$) is sufficient to come by appropriate results (Virella et al., 2005).

As indicated in Fig. 3.1B, considering shell deformability in the mass-spring model results in breaking the impulsive mass into two masses: rigid-impulsive and flexible impulsive mass. It should be noted that the tank flexibility makes an insignificant change to convective hydrodynamic pressures. Hence, they are usually assumed to be the same as rigid tanks. While, as shown in Fig. 3.2, the distribution of the impulsive pressure along the flexible cylinder's height differs from rigid ones. However, the difference is not considered for relatively broad tanks, especially for those with thicker shells (Buratti & Tavano, 2014; Haroun & Housner, 1981). In other words, only in thin tanks and considerably slender ones, the tank shell deformability is dominant (Jaiswal et al., 2007).

In addition to the mass-spring analogy, extensive studies have been conducted on the numerical evaluation of hydrodynamic effects in tanks. Some of the abovementioned studies are based on the boundary element method (Firouz-Abadi et al., 2008) some others are based on coupled boundary element-finite element (FE) technics (Cho & Cho, 2007), and the rest on finite-element based analyses (Goudarzi et al., 2010; Merino et al., 2020; Ozsarac et al., 2021).

Hydrodynamic actions create an overturning moment on the bottom plate. As a result of the overturning moment, the tank tends to have a

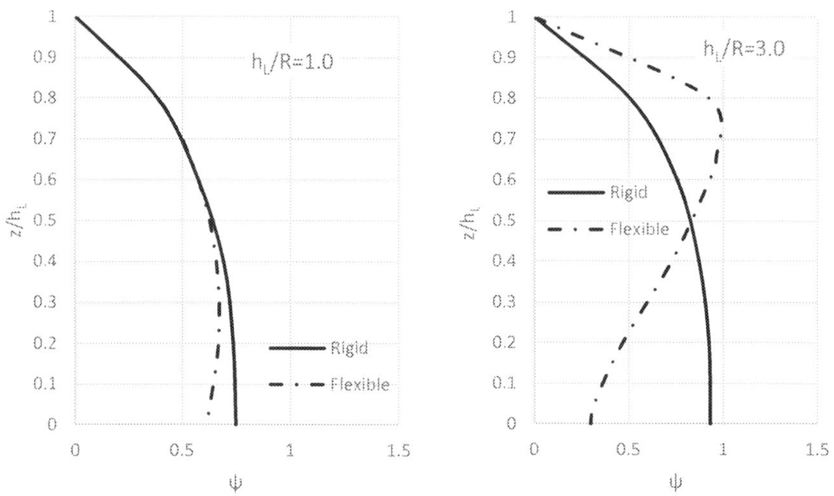


**Figure 3.2** Comparison of pressure distribution for rigid and flexible tanks of different aspect ratios. *After Veletsos, A.S., & Shivakumar, P. (1997). Chapter 15—Tanks containing liquids or solids. In:* Computer analysis and design of earthquake resistant structures. *London: Computational Mechanics Publications.*

rocking response. The response of unanchored tanks and anchored ones to the hydrodynamic overturning moment is considerably different.

### 3.2.2 Response of unanchored tanks

Subjecting to strong ground motions, unanchored tanks may experience a partial base uplifting. The tank uplift exerts significant stresses on base plates. In addition, the reduction of shell-to-foundation contact increases axial compression in nonuplifted regions (Malhotra, 1998). As the uplifting response of the tank is geometrically nonlinear, modeling the tank uplift is rather complicated. Assessing the seismic response of the uplifting tanks has attracted several researchers in recent decades. The early studies on the subject have taken place in the 1970s. Perhaps Wozniak and Mitchel (1978) are the first researchers who provided an analytical model for tank uplift. They considered the uplift of a strip of baseplate neglecting the effect of membrane action. Malhotra and Veletsos (1994) presented a beam model for the analysis of the uplifting behavior of tanks. They modeled the baseplate as semiinfinite beams of constant width resting on a rigid foundation (See Fig. 3.3). They also considered the plasticization at the uplifting end of the beam. Malhotra then generalized the beam model to a flexibly supported tank as illustrated in Fig. 3.3 (Malhotra, 1995). The drawback of the beam model is that the shape of the semiinfinite beam is not sufficiently compatible with the shape of the base plate. Accordingly, Ahari et al. (2009) presented the tapered beam model as shown in Fig. 3.4. However, their solution technique encounters numerical difficulties for the small uplift lengths. In other words, their solution may encounter chaotic estimations around the exact result. The problem in their solution arises from the drastic changes in the second-order shortening of the baseplate due to tank uplift (Ahari et al., 2009).

In addition to the above analytical studies, valuable experimental investigations were conducted by Clough (1977) and Cambra (1982, 1983) they presented empirical relations between the overturning moment and the tank uplift. It is worth mentioning that the modified Cambra's model is employed for the New Zealand seismic design specification for tanks (NZSEE, 2009).

As the uplifting responses of tanks are remarkably nonlinear from the material and geometric point of view, uplift modeling is inherently complicated. Therefore, the conventional analytical models are associated with various levels of simplifications. Thus, they may be incapable of estimating the actual performance of an uplifting tank precisely (Miladi & Razzaghi, 2019). The

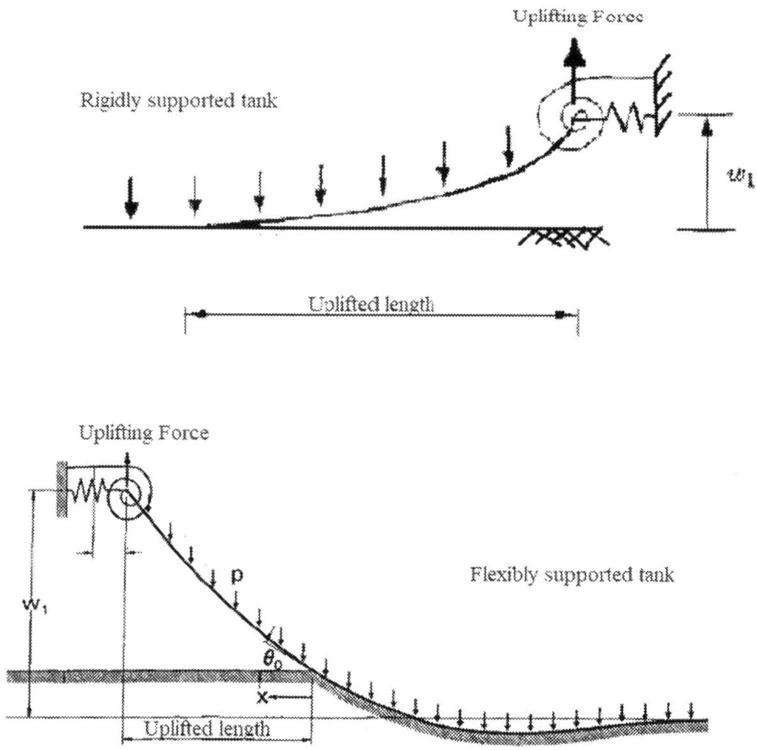


**Figure 3.3** Beam model for rigidly and flexibly supported tank uplift.

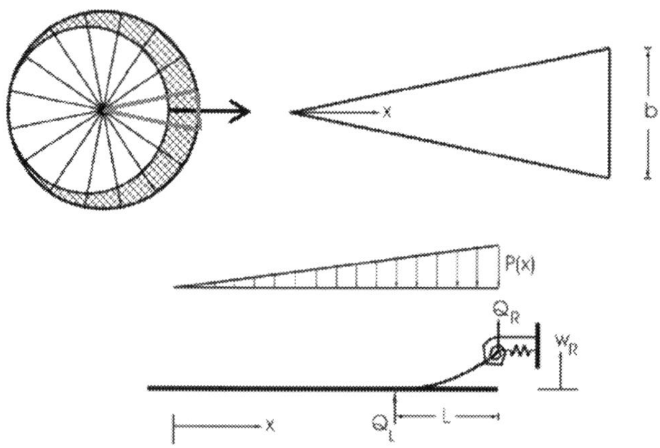


**Figure 3.4** Tapered beam model according to (Ahari et al., 2009).

development in numerical techniques following the advancement of computer hardware made notable progress in the FE modeling of the tank uplift. Recently, several studies have been performed on the seismic behavior of uplifting tanks using FE techniques (Bakalis & Karamanos, 2021; Razzaghi & Eshghi, 2008a, 2008b; Taniguchi & Katayama, 2016; Zhou & Zhao, 2021).

### 3.2.3 Response of anchored tanks

Cylindrical anchored tanks generally exhibit a cantilever-type behavior due to horizontal seismic accelerations (Buratti & Tavano, 2014). Subjecting to strong ground motions, mechanical anchors prevent the tank base from uplifting behavior. Hence, axial stresses are created in mechanical anchors. In such a circumstance, the overturning moment is transmitted to the foundation and the global tank-foundation overturning is credible. On the other hand, pulling off or stretching the mechanical anchors and shell failure around the anchor chair junction may occur in vulnerable tanks (Malhotra, 1998). The failure of the anchorage system may cause a sudden complicated uplifting behavior. As practicing engineers generally do not consider tank uplift in the calculation procedure of anchored tanks, significant failure of anchor bolts could lead to catastrophic consequences. The rupture of the baseplate, the fracture of the shell-to-baseplate junction, and the breaking of the rigid appurtenances attached to tanks are some of the potential aftermaths of anchor-bolt failure in such tanks. However, some researchers and practicing engineers believe that providing the controlled ability for limited uplift response in tanks may improve their overall seismic performance (Najmabad et al., 2021). The logic behind such a viewpoint is dissipating the seismic energy via the nonlinear behavior of anchor bolts alongside the small amount of the base uplift.

## 3.3 Typical failure modes

During recent decades a considerable number of tanks have suffered different levels of damage during devastating earthquakes. Diverse failure modes have been observed in damaged tanks following seismic events (Brunesi et al., 2015; Eshghi & Razzaghi, 2007; Fischer et al., 2016; Manos & Clough, 1985). Table 3.1 presents the list of most frequent failure modes (but not necessarily all of them) as well as their reasons and consequences.

**Table 3.1** The most frequent failure modes of tanks following the occurred earthquakes.

| Failure mode | Frequent reason(s) | Probable consequence(s) |
|---|---|---|
| Elephant's foot buckling | – Excessive compression in bottom parts of thicker shells<br>– Tank-to-foundation collision of uplifted tanks | – Welding failure<br>– Failure of shell-to-pipe junction<br>– Loss of content |
| Diamond-shaped buckling | – Excessive shell compression in thinner shells | Welding failure<br>Failure of shell-to-pipe junction<br>Shell rupture<br>Loss of content |
| Shell yielding | – Excessive hoop stress | – Shell wrinkling<br>– Welding failure |
| Local shell wrinkling/rupture | – Stress concentration around connections<br>– Excessive uplift | – Shell rupture<br>– Loss of content |
| Welding failure | – Welding defects<br>– Excessive tank uplift (for bottom plate-to-shell junction welding)<br>– Shell excessive deformation (e.g., due to buckling) | – Loss of content<br>– Tank collapse |
| Failure of the bottom plate | – Excessive tank uplift | – Loss of content |
| Pipe failure | – The considerable relative displacement of tank and pipe mainly because of tank uplift | – Loss of content |
| Yielding/Rupture of anchors | – Insufficient number of anchors<br>– Insufficient anchor cross-sectional area<br>– Anchor corrosion | – Tank uplift and its consequences |
| Anchor pull-out | – Insufficient bonding strength<br>– Anchor corrosion<br>– Degradation of foundation material due to environmental attacks | – Tank uplift and its consequences<br>– Structural failure of foundation |
| Content leakage | – Pipe failure<br>– Local shell rupture<br>– Rupture of the bottom plate<br>– Welding failure<br>– Liquid sloshing<br>– Failure of roof sealing (in floating roofs) | – Environmental pollutions<br>– Fire and/or explosions |

(*Continued*)

**Table 3.1** (Continued)

| Failure mode | Frequent reason(s) | Probable consequence(s) |
|---|---|---|
| Excessive Foundation settlement | – Geotechnical instability (e.g., liquefaction)<br>– Tank-to-foundation collision of uplifted tanks | – Failure of pipes or appurtenances |
| Structural failure of foundation | – Tank-to-foundation collision of uplifted tanks<br>– Anchor pull-out | – N/A |
| Roof wrinkling (mainly in fixed roofs) | – Hydrodynamic effects<br>– Tank uplift | – N/A |
| Failure of floating roof sealing | – Liquid sloshing<br>– Roof rocking response | – Leakage of content and its consequences |
| Floating roof submerging | – Liquid sloshing<br>– Roof rocking response | – Leakage of content and its consequences |

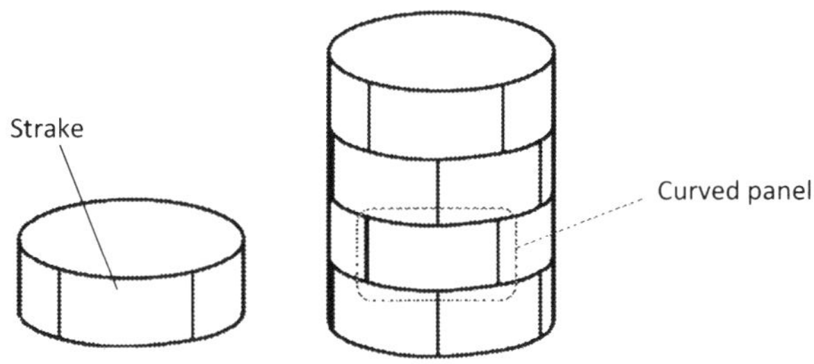


**Figure 3.5** Typical structural configuration of steel tank shells.

As indicated in Fig. 3.5 tank shells consist of many strakes that are fabricated by joining relatively thin curved panels (Teng & Rotter, 2006). Based on the economic aspects, the thickness of the shell usually decreases from bottom to top strake. Hence, as steel tanks are thin-walled structures, shell buckling is one of their most probable failure modes. Tank shells may buckle in elastic and/or inelastic forms. The elastic buckling (so-called diamond-shaped buckling) takes place in relatively thinner shells (Miladi et al., 2020; Virella et al., 2006). In thinner tanks such as stainless steel ones which are more common in food and beverage industries, it may happen in the lower parts of the tank shell (See Fig. 3.6); but in relatively thicker tanks, such as prevalent oil tanks the elastic buckling may

**Figure 3.6** Diamond-shaped buckling of a steel tank during the Silakhor Earthquake of 2006 in Iran.

occur in upper strakes (Razzaghi, 2007). On the other hand, the inelastic (elephant's foot) buckling which is characterized by an outward bulge, occurs in thicker cylindrical shells (See Fig. 3.7). In other words, it can usually take place in the lower strakes of thicker tanks (Hamdan, 2000; Razzaghi, 2007). However, as indicated in Fig. 3.8, as an infrequent case, the inelastic buckling has been observed in the middle strakes known as the elephant's knee-buckling (FEMA, 2011). Further information about the shell buckling is presented in section 3.3.

The excessive deformation of the tank shell may lead to overstressing in the attached pipes or shell-to-pipe junction. Unanchored tanks are notably more susceptible to such a failure mode because of the uplift behavior. Fig. 3.7 illustrates the typical failure of pipes in unanchored tanks during the Bam Earthquake of 2003 in Iran. As indicated in Fig. 3.8, damage to the tank piping system can lead to the content spill and consequently environmental pollution, fire, and explosions. The sudden loss of content may also cause damage because of the vacuum created in the upper parts of the tank (González et al., 2013).

Generally, the tank bottom plate is a thin plate whose out-of-plane deformation may lead to its inelastic behavior. The shell-to-baseplate welding failure and the content loss are aftermaths of such a performance. On the other hand, the tank-to-foundation collision may cause damage to the foundation (See Fig. 3.9).

**Figure 3.7** Pipe failure in an uplifted tank following the Bam Earthquake of 2003 in Iran.

In anchored tanks, the overturning moment may apply considerable tension on mechanical anchors, leading to anchorage inelastic elongation or anchor bolt pull-out. Fig. 3.10 illustrates the residual elongation of an anchor bolt. It should be noted that an anchored tank may experience uplift following the failure of its anchorage system. Since such a tank is not necessarily designed for uplifting behavior it may

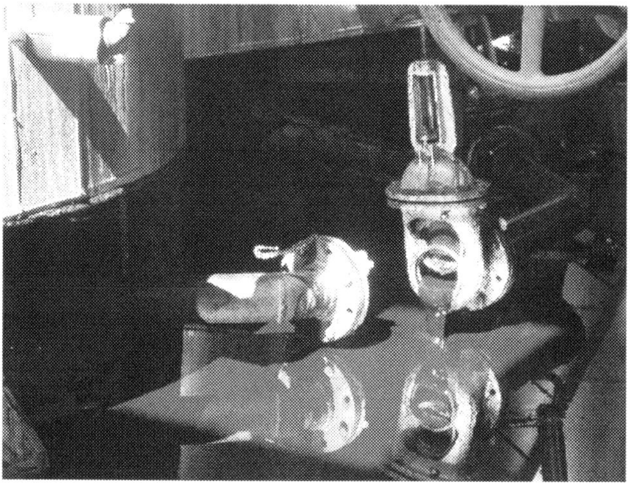

**Figure 3.8** Liquid spill following the pipe failure during the Silakhor Earthquake of 2006 in Iran.

**Figure 3.9** Damage to the foundation due to tank-to-foundation collision during the Silakhor Earthquake of 2006 in Iran.

encounter catastrophic failure mechanisms as an aftermath of damage to mechanical anchors.

As indicated in Fig. 3.11, the sloshing liquid may apply hydrodynamic forces on fixed roofs and cause roof wrinkling. It can also lead to damage to the floating roof sealing apparatus and make the floating roof

**Figure 3.10** Residual elongation of an anchor bolt.

**Figure 3.11** Roof wrinkling as a result of the impact of sloshing liquid.

**Figure 3.12** Oil spill because of sloshing behavior of the tank content during the Bam Earthquake of 2003 in Iran.

submerged. Moreover, as shown in Fig. 3.12 the sloshing liquid can cause liquid to spill from the roof openings (e.g., manholes and exhaust valves).

The tank failure modes due to the earthquakes are not limited to the abovementioned ones. Several other potential damages such as foundation settlement, shell rupture, and tank-foundation overturning may happen during a devastating earthquake. However, the above failure modes are the most frequently observed damages following the past earthquakes.

## 3.4 Shell buckling

Since shell buckling is one of the most frequent failure modes in steel tanks, it has attracted the attention of several researchers in recent decades. The buckling of steel tank shells depends on several parameters such as geometrical specifications, boundary conditions (e.g., anchorage system and roof type), the mechanical properties of shell material, loading conditions, shell imperfections, and structural openings (Brunesi &

Nascimbene, 2018; Miladi & Razzaghi, 2014). The seismic design codes of practice provide diverse criteria to prevent steel tanks from different types of buckling. In addition, researchers have presented analytical and empirical relations and numerical techniques to estimate the buckling capacity of tanks.

### 3.4.1 Analytical solutions

The early studies on buckling of the cylindrical shells were independently conducted by Timoshenko (1910), Lorenz (1908), and Southwell (1914), which resulted in the elastic buckling stress of a perfect cylinder under pure axial compression. The abovementioned buckling stress is known as the classical buckling stress and for a steel cylindrical shell can be expressed as follows:

$$\sigma_{cl} = 0.605E\frac{t}{R} \tag{3.5}$$

where $E$ is Young's modulus, $t$ is the shell thickness and $R$ is the cylinder's radius. Subsequent studies revealed that in practice, the buckling strength is considerably less than the classical stress (Harris et al., 1957; Teng & Rotter, 2006). Such observations paved the way for further extensive studies showing the significant impact of imperfections on cylindrical shell buckling (Hutchinson & Amazigo, 1967; Song et al., 2004; Teng & Rotter, 1992).

Although classical buckling deals with perfect shells, it is the basis of many practical relations for the elastic buckling of cylindrical shells, even in seismic design codes. For instance, the European Convention for Constructional Steelwork (ECCS, 1988) suggests a modified form of the classical buckling for estimating the elastic buckling capacity of cylinders as follows:

$$\sigma_{c} = \alpha\left(0.605E\frac{t}{R}\right) \tag{3.6}$$

where $\alpha$ is a modification factor with respect to imperfections. Besides, the elastic buckling criteria in Eurocode 8 and NZSEE are principally further developments of the classical buckling (Buratti & Tavano, 2014).

On the other hand, inelastic buckling is one of the frequent modes of buckling which may occur in thicker shells. It usually happens in the lower parts of tanks near the base and sometimes appears in positions the shell thickness changes. In liquid storage tanks, the above zones are prone to local bending. In the lower parts, the bending moment appears because of the shell's cantilever action. However, in thickness changing zones, the difference in membrane deflection state and possible eccentricity of the thrust line

(Teng & Rotter, 2006). The axial compression can amplify the local bending moment and consequently yield the tank shell. Such behavior leads to inelastic buckling in the abovementioned zones known as elephant's foot- or elephant's knee-buckling according to the location of the buckled area. Rotter's formula is one of the most significant relations for estimating the inelastic buckling strength of steel cylindrical tanks. Based on the Rotter's relation the inelastic buckling strength is dependent on the yield stress, internal pressure, Young's modulus, and thickness-to-radius ratio as follows:

$$\sigma = \sigma_{cl}\left(1 - \left(\frac{PR}{f_y t}\right)^2\right)\left(1 - \frac{1}{1.12 + r^{1.5}}\right)\left(\frac{r + \frac{f_y}{250}}{r + 1}\right) \tag{3.7}$$

where $r = R/(400t)$, $P$ is the total internal pressure and $f_y$ is the yielding stress of steel. It should be noted that the Rotter's relation is the basis for calculating the buckling strength in NZSEE and Eurocode8.

### 3.4.2 Dynamic buckling assessment

As the buckling of cylindrical tanks during an earthquake is a dynamic phenomenon, it is highly dependent on the characteristics of the strong ground motion and the dynamic properties of the tank. The most significant advantage of analytical relations is their simplicity and ease of use. However, they do not take the dynamic nature of seismic buckling into account. To consider the dynamic characteristics associated with the buckling of a tank during a particular earthquake, the Budiansky-Roth (Budiansky & Roth, 1962) procedure is usually employed. The procedure was not mainly developed for liquid storage tanks but was employed by several researchers for dynamic buckling of tank shells (Maheri & Abdollahi, 2013; (Miladi & Razzaghi, 2019); Virella et al., 2006). It is based on the incremental dynamic analysis (IDA). In other words, the tank shall be repeatedly subjected to nonlinear response history analyses using the increasing levels of a single earthquake record. Accordingly, the Budiansky-Roth approach should be associated with a nonlinear numerical technique (generally FEs). The radial deformation ($U_{rad}$) of the tank shell shall be measured for every level of the selected strong ground motion. Note that $U_{rad}$ shall be measured in every node which may be prone to buckling. Buckling happens when a sudden change is created in a particular node of a tank shell, as shown in Fig. 3.13. The minimum peak ground acceleration (PGA) corresponding to such a jump in the tank response is known as the critical peak ground acceleration ($PGA_{cr}$). Usually, to eliminate the human mistakes in the visual recognition of $PGA_{cr}$, the pseudo-equilibrium path is

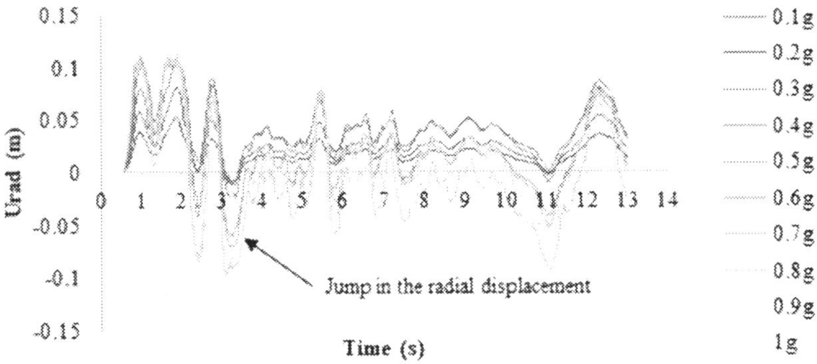


**Figure 3.13** Variation of $U_{rad}$ response history with PGA.

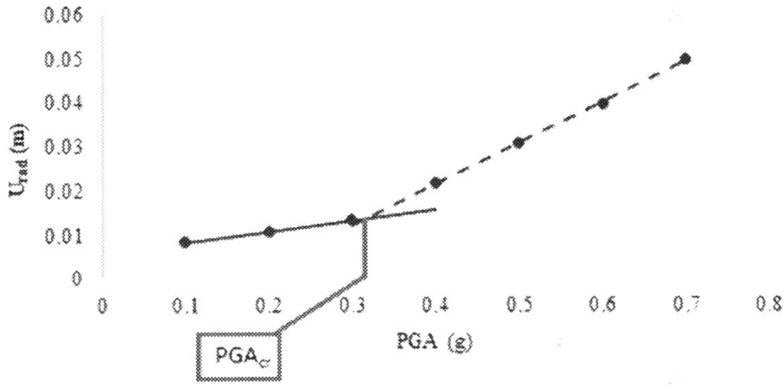


**Figure 3.14** Determination of the $PGA_{cr}$ using pseudo-equilibrium path.

employed. The pseudo-equilibrium path is a plot of $U_{rad}$ versus PGA for a particular shell node. Linear regression is usually utilized to make the pseudo-equilibrium path a bilinear function. The PGA corresponding to the intersection of the lines in the pseudo-equilibrium path is $PGA_{cr}$ (See Fig. 3.14).

## 3.5 Factors affecting the seismic performance

### 3.5.1 Geometrical specifications

The seismic performances of tanks depend on geometric configuration. Some researchers relate the seismic vulnerability of tanks to their height-to-diameter ratio (H/D), especially in unanchored tanks. O'Rourke and

So (2000) have prepared a dataset of damaged tanks during the occurred devastating earthquakes in the United States. To investigate the vulnerability of tanks, they categorized them into two classes $H/D < 0.7$ and $H/D \geq 0.7$. They revealed that the failure during an earthquake is more probable in the tanks of $H/D \geq 0.7$. Razzaghi (2007) categorized the tanks into three classes broad ($H/D < 0.5$), moderate ($0.5 \leq H/D < 0.6$), and tall ($H/D \geq 0.6$), and stated that the seismic performances of tall tanks significantly differ from broad ones. By the way, Razzaghi and Eshghi (2015) showed that H/D is a significant source of uncertainty in the seismic performance assessment of tanks.

### 3.5.2 The relative amount of content

The relative amount of content (%full) plays a significant role in the seismic performance of tanks. The dependency of the tank seismic performance to %full has two reasons. The first reason is that by increasing the relative amount of content, the impulsive and convective mass of a particular tank increases, and thus the earthquake-induced load rises. The second one is that the distribution of the hydrodynamic pressure exerted on the tank shell is notably dependent on the content height-to-diameter ratio ($h_L/D$). Razzaghi and Eshghi (2008a) showed that in unanchored broad tanks, the amount of uplift and the shell axial stress decreases by reducing ($h_L/D$). However, their numerical analyses showed that in unanchored tall tanks the reduction in ($h_L/D$) does not necessarily lead to a reduction in their seismic responses (See Fig. 3.15).

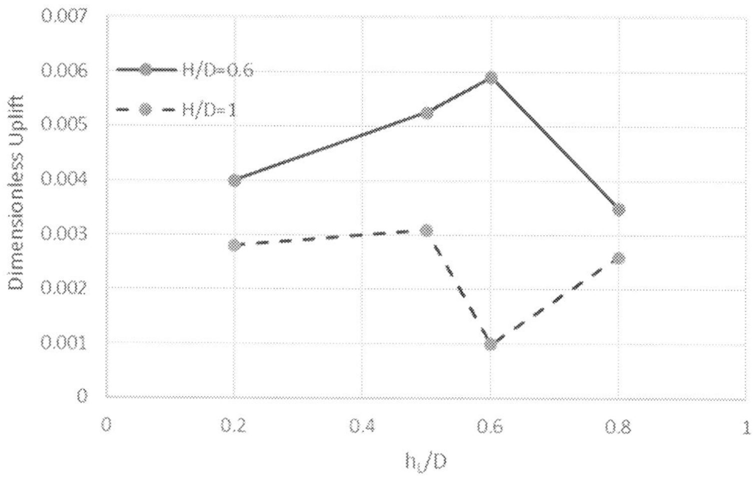


**Figure 3.15** Variation of tank uplift with $h_L/D$ (Razzaghi & Eshghi, 2008a).

Such a result is consistent with uplift-induced damages observed in a tall tank of $h_L/D < 0.5$ following the Bam earthquake of 2003 in Iran (Eshghi & Razzaghi, 2005). It seems that the reason for such a complicated behavior is that in tall tanks, particularly those with lower diameters, the seismic response is mostly controlled by the dynamic characteristics of the tank structure than the tank-liquid system (Razzaghi, 2007). For this reason, the Iranian guideline for the seismic evaluation and retrofit of petroleum facilities (MOP, 2016b) states that the reduction in %full as a damage prevention strategy requires accurate and precise investigations. O'Rourke and So (2000), showed that the probability of failure in tanks with %full < 50% is remarkably low. Razzaghi and Eshghi (2015), revealed that the potential for damage to unanchored tanks during an earthquake is remarkably dependent to %full.

## 3.5.3 Strong ground motion characteristics

Generally, the seismic performances of structures are remarkably dependent on the characteristics of earthquakes (e.g., frequency content and the duration of strong ground motion). Based on the structural configuration and specifications, structures may be more susceptible to some of the characteristics of earthquakes (Mosleh et al., 2016b). As the site-to-source distance plays a significant role in strong ground motion characteristics, the structural responses may notably differ due to near- and far-field earthquakes (Alonso-Rodríguez & Miranda, 2015; Mosleh et al., 2016a). Regarding liquid storage tanks, Razzaghi and Eshghi (2004) conducted a brief numerical study on the seismic behavior of tanks during near-fault earthquakes and revealed that steel tanks might be susceptible to near-fault excitation. Razzaghi (2007) prepared a dataset of the performance of about 50 steel tanks during devastating earthquakes in Iran and showed that most of the damaged tanks were located near the seismic sources. He related such an observation to the probable susceptibility of tanks to near-fault ground motions. Kalogerakou et al. (2017) investigated the hydrodynamic effects on tanks and showed that the second sloshing mode is not negligible for the tanks subject to near-fault earthquakes. Miladi et al. (2020) investigated the buckling and uplift of unanchored tanks during near- and far-field earthquakes. As indicated in Fig. 3.16, they showed that in most tank-PGA cases, the average tank uplift due to near-field earthquakes was higher than that of far-field ones. On the other hand, for a particular tank, the buckling happened in lower PGAs subjected to near field earthquakes than in far ones (See Fig. 3.17). In other words,

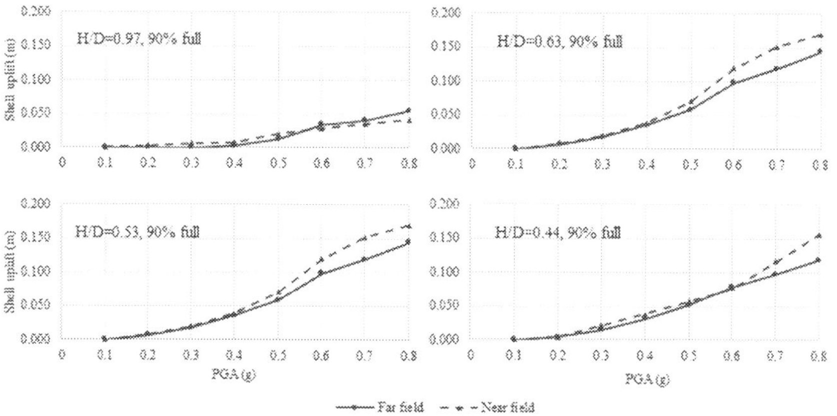


**Figure 3.16** Comparison of average uplift due to near- and far-field earthquakes in different tanks. *After Miladi, S., Razzaghi, M. S., & Ghasemi, S. H. (2020). Seismic performance of imperfect unanchored tanks. Proceedings of the Institution of Civil Engineers-Structures and Buildings, 1—10.*

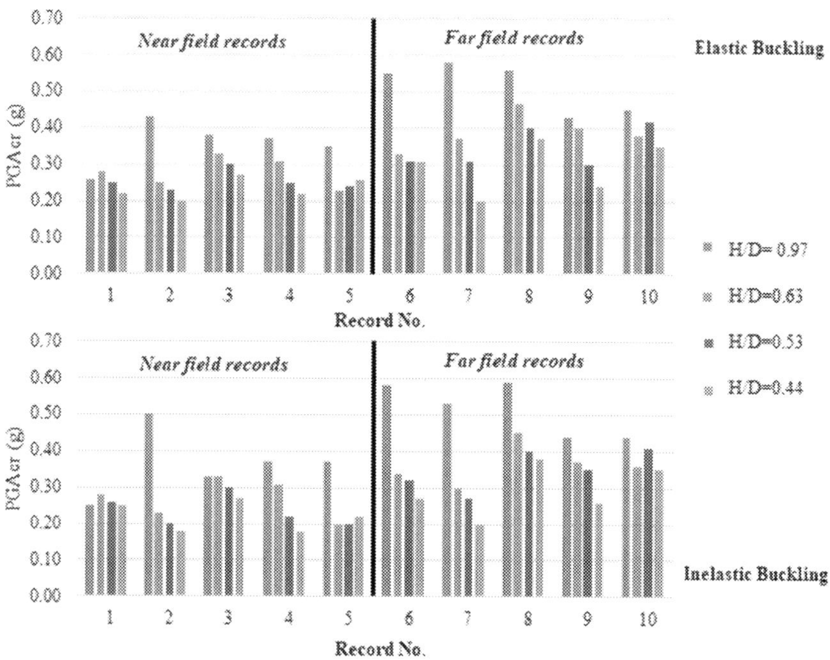


**Figure 3.17** Critical PGAs due to near- and far-field earthquakes in different tanks. *After Miladi, S., Razzaghi, M. S., & Ghasemi, S. H. (2020). Seismic performance of imperfect unanchored tanks.* Proceedings of the Institution of Civil Engineers-Structures and Buildings, *1—10.*

subjecting to near-fault strong ground motion, unanchored tanks are mostly more vulnerable. They related such performance to the frequency content of near-field earthquakes and demonstrated that as near-field ground motions are associated with high frequencies, they may generate significant dynamic responses in the impulsive mode of tanks.

### 3.5.4 Fabrication quality and imperfection

Since the steel cylindrical tanks are thin-walled structures, they are significantly prone to geometrical imperfections. Such imperfections can be classified into two categories: inherent and fabrication induced imperfections. While fabrication-induced imperfections are localized around the weld lines in the junctions of shell segments, inherent imperfections are deviations from nominal geometry, which are randomly distributed on tank structures. The amplitude of fabrication-induced imperfections is strongly related to the quality of construction (Teng & Rotter, 2006). The buckling strength of axially loaded cylindrical shells considerably depends on imperfection amplitude (Rotter & Teng, 1989; Teng & Rotter, 2006). Since seismic loads are mainly transversal, such an imperfection-dependency may not be necessarily valid for the earthquake-induced buckling. Miladi et al. (2020) conducted a series of nonlinear response history analyses to investigate the effect of inherent and fabrication-induced imperfections. They considered several random imperfection patterns on tanks using probabilistic approaches. The results of their study revealed that for imperfection amplitude-to-thickness ratios (w/t)s less than or equal to 3; the critical PGA is not significantly susceptible to the pattern and amplitude of inherent imperfections. Nevertheless, the dependency of the dynamic buckling strength on fabrication-induced imperfections may not be negligible, especially for poorly fabricated tanks. They have also shown that the tank uplift in unanchored tanks is not dependent on geometric imperfections.

### 3.5.5 Corrosion and maintenance

Steel liquid storage tanks especially those located in chemical and petrochemical plants are permanently prone to corrosion. The amount of corrosion is often defined by the yearly corrosion rate. It depends on several parameters such as the chemical specifications of the tank content, the level of protective techniques, and maintenance. For instance, the corrosion rate for tanks containing crude oil and gasoil are approximately 0.3

and 0.5 mm/year respectively (Martinez, 2013). On the other hand, the corrosion rate is not the same in different parts of the tank. It mostly takes place at the internal face of the tank structure, and the corrosion of the outside layer is notably less (Manshadi & Maheri, 2010; Zagórski et al., 2004). The lower parts of tanks near their base are the most susceptible regions to local acidification-induced corroding. The reason is the sedimentation of corrosive material such as hydrogen sulfide. Moreover, the upper parts of the tank shell, which are prone to the daily changes in the liquid level, are the second most susceptible parts to corrosion, while the other parts have less potential for corroding (Manshadi & Maheri, 2010).

The excessive corrosion due to the weak maintenance and deploying inefficient protective strategies can significantly reduce the service life of liquid storage tanks (Medvedeva & Tiam, 1998). Many of the tank failure modes are consequences of steel corrosion. It causes about 20% of the loss in hydrocarbon material due to spillage (Chang & Lin, 2006; Maheri & Abdollahi, 2013). In addition, corrosion may considerably increase the level of the seismic vulnerability of tanks. However, so far, only a few research articles are available on the seismic performance of corroded tanks. Tavakoli Joorabi and Razzaghi (2019) showed that tank corrosion might affect the retrofitting strategy. They stated that neglecting the steel corrosion in the retrofitting design procedure may notably reduce the reliability of the retrofitted structure.

As indicated in Fig. 3.18, steel corrosion may increase the uplift in unanchored tanks, especially for higher PGAs. It is worth mentioning that the curves illustrated in Fig. 3.18, present the average of IDA curves for ten earthquake records. Since the corrosion in the bottom plate reduces

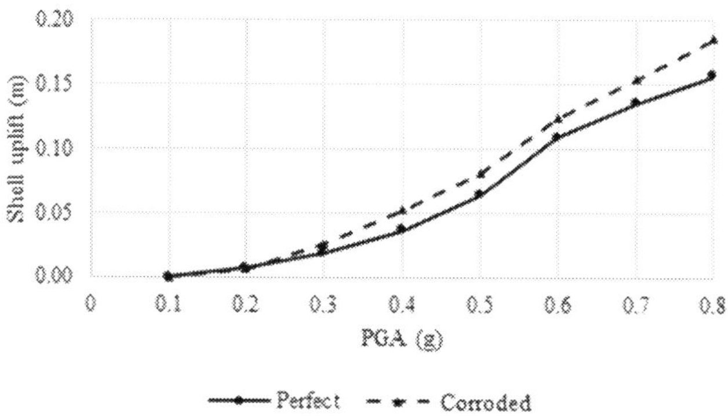


**Figure 3.18** The IDA curve for the uplift of a broad tank with and without considering corrosion.

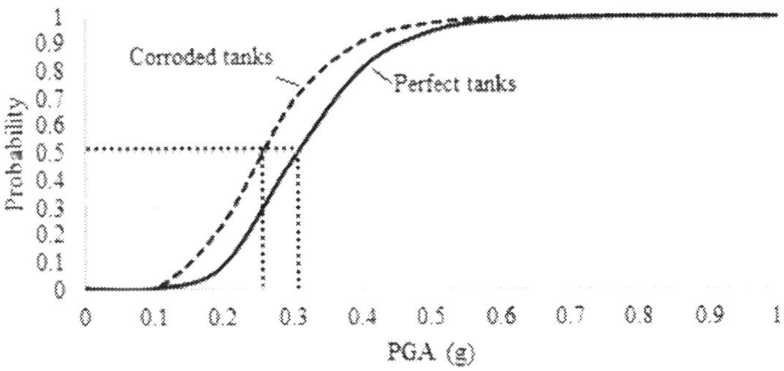


**Figure 3.19** Buckling probability of tanks with and without considering (Corrosion rate 0.5 mm/year, after 5 years).

its thickness, it may encounter failure during the tank uplift. On the other hand, the thinning of tank shells may increase the potential for shell buckling in corroded tanks. Mahri and Abdollahi (2013) showed that by thinning tank shells caused by corrosion, the critical buckling of the tank decreases. Fig. 3.19 compares the probability of reaching (or exceeding) buckling damage state in a series of the tank with and without considering the corrosion in different levels of PGA. As indicated in Fig. 3.19, the probability of the inelastic buckling of tanks considerably increases by the corrosion. It should be noted that Fig. 3.19 is based on the assumption of constant thinning along the shell axis which is the worst scenario of corrosion.

## 3.6 Seismic design codes

During an earthquake, a particular liquid storage tank encounters a considerable fluid-structure interaction. Furthermore, Tanks generally have lower redundancy and lesser ductility to buildings and also are different in damping levels. Therefore, their seismic analysis and design specifications differ from ordinary buildings. On the other hand, liquid storage tanks have more significant failure consequences than ordinary buildings. Hence the design performance targets for such structures differ from buildings.

Hence, specific codes and standards are required for the seismic design of liquid storage tanks. Accordingly, several national codes and standards are provided for the seismic analysis and design of tanks. Table 3.2

**Table 3.2** Some of the available codes and standards for seismic loading, analysis, and design of steel tanks.

| Code | Mechanical analogy | Seismic force level |
|---|---|---|
| NZSEE | Flexible shell | Strength design |
| API 650 | Rigid shell | Allowable stress design |
| 038 | Rigid shell | Allowable stress design[a] |
| AWWA D100 | Rigid shell | Allowable stress design |
| ASCE-7 | Rigid shell | Strength design |
| Eurocode 8 | Rigid shell | Strength design |

[a]The code is generally based on the strength design but for steel tanks, the allowable stress design is recommended.

presented a list of some available codes and standards for the seismic analysis and design of steel tanks.

The Annex E of API 650, has included the tank seismic design provisions for several decades. In recent decades, it has been widely used internationally for the seismic design of welded steel oil tanks. Since 2008 Appendix E of API 650 has linked to ASCE 7.

The ASCE 7 is a standard that is internationally used for minimum design load and associated criteria for various types of structures. Its chapter 15 includes seismic design requirements for nonbuilding structures such as liquid storage tanks.

The early edition of NZSEE was published as *The Recommendations for Seismic Design of Storage Tanks* in 1986. Since then it was widely used in New Zealand and some other countries. In 2009, NZSEE published its latest edition Seismic Design of Storage tanks: 2009. The revised version introduced significant changes to the original one. For instance, the design basis of the 1986 edition did not allow assumed any plasticization or damage to tanks under the design earthquake level. While, the 2009 revision allows some ductility for particular types of tanks (NZSEE, 2009).

Part 4 of Eurocode 8, presents the seismic design provisions for liquid storage tanks as well as silos and pipelines. Its tank seismic design provisions have some similarities to NZSEE. For example, both of them employed Rotter's relation as the Elephant's foot buckling criteria.

In Iran, the first edition of the Iranian seismic design code for oil industries, known as the 038 code, was published in 2008. Its chapter 12 includes the seismic design provisions for liquid storage tanks. The second edition of 038 was published in 2010 and considered mandatory to apply for every oil-related project in Iran. In 2016, the third edition of 038 (MOP, 2016a) was published, including significant changes to chapter 12.

It is worth mentioning that the third edition was recommended by the Iranian seismic code (standard No. 2800) to be applied for all other non-building components than oil-related ones.

Despite some similarities among the seismic codes of practices, their design objectives and procedures are rather different. This section aims to present the most significant differences and similarities of the tank seismic design codes.

### 3.6.1 Seismic performance target

Generally, every seismic design procedure shall provide for structural integrity. In addition, for liquid storage facilities, the content spilling may lead to secondary impacts such as environmental pollution and explosions. Therefore, their seismic design procedures shall assure containment of their content (ASCE, 2011). On the other hand, as a component of a process industry or a water distribution system, storage tanks should maintain their specific functionality. In other words, the background of seismic performance objectives of every code of practice is assuring structural integrity, containment, and functionality. However, the definition of seismic performance target may vary from code to code. Table 3.3 lists the description of the seismic performance targets in different codes.

### 3.6.2 Mechanical analogy

All the above codes and standards recommend the quasistatic approach for calculating the seismic actions on liquid storage tanks. To do so, they suggest a mass-spring analogy to simplify the tank-liquid behavior. It is worth mentioning, NZSEE employs a flexible tank model based on (Haroun & Housner,

**Table 3.3** Description of the seismic performance target in different codes.

| Code | Seismic performance target |
|---|---|
| NZSEE | Preventing people from injury or loss of amenity caused by the content spill or tank damage. Protecting the environment and other property from the consequences of the content spill. |
| API 650 | Life protection and preventing catastrophic collapse. |
| 038 | Minimizing the probability of operation and function interruption in nonsevere earthquakes.<br>Minimizing the probability of extensive loss during a strong earthquake. |
| ASCE 7 | Providing sufficient ductility, strength, and stiffness to resist the effects of earthquakes. |

1981) for flexible tanks, while most of the other ones are based on the rigid tank assumption. However, the latter codes and standards ignore the shell flexibility only for calculating the convective and impulsive masses and include it in the evaluation of the impulsive natural period (Jaiswal et al., 2007). The impulsive and convective hydrodynamic loads can be calculated by multiplying the corresponding masses by seismic action coefficients. The impulsive and convective actions are usually combined by the square root of the sum of squares (SRSS) method. However, NZSEE takes the absolute some of the rigid-impulsive and flexible-impulsive actions to calculate the effect of impulsive pressure. The seismic action coefficients are functions of the impulsive and convective natural periods. Codes and standards present different relations for calculating the natural periods corresponding to impulsive and convective portions (See Table 3.4). Although they are apparently different, they result in approximately the same values for a particular case. The seismic action coefficients generally depend on the tank importance, ductility, damping, and seismic hazard level.

The seismic action coefficients generally depend on the tank importance, ductility, damping, and seismic hazard level. The state of these parameters in codes and standards is presented herein:

– Tank importance:

The level of design base earthquake depends on the level of risk associated with the seismic performance of tanks. The level of seismic risk in a particular tank is associated with its systemic function and hazard to the public and environment. Seismic codes utilize different codes for determining the tank's importance. API 650 classifies tanks into three categories so-called seismic use group (SUG) based on their function and their content's hazard to the public

**Table 3.4** Natural period relations in different codes.

| | NZSEE (2009) | API 650 (API, 2020) | 038 (MOP, 2016b) |
|---|---|---|---|
| Impulsive natural period | $T_i = \frac{5.61\pi h_l}{k_h}\sqrt{\frac{\gamma}{Eg}}$ | $T_i = \frac{1}{\sqrt{2000}}\left(\frac{C_i h_l}{\sqrt{\frac{t_e}{D}}}\right)\sqrt{\frac{\rho}{E}}$ | $T_i = (C_i^* h_l)\sqrt{\frac{\rho D}{2t_e E}}$ |
| Period of vertical motion | $T_v = \frac{5.61\pi h_l}{k_v}\sqrt{\frac{\gamma}{Eg}}$ | NA | NA |
| Convective period | $T_{ci} = \frac{2\pi\sqrt{\frac{R}{g}}}{\sqrt{\lambda_i \tanh\left(\lambda_i \frac{h_l}{R}\right)}}$ | $T_c = 1.8k_s\sqrt{D}$ | $T_c = 2\pi\sqrt{\frac{D_1}{g}}$ |

$C_i$ and $C_i^*$, Unitless coefficients depending $\frac{\text{onh}_l}{\text{D}}$; $D_1$, $\frac{D}{3.67\tanh\left(\frac{3.67h_l}{D}\right)}$; $k_h$, Period coefficient according to (Haroun & Housner, 1981); $k_s$, $\frac{0.578}{\sqrt{\tanh\left(\frac{3.68h_l}{D}\right)}}$; $k_v$, Period coefficient of the breathing mode according to Veletsos (1984); $\lambda_i$, The *i*th root of the Bessel's function ($\lambda_1 = 1.841$, $\lambda_2$, 5.331and $\lambda_3$, 8.536); $\gamma$, Unit weight of liquid.

(See Table 3.5). It should be noted that the highest and lowest importance belongs to SUG III and SUG I respectively. As tabulated in Table 3.6, chapter twelve of 038 defines four risk and function categories. The lowest importance belongs to the first category which is not required to design for withstanding an earthquake. The NZSEE defines four consequences (life safety, environmental exposure, community or national significance, and the adjacent property value) for tank damage, among which the importance level shall be

**Table 3.5** Tank importance according to API 650.

| Category | Description | Importance factor (I) |
|---|---|---|
| SUG I | Tanks in an area isolated from public access with secondary spill prevention, or tanks without secondary spill prevention that are sufficiently removed from areas of public access to minimize the hazard. | 1.0 |
| SUG II | Tanks that shall continue to function, following an earthquake, or those tanks that contain moderate-risk content to the public if released. | 1.25 |
| SUG III | Tanks with an essential function for public safety or those tanks that store high-risk content to the public if released and lack secondary protection. | 1.5 |

**Table 3.6** Tank importance categories according to 038.

| Category | Description | Importance level | Importance factor (I) |
|---|---|---|---|
| I | Temporary[a] tanks whose failure does not lead to life loss. | Low | NA |
| II | Those tanks whose failure operation stoppage does not lead to life loss or considerable economic impact. | Moderate | 1.0 |
| III | Tanks that are allowed to suffer an insignificant amount of damage in such a way may bring back into operation in a short period. | High | 1.25 |
| IV | Tanks whose function or content are essential in postearthquake, or tanks with considerably harmful content to public or environment. | Extremely high | 1.5 |

[a]The expression "Temporary" refers to tanks whose service life is less than two years.

determined from the most severe. Four importance levels corresponding to the overall risk level (negligible/slight, moderate, serious, and extreme) are defined by NZSEE. Every importance level corresponds to a specific return period factor, ranging from 0.5 to 1.8. The return period factor directly changes the elastic site hazard spectrum.

– Ductility and performance factor:

Liquid storage tanks generally have low ductility and redundancy, and thus, seismic design codes allow them to pose insignificant inelastic behavior during the design earthquake. In the elastic design procedures, performance or ductility factors (also known as response modification factors) are employed for considering the inelastic behavior of structures. In other words, they employ to reduce the design seismic loads. The tank design codes take different strategies to reduce the elastic design seismic loads. For instance, API 650, ASCE-7, and 038 divide the elastic response to the modifications factors that are greater than 1. As indicated in Table 3.7, API 650 uses impulsive and convective design response modification factors ($R_{wi}$ and $R_{wc}$) for the allowable stress design (ASD) method. Chapter 12 of the 038 code, introduce separate impulsive and convective response modification factors for ASD and strength design (SD). In addition, ASCE-7 utilizes modification factors SD. The NZSEE approach to response modification is considerably different. It provides displacement ductility factors ($\mu$) for different types of tanks, then employs a modification factor, $k_f(\mu, \xi)$ which depends on the ductility factor ($\mu$) and damping level ($\xi$) (See Table 3.8). As indicated in Table 3.8, the modification factor is less than 1.0 for every case except for $\xi = 0.5$ and $\xi = 1$ when $\mu = 1.25$. However, for most practical circumstances, the damping ratio of steel tanks would be higher than 2%. It should be mentioned that $\mu = 1.25$ refers to nonductile

**Table 3.7** Response modification factors in different codes.

| Code | Tank type | Response modification factor | |
|---|---|---|---|
| | | Impulsive | Convective |
| API 650 | Mechanically anchored | | 2 |
| | Self-anchored (unanchored) | 3.5 | 2 |
| ASCE-7 | Mechanically anchored | 3 | 1.5 |
| | Self-anchored (unanchored) | 2.5 | 1.5 |
| 038 (ASD) | Mechanically anchored | 4 | 2 |
| | Self-anchored (unanchored) | 3.5 | 2 |
| 038 (SD) | Mechanically anchored | 3 | 1.5 |
| | Self-anchored (unanchored) | 2.5 | 1.5 |

**Table 3.8** Modification factors in NZSEE.

| μ | $k_f(\mu, \xi)$ | | | | | | | |
|---|---|---|---|---|---|---|---|---|
| | $\xi = 0.5$ | $\xi = 1$ | $\xi = 2$ | $\xi = 5$ | $\xi = 10$ | $\xi = 15$ | $\xi = 20$ | $\xi = 30$ |
| 1.25 | 1.08 | 1.04 | 0.96 | 0.82 | 0.67 | 0.58 | 0.52 | 0.44 |
| 2 | 0.91 | 0.89 | 0.84 | 0.74 | 0.63 | 0.55 | 0.5 | 0.43 |

unanchored or anchored tanks. Besides, in unanchored tanks with limited ductile behavior or those anchored by ductile bolts, $\mu = 2$.

– Damping:

Seismic design codes generally consider different damping ratios for impulsive and convective motion. All of them specify 0.5% damping for convective motion. Such a prescription is based on the assumption that the tank content is inviscid. It should be noted that only a few studies have focused on the influence of fluid viscosity on the damping ratio of tanks. However, for most fluids, the tank content viscosity does not make a notable change to the seismic response (NZSEE, 2009). Many codes and standards prescribe 5% damping for the impulsive mode of steel tanks. However, NZSEE considers another strategy by taking the effects of foundation radiation, soil hysteretic damping, and tank material damping. The NZSEE's procedure for calculating the damping ratio for the horizontal and vertical impulsive modes is based on (Veletsos & Shivakumar, 1997) and (Veletsos & Tang, 1986) respectively. To do so, it provides different graphs for the vertical and horizontal impulsive modes. They are plots of the damping ratio versus the shear wave velocity of the site for various shell thickness-to-radius ratios.

## 3.6.3 Vertical seismic effects

The effect of the vertical component of earthquakes on steel tanks is not negligible. It is mainly effective on mechanical anchors and the shell axial and hoop stresses. Hence, seismic codes generally take the vertical seismic component into the account.

As indicated in Table 3.4, NZSEE presents a relation for calculating the vertical motion (also known as breathing) period. The Vertical seismic loads can then be calculated by employing the spectral acceleration corresponding to the breathing period. API 650 and 038 have not presented any relation for the period of breathing mode. API 650 takes the vertical seismic acceleration as $Av = 0.47S_{DS}$, where $S_{DS}$ is the 5% damped, spectral response acceleration at the design earthquake level at short periods based on ASCE 7 specification.

Provided that Purchaser specifies the vertical ground motion acceleration instead of $S_{DS}$, 70% of that acceleration shall be taken as *Av* according to API 650. On the other hand, Chapter 12, Risk Analysis in Various Countries of the 038 code specifies *Av* as $0.2S_{DS}I$.

### 3.6.4 Anchorage criteria

One of the crucial steps in the seismic design of steel tanks is deciding to provide mechanical anchors. Because of economic aspects and ease of construction most of the older tanks especially precode ones were unanchored (Razzaghi & Eshghi, 2015). Catastrophic failure of such tanks during devastating earthquakes as a result of unacceptable uplift attracted the attention of researchers to stability aspects. Since then seismic design codes have gradually presented tank anchorage criteria. Current editions of the seismic design codes and standards generally provide a conservative seismic design for uplifting tanks (Ormeño et al., 2015).

The recent editions of API 650 have presented an anchorage ratio, *J*, to estimate the overall tank stability to overturning moment:

$$J = \frac{M_{rw}}{D^2\left[w_t(1 - 0.4A_v) + w_a - F_p w_{int}\right]} \tag{3.8}$$

where $M_{rw}$ is the overturning moment, $w_t$ are the tank and roof weigh acting on tank base, and $w_a$ is the force resisting uplift in the annular region. It is worth mentioning that the term $F_p w_{int}$ has inserted into the Eq. (3.8) since the 2020 edition for taking the uplifting effect of internal pressure in pressurized tanks into account. The anchorage criteria of API 650 are then prescribed based on the range of the J ratio as listed in Table 3.9. ASCE-7 ignores the effect of vertical seismic action and internal pressure on J-ratio in its 2016 edition as follows:

$$J = \frac{M_{rw}}{D^2(w_t + w_a)} \tag{3.9}$$

On the other hand, 038 considers the effect of vertical seismic actions but ignores the effect of internal pressure on the pressurized tank. The anchorage ratio of the 038 code is as follows:

$$J = \frac{M_{rw}}{\frac{\pi D^2}{2}\left[w_t(1 - A_v) + w_a\right]} \tag{3.10}$$

**Table 3.9** Anchorage criteria according to different codes and standards.

| J range | Code/standard | | |
|---|---|---|---|
| | **API650** | **ASCE-7** | **038** |
| $J \leq 0.785$ | The tank is self-anchored with no calculated uplift under the design loads | The tank is self-anchored with no uplift under the design seismic loads. | NA |
| $0.785 < J \leq 1.54$ | The tank is self-anchored but uplifting. It is stable for the design loads if the shell compression requirements are satisfied. | The tank is self-anchored but uplifting. It is stable if the shell compression requirements are satisfied. | NA |
| $J > 1.54$ | The tank is unstable and is not self-anchored for the design level of seismic loads.<br>The annular ring shall be modified if L < 0.035D is not controlled or mechanical anchorage shall be provided. | The tank shall be anchored for the design level of seismic loads | NA |
| $J \leq 0.5$ | NA | NA | The tank is self-anchored with no uplift under the design seismic loads. |
| $0.5 < J \leq 1.0$ | NA | NA | The tank is self-anchored but uplifting. It is stable if the shell compression requirements are satisfied. |
| $J > 1.0$ | NA | NA | The tank shall be anchored for the design level of seismic loads |

*Note*: For API 650L is the required minimum width of a thickened bottom annular ring.

The anchorage criteria of 038 are tabulated in Table 3.9. Note in this regard that 038 provides some additional requirements for $0.5 < J \leq 1.0$. For instance, the flexibility requirements of the tank pipes should be satisfied. Otherwise, mechanical anchors shall be provided.

The NZSEE's criteria for the tank anchorage are considerably different. According to NZSEE, the tank designer shall consider mechanical anchors for the tank if:

$$\frac{h_l}{R} > \frac{2}{C_d\left(T_f\right)}, \quad \text{and} \quad R < [C(0)]^2 \frac{h_l}{R}$$

where $C_d\left(T_f\right)$ is the horizontal design action coefficient for the first horizontal impulsive mode considering the foundation flexibility and $\mu = 1.25$, $C(0)$ the site hazard coefficient for $T = 0$, and $R = D/2$.

### 3.6.5 Freeboard requirement

As the sloshing wave displacement may cause damage to fixed roofs or lead to overtopping, sufficient freeboard shall be considered for liquid storage tanks. Accordingly, the required freeboard is directly dependent on convective wave height. As tabulated in Table 3.10, based on API 650, ASCE-7, and 038, the required freeboard is related to the seismicity and seismic usage or risk category. It is worth mentioning that their freeboard relations include adjusting factors to convert 5% damped spectral acceleration to 0.5% one. In other words, the designer shall utilize a 5% damped spectral acceleration for calculating the required freeboard using the above codes or standards. They also suggest 70% of the sloshing wave height as the maximum required freeboard for tanks other than critical or essential ones. However, according to NZSEE, the freeboard should be equal to the maximum vertical displacement of the convective liquid. To this end, it presents a single equation for the sloshing wave height for every tank-site-content case using the SRSS of the first two convective mode components as follows:

$$d_{\max} = R\sqrt{[0.84C_d(T_1)]^2 + [0.07C_d(T_2)]^2} \tag{3.11}$$

where $C_d(T_1)$ and $C_d(T_2)$ are the first and the second convective mode horizontal design action coefficient for a 0.5% damped system of $\mu = 1$.

It should be noted that the above codes and standards have made some exceptions in their freeboard requirements (See Table 3.11).

**Table 3.10** The freeboard requirement in various codes or standards.

| Code | $S_{DS}$ values | Seismic Usage group/risk category | | |
|---|---|---|---|---|
| **API650** | | **SUG I** | **SUG II** | **SUG III** |
| | $S_{DS} < 0.33$ | Not required | Not required | $(0.42D)(KS_{D1}I)(1/T_c)$, for $T_c \leq T_L$<br>$(0.42D)(KS_{D1}I)(T_L/T_c^2)$, for $T_c > T_L$ |
| | $S_{DS} \geq 0.33$ | Not required | $(0.7)(0.42D)(KS_{D1}I)(1/T_c)$, for $T_c \leq 4$<br>$(0.7)(0.42D)(KS_{D1}I)(4/T_c^2)$, for $T_c > 4$ | $(0.42D)(KS_{D1}I)(1/T_c)$, for $T_c \leq T_L$<br>$(0.42D)(KS_{D1}I)(T_L/T_c^2)$, for $T_c > T_L$ |
| ASCE-7 | | I & II | III | IV |
| | $S_{DS} < 0.33$ | Not required | Not required | $(0.42D)(1.5S_{D1}I)(1/T_c)$, for $T_c \leq T_L$<br>$(0.42D)(1.5S_{D1}I)(T_L/T_c^2)$, for $T_c > T_L$ |
| | $S_{DS} \geq 0.33$ | Not required | $(0.7)(0.42D)(1.5S_{D1}I)(1/T_c)$, for $T_c \leq T_L$<br>$(0.7)(0.42D)(1.5S_{D1}I)(T_L/T_c^2)$, for $T_c > T_L$ | $(0.42D)(1.5S_{D1}I)(1/T_c)$, for $T_c \leq T_L$<br>$(0.42D)(1.5S_{D1}I)(T_L/T_c^2)$, for $T_c > T_L$ |
| 038 | | II | III | IV |
| | $S_{DS} < 0.33$ | Not required | Not required | $(0.63D)(S_{D1}I/T_c)$, for $T_c \leq 4$<br>$(2.5D)(S_{D1}I/T_c^2)$, for $T_c > 4$ |
| | $S_{DS} \geq 0.33$ | Not required | $(0.7)(0.63D)(S_{D1}I/T_c)$, for $T_c \leq 4$<br>$(0.7)(2.5D)(S_{D1}I/T_c^2)$, for $T_c > 4$ | $(0.63D)(S_{D1}I/T_c)$, for $T_c \leq 4$<br>$(2.5D)(S_{D1}I/T_c^2)$, for $T_c > 4$ |

$K$, Adjusting coefficient to convert spectral acceleration from 5% to 0.5% damping ($K$, 1.5 unless otherwise specified); $S_{D1}$, 5% damped spectral acceleration parameter of design earthquake corresponding to the period of one second; $T_L$, Transition point for long-period strong ground motion.

**Table 3.11** Freeboard exceptions in various codes or standards.

| Code/ standard | Freeboard exception(s) |
|---|---|
| API 650 | – Secondary components are provided to prevent the product spill<br>– The roof and its supporting structure are designed for sloshing wave pressure. |
| ASCE-7 | – Open-top tanks whose content is not hazardous if the site-specific content spill prevention, control, and countermeasure plan are available.<br>– The roof and its supporting structure are designed for sloshing wave pressure. |
| 038 | Same as API 650 |
| NZSEE | – The roof and its supporting structure are designed for sloshing wave pressure. |

## 3.7 Fragility based seismic performance assessment

Significant uncertainties associated with the seismic performance of tanks necessitate providing a probabilistic framework for the seismic safety assessment. A fundamental requirement for such an assessment is the ability to quantify the potential for structural damage as a function of the severity of seismic hazards (so-called an intensity measure) (Manjily et al., 2021). Fragility curves provide such a platform through a conditional probabilistic function as follows:

$$F = P\left[\left(\frac{S_d}{S_c} \geq 1|, IM\right)\right] \tag{3.12}$$

where F is the fragility function, $S_d$ and $S_c$ are the structural demand and capacity of the tank respectively and $IM$ is the ground motion intensity measure. A cumulative lognormal distribution is mathematically convenient for specifying the uncertainties associated with the seismic performances of structures (Hancilar et al., 2013; Razzaghi & Eshghi, 2015). Hence, usually, it is assumed that the fragility function can be represented by the cumulative standard lognormal function (Kircher et al., 1997; Shinozuka et al., 2000) as follows:

$$F = \Phi\left[\frac{1}{\beta}\ln\left(\frac{S_d}{S_c}\right)\right] \tag{3.13}$$

where $\Phi[.]$ is the cumulative standard normal distribution function and $\beta$ is the logarithmic standard deviation of the variables. To develop a fragility function for a particular type of structure a dataset of seismic performance is required. Accordingly, fragility curves can be classified into four categories based on their data collection methodology: judgmental, empirical, analytical, and hybrid (Razzaghi et al., 2018; Tafti et al., 2020). In judgmental fragility curves, the dataset is prepared based on expert opinion, while the empirical fragility curves are based on observational data collected following the occurred earthquakes. Meanwhile, the analytical fragility curves are developed based on the results of analyses (usually nonlinear response history analysis). It should be noted that in a hybrid approach, the database is prepared by employing a combination of different methods. Fragility curves are widely used in seismic risk analysis of structures (Bakalis et al., 2017). They can also be utilized in the prioritization of structures in retrofitting projects. Therefore, several researchers have developed different fragility curves in recent decades. A summary of available fragility curves for on-ground steel tanks is presented in Table 3.12

**Table 3.12** A summary of the most significant fragility curves for steel tanks.

| Reference | Category | Description | IM |
|---|---|---|---|
| NIBS (1999) | Judgmental | Separate expert opinion-based fragility curves for anchored and unanchored tanks. | PGA |
| O'Rourke and So (2000) | Empirical | Empirical fragility curves in terms of H/D and %full for pre1995 US seismic events. | PGA |
| Berahman and Behnamfar (2007) | Hybrid | Bayesian-based fragility curves for unanchored tanks using historical data and ALA database. | PGA |
| Razzaghi (2007) | Judgmental-empirical-analytical | Separate judgmental, empirical, and analytical fragility curves for unanchored tanks in terms of H/D and %Full. | PGA |
| Berahman a Behnamfar (2009) | Hybrid | Fragility curves for elephant's foot buckling and welding failure of shell-to-bottom plate junction using numerical analysis and Bayesian updating technique. | $Sa(T_i)$ |

(*Continued*)

**Table 3.12** (Continued)

| Reference | Category | Description | IM |
|---|---|---|---|
| Buratti and Tavano (2014) | Analytical | Analytical fragility curves for shell buckling using incremental dynamic analysis. | PGA, PGV, PGD, PSA |
| Razzaghi and Eshghi (2015) | Analytical-empirical | Analytical fragility curves for precode unanchored tanks in terms of H/D and %full as well as an empirical fragility curve using data collected following three major earthquakes in Iran. | PGA |
| D'Amico and Buratti (2019) | Empirical | Empirical fragility curves based on observed seismic performance of tanks and Bayesian approach. | PGA |
| Phan et al. (2019) | Analytical | Analytical fragility curves for shell buckling and shell-to-bottom plate rotation using pushover analysis of simplified models. | $Sa(T_i)$ |

*ALA*, American lifeline alliance; *PGD*, Peak ground displacement; *PGV*, Peak ground velocity; *PSA*, Pseudo spectral acceleration; *Sa(Ti)*, Spectral acceleration corresponding to the period of impulsive mass.

The earliest fragility curves for anchored and unanchored cylindrical tanks are provided by HAZUS (NIBS, 1999). HAZUS defined five qualitative damage states ranging from no damage to the complete collapse of tanks (see Table 3.13). O'Rourke and So (2000) developed a series of empirical fragility curves for on-ground tanks. Their database contained the observed performance of 424 tanks following nine pre1995 seismic events in the United States. They developed separate fragility curves for the tanks of $H/D \geq 0.7$ and those with $H/D < 0.7$. Furthermore, they provide different fragility curves for the tanks with the relative amount of contents less than 50% and greater than 50% respectively. It should be mentioned that O'Rouke and So have adopted HAZUS definitions of the tank damage states. The earliest analytical fragility curves of unanchored tanks were developed by Razzaghi (2007) by employing nonlinear response history analysis utilizing FE models. Since then several analytical fragility curves were developed for different classes of steel tanks (Buratti & Tavano, 2014; Phan et al., 2019; Razzaghi & Eshghi, 2015).

Fig. 3.20 illustrates three of the above fragility curves for different levels of damage. As indicated in Fig. 3.20, there is a notable difference

**Table 3.13** Definition of the tank damage states in HAZUS (NIBS, 1999).

| Damage state | Severity | Description |
|---|---|---|
| DS1 | None | No damage |
| Ds2 | Slight | The tank is functional but suffering minor damage without loss of its contents. Localized wrinkles in structure. Minor damage to the tank roof. |
| Ds3 | Moderate | Considerable damage with only minor loss of content. Elephant's foot buckling without loss of tank content. |
| Ds4 | Extensive | The tank going out of service because of severe damage. Elephant's foot buckling with a loss of content |
| Ds5 | Complete collapse | Tank collapsing and losing all of its content |

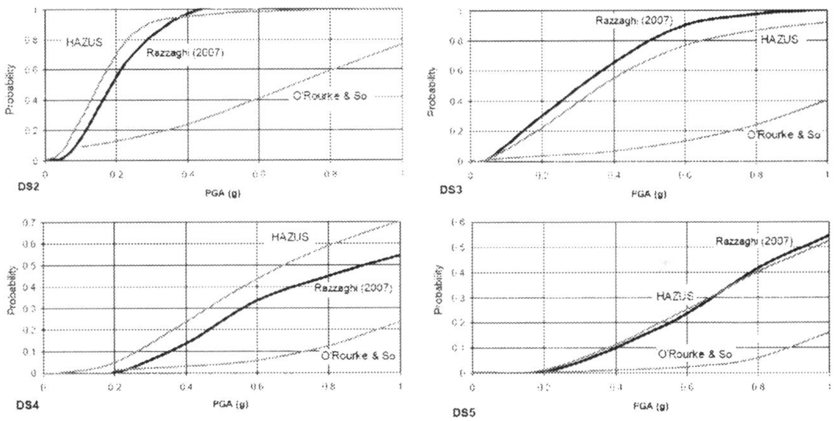


**Figure 3.20** Comparison of available fragility curves.

among the illustrated fragility curves. The main reason for such a difference is significant uncertainties associated with the seismic performance assessment of liquid storage tanks (Razzaghi & Eshghi, 2015).

## 3.8 New horizons for further developments

Nonlinear response history analyses are the most realistic analysis methods for simulating the seismic behavior of structures. However, they usually claim a considerable amount of computer memory and analysis time. On the other hand, they may encounter convergence difficulties, especially for highly nonlinear problems. Hence, during recent decades nonlinear response history analysis of liquid storage tanks was associated with different levels of simplifications. For instance, in most of the research programs, it was assumed that the tanks are perfect. The extensive progress in nonlinear FE techniques alongside the advancement of computer hardware and software provide the basis for more comprehensive and complicated analyzes. Such a condition is a crucial opportunity to conduct sample-based probabilistic approaches. It is beneficial for modeling tank imperfections or various corrosion patterns. For example, by deploying an appropriate probabilistic model, one can generate a considerable number of random variables (e.g., imperfection pattern, material properties, etc.). Analyzing an enormous number of sample-based simulated tanks paves the path to reliable and comprehensive conclusions on the factors affecting the tank's seismic performance.

## 3.9 Conclusions

This chapter has outlined an overview of the field of seismic performance assessment and design of liquid storage tanks. A review of current knowledge on different aspects of the subject has been presented. The chapter has addressed the most frequent failure modes of steel cylindrical tanks during earthquakes, alongside their reasons and associated consequences.

The seismic performance assessment of tanks is associated with considerable complexities. The overall seismic behavior of tanks is complicated and includes different types of nonlinearities. Besides, diverse parameters may influence the response of tanks to the strong ground motions. Some of them, such as inherent imperfections, have random characteristics and are not easy to model. However, recent advancements in numerical analysis techniques, particularly FE methods, provide an appropriate basis to reduce the uncertainties associated with modeling.

Different design codes and standards are in current use. Although they are the same in some aspects, they differ in many assumptions and design criteria. Therefore, the safety level of a particular tank as designed by different codes of practice may vary markedly. Also, Because of the evolution of design specifications in consecutive editions of a particular code, the tanks designed by its different editions are not the same in safety level.

Based on the above, the field of seismic performance assessments of liquid storage tanks is strongly active. Different aspects of the subject still need a considerable body of research programs.

## References


Ahari, M. N., Eshghi, S., & Ashtiany, M. G. (2009). The tapered beam model for bottom plate uplift analysis of unanchored cylindrical steel storage tanks. *Engineering Structures*, *31*(3), 623–632.

Alonso-Rodríguez, A., & Miranda, E. (2015). Assessment of building behavior under near-fault pulse-like ground motions through simplified models. *Soil Dynamics and Earthquake Engineering*, *79*, 47–58.

API (2020). Welded tanks for oil storage, API Standard 650, American Petroleum Institute.

ASCE (2011). Guidelines for seismic evaluation and design of petrochemical facilities. American Society of Civil Engineers. Task Committee on Seismic Evaluation and Design of Petrochemical Facilities, USA.

Bakalis, K., & Karamanos, S. A. (2021). Uplift mechanics of unanchored liquid storage tanks subjected to lateral earthquake loading. *Thin-Walled Structures*, *158*, 107145.

Bakalis, K., Vamvatsikos, D., & Fragiadakis, M. (2017). Seismic risk assessment of liquid storage tanks via a nonlinear surrogate model. *Earthquake Engineering & Structural Dynamics*, *46*(15), 2851–2868.

Berahman, F., & Behnamfar, F. (2007). Seismic fragility curves for un-anchored on-grade steel storage tanks: Bayesian approach. *Journal of Earthquake Engineering*, *11*(2), 166–192.

Berahman, F., & Behnamfar, F. (2009). Probabilistic seismic demand model and fragility estimates for critical failure modes of un-anchored steel storage tanks in petroleum complexes. *Probabilistic Engineering Mechanics*, *24*(4), 527–536.

Brunesi, E., & Nascimbene, R. (2018). Effects of structural openings on the buckling strength of cylindrical shells. *Advances in Structural Engineering*, *21*(16), 2466–2482.

Brunesi, E., Nascimbene, R., Pagani, M., & Beilic, D. (2015). Seismic performance of storage steel tanks during the May 2012 Emilia, Italy, earthquakes. *Journal of Performance of Constructed Facilities*, *29*(5), 04014137.

Budiansky, B., & Roth, R. S. (1962). Axisymmetric dynamic buckling of clamped shallow spherical shells. NASA TN-D 1510, pp. 597–606.

Buratti, N., & Tavano, M. (2014). Dynamic buckling and seismic fragility of anchored steel tanks by the added mass method. *Earthquake Engineering & Structural Dynamics*, *43* (1), 1–21.

Cambra, F. J. (1982). Earthquake response considerations of broad liquid storage tanks. NASA STI/Recon Technical Report No. 84, 13393.

Cambra, F. J. (1983). A study of liquid storage tank seismic uplift behavior (No. CONF-830607-). New Orleans, LA: Chevron USA, Inc.

Chang, J. I., & Lin, C. C. (2006). A study of storage tank accidents. *Journal of Loss Prevention in the Process Industries*, *19*(1), 51–59.

Cho, K. H., & Cho, S. Y. (2007). Seismic response of cylindrical steel tanks considering fluid-structure interaction. *Steel Structures*, 7(2), 147–152.

Clough, D. P. (1977). *Experimental evaluation of seismic design methods for broad cylindrical tanks*. Berkeley, CA: Earthquake Engineering Research Center, University of California.

D'Amico, M., & Buratti, N. (2019). Observational seismic fragility curves for steel cylindrical tanks. *Journal of Pressure Vessel Technology*, *141*(1).

Eshghi, S., & Razzaghi, M. S. (2005). Performance of industrial facilities in the 2003 Bam, Iran, earthquake. *Earthquake Spectra, 21*, 395–410.

Eshghi, S., & Razzaghi, M. S. (2007). Performance of cylindrical liquid storage tanks in Silakhor, Iran earthquake of March 31, 2006. *Bulletin of the New Zealand Society for Earthquake Engineering, 40*(4), 173–182.

European Convention for Constructional Steelwork (ECCS). (1988). *Buckling of steel shells—european recommendations* (4th ed.). Brussels, Belgium: ECCS.

FEMA (2011). Reducing the risks of nonstructural earthquake damage—a practical guide. FEMA E-74, Washington, DC.

Firouz-Abadi, R. D., Haddadpour, H., Noorian, M. A., & Ghasemi, M. (2008). A 3D BEM model for liquid sloshing in baffled tanks. *International Journal for Numerical Methods in Engineering, 76*(9), 1419–1433.

Fischer, E. C., Liu, J., & Varma, A. H. (2016). Investigation of cylindrical steel tank damage at wineries during earthquakes: lessons learned and mitigation opportunities. *Practice Periodical on Structural Design and Construction, 21*(3), 04016004.

Goudarzi, M. A., Sabbagh-Yazdi, S. R., & Marx, W. (2010). Seismic analysis of hydrodynamic sloshing force on storage tank roofs. *Earthquake Spectra, 26*(1), 131–152.

González, E., Almazán, J., Beltrán, J., Herrera, R., & Sandoval, V. (2013). Performance of stainless steel winery tanks during the 02/27/2010 Maule earthquake. *Engineering Structures, 56*, 1402–1418.

Hamdan, F. H. (2000). Seismic behaviour of cylindrical steel liquid storage tanks. *Journal of Constructional Steel Research, 53*(3), 307–333.

Hancilar, U., Taucer, F., & Corbane, C. (2013). Empirical fragility functions based on remote sensing and field data after the 12 January 2010 Haiti earthquake. *Earthquake Spectra, 29*(4), 1275–1310.

Haroun, M. A., & Housner, G. W. (1981). Earthquake response of deformable liquid storage tanks. *Journal of Applied Mechanics, 48*(2), 411–418.

Harris, L. A., Suer, H. S., Skene, W. T., & Benjamin, R. J. (1957). The stability of thin-walled unstiffened circular cylinders under axial compression including the effects of internal pressure. *Journal of the Aeronautical Sciences, 24*(8), 587–596.

Housner, G. W. (1954). *Earthquake pressures on fluid containers*. Pasadena, CA: California Institute of Technology.

Hutchinson, J. W., & Amazigo, J. C. (1967). Imperfection-sensitivity of eccentrically stiffened cylindrical shells. *AIAA Journal, 5*(3), 392–401.

Jacobsen, L. S. (1949). Impulsive hydrodynamics of fluid inside a cylindrical tank and of fluid surrounding a cylindrical pier. *Bulletin of the Seismological Society of America., 39*(3), 189–203.

Jaiswal, O. R., Rai, D. C., & Jain, S. K. (2005). Review of code provisions on design seismic forces for liquid storage tanks. *Report IITK-GSDMA Project on Building Codes, IITK-GSDMA-EQ01-V1. 0.*

Jaiswal, O. R., Rai, D. C., & Jain, S. K. (2007). Review of seismic codes on liquid-containing tanks. *Earthquake Spectra, 23*(1), 239–260.

Kalogerakou, M. E., Maniatakis, C. A., Spyrakos, C. C., & Psarropoulos, P. N. (2017). Seismic response of liquid-containing tanks with emphasis on the hydrodynamic response and near-fault phenomena. *Engineering Structures, 153*, 383–403.

Kildashti, K., Mirzadeh, N., & Samali, B. (2018). Seismic vulnerability assessment of a case study anchored liquid storage tank by considering fixed and flexible base restraints. *Thin-Walled Structures, 123*, 382–394.

Kircher, C. A., Nassar, A. A., Kustu, O., & Holmes, W. T. (1997). Development of building damage functions for earthquake loss estimation. *Earthquake Spectra, 13*(4), 663–682.

Lorenz, R. (1908). Buckling of a cylindrical shell under axial compression. *Zeitschrift des Vereines Deutscher Ingenieure, 52*(1908), 1766. (in German).

Maheri, M. R., & Abdollahi, A. (2013). The effects of long term uniform corrosion on the buckling of ground based steel tanks under seismic loading. *Thin-Walled Structures, 62*, 1–9.

Malhotra, P. K. (1995). Base uplifting analysis of flexibly supported liquid-storage tanks. *Earthquake Engineering & Structural Dynamics, 24*(12), 1591–1607.

Malhotra, P. K. (1998). Seismic strengthening of liquid-storage tanks with energy-dissipating anchors. *Journal of Structural Engineering, 124*(4), 405–414.

Malhotra, P. K., & Veletsos, A. S. (1994). Uplifting response of unanchored liquid-storage tanks. *Journal of Structural Engineering, 120*(12), 3525–3547.

Manjily, F., Mosleh, A., & Razzaghi, M. S. (2021). Seismic fragility of concrete bridges designed to current and older Aashto specifications. *Proceedings of the Institution of Civil Engineers-Structures and Buildings*.

Manos, G. C., & Clough, R. W. (1985). Tank damage during the May 1983 Coalinga earthquake. *Earthquake Engineering & Structural Dynamics, 13*(4), 449–466.

Manshadi, S. D., & Maheri, M. R. (2010). The effects of long-term corrosion on the dynamic characteristics of ground based cylindrical liquid storage tanks. *Thin-Walled Structures, 48*(12), 888–896.

Martinez, S. (2013). Estimating internal corrosion rate and internal inspection interval of aboveground hydrocarbon storage tanks. *Goriva i maziva: časopis za tribologiju, tehniku podmazivanja i primjenu tekućih i plinovitih goriva i inžinjerstvo izgaranja, 52*(2), 134–143.

Medvedeva, M. L., & Tiam, T. D. (1998). Classification of corrosion damage in steel storage tanks. *Chemical and Petroleum Engineering, 34*(9), 620–622.

Merino, R. J., Brunesi, E., & Nascimbene, R. (2020). Probabilistic evaluation of earthquake-induced sloshing wave height in above-ground liquid storage tanks. *Engineering Structures, 202*, 109870.

Miladi, S., & Razzaghi, M. S. (2014). A parametric study on inelastic buckling in steel cylindrical shells with circular cutouts. *International Journal of Advanced Structural Engineering (IJASE), 6*(1), 47.

Miladi, S., & Razzaghi, M. S. (2019). Failure analysis of an un-anchored steel oil tank damaged during the Silakhor earthquake of 2006 in Iran. *Engineering Failure Analysis, 96*, 31–43.

Miladi, S., Razzaghi, M. S., & Ghasemi, S. H. (2020). Seismic performance of imperfect unanchored tanks. *Proceedings of the Institution of Civil Engineers-Structures and Buildings*, 1–10.

MOP (2016a). Iranian seismic code for petroleum facilities (Publication No. 038), Ministry of Petroleum, Tehran, Iran.

MOP (2016b). Guidelines for seismic evaluation and retrofit of petroleum (Publication No. 041), Ministry of Petroleum, Tehran, Iran (In Persian).

Mosleh, A., Razzaghi, M. S., Jara, J., & Varum, H. (2016a). Seismic fragility analysis of typical pre-1990 bridges due to near-and far-field ground motions. *International Journal of Advanced Structural Engineering (IJASE), 8*(1), 1–9.

Mosleh, A., Razzaghi, M. S., Jara, J., & Varum, H. (2016b). Development of fragility curves for RC bridges subjected to reverse and strike-slip seismic sources.

Najmabad, S. J., Razzaghi, M. S., & Alahi, F. N. (2021). Retrofitting of a damaged liquid storage tank using steel and SMA anchor bolts. *International Journal of Pressure Vessels and Piping, 194*, 104530.

NIBS (1999). *Earthquake loss methodology*, HAZUS 99, Technical manual, Vol. II, Washington, DC.

NZSEE (2009). *Seismic design of storage tanks*. New Zealand: New Zealand Society for Earthquake Engineering.

Ormeño, M., Larkin, T., & Chouw, N. (2015). The effect of seismic uplift on the shell stresses of liquid-storage tanks. *Earthquake Engineering & Structural Dynamics*, *44*(12), 1979–1996.

O'Rourke, M. J., & So, P. (2000). Seismic fragility curves for on-grade steel tanks. *Earthquake Spectra*, *16*(4), 801–815.

Ozsarac, V., Brunesi, E., & Nascimbene, R. (2021). Earthquake-induced nonlinear sloshing response of above-ground steel tanks with damped or undamped floating roof. *Soil Dynamics and Earthquake Engineering*, *144*, 106673.

Phan, H. N., Paolacci, F., & Alessandri, S. (2019). Enhanced seismic fragility analysis of unanchored steel storage tanks accounting for uncertain modeling parameters. *Journal of Pressure Vessel Technology*, *141*(1).

Phan, H. N., Paolacci, F., Corritore, D., Akbas, B., Uckan, E., & Shen, J. J. (2016). Seismic vulnerability mitigation of liquefied gas tanks using concave sliding bearings. *Bulletin of Earthquake Engineering*, *14*(11), 3283–3299.

Razzaghi, M. S. (2007). Development of seismic fragility curves for on-ground cylindrical oil-storage tanks. Ph. D. Dissertation, International Institute of Earthquake Engineering and Seismology, Tehran, Iran.

Razzaghi, M. S., & Eshghi, S. (2004). Behavior of steel oil tanks due to near-fault ground motion. In *13th World conference on earthquake engineering*, August, Vancouver, BC, Canada (pp. 1–6).

Razzaghi, M. S., & Eshghi, S. (2008a). A parametric study on seismic behavior of unanchored oil storage tanks using numerical analysis. *Journal of Seismology and Earthquake Engineering (JSEE)*, *9*(4). (In Persian).

Razzaghi, M. S., & Eshghi, S. (2008b). Development of analytical fragility curves for cylindrical steel oil tanks. In *Proccedings of the 14th world conference on earthquake engineering*.

Razzaghi, M. S., & Eshghi, S. (2015). Probabilistic seismic safety evaluation of precode cylindrical oil tanks. *Journal of Performance of Constructed Facilities*, *29*(6), 04014170.

Razzaghi, M. S., Safarkhanlou, M., Mosleh, A., & Hosseini, P. (2018). Fragility assessment of RC bridges using numerical analysis and artificial neural networks. *Earthquakes and Structures*, *15*(4), 431–441.

Rotter, J. M., & Teng, J. G. (1989). Elastic stability of cylindrical shells with weld depressions. *Journal of Structural Engineering*, *115*(5), 1244–1263.

Shinozuka, M., Feng, M. Q., Lee, J., & Naganuma, T. (2000). Statistical analysis of fragility curves. *Journal of Engineering Mechanics*, *126*(12), 1224–1231.

Song, C. Y., Teng, J. G., & Rotter, J. M. (2004). Imperfection sensitivity of thin elastic cylindrical shells subject to partial axial compression. *International Journal of Solids and Structures*, *41*(24–25), 7155–7180.

Southwell, R. V. (1914). On the general theory of elastic stability. *Philosophical Transactions of the Royal Society of London. Series A, Containing Papers of a Mathematical or Physical Character*, *213*(497–508), 187–244, 213, 187.

Tafti, M. F., Hosseini, K. A., & Mansouri, B. (2020). Generation of new fragility curves for common types of buildings in Iran. *Bulletin of Earthquake Engineering*, *18*(7), 3079–3099.

Taniguchi, T., & Katayama, Y. (2016). Masses of fluid for cylindrical tanks in rock with partial uplift of bottom plate. *Journal of Pressure Vessel Technology*, *138*(5).

Tavakoli Joorabi, A., & Razzaghi, M. S. (2019). Seismic fragility analysis of retrofitted steel tanks considering corrosion. *Proceedings of the Institution of Civil Engineers-Structures and Buildings*, *172*(10), 712–720.

Teng, J. G., & Rotter, J. M. (1992). Buckling of pressurized axisymmetrically imperfect cylinders under axial loads. *Journal of Engineering Mechanics*, *118*(2), 229–247.

Teng, J. G., & Rotter, J. M. (Eds.), (2006). *Buckling of thin metal shells*. CRC Press.

Timoshenko, S. (1910). Some stability problems of the theory of elasticity. *Magazine for Mathematics and Physics, 58*. (in German).

Veletsos, A. S. (1974). Seismic effects in flexible liquid storage tanks. In *Proceedings of the 5th world conference on earthquake engineering* (Vol. 1, pp. 630–639).

Veletsos, A. S. (1984). *Seismic response and design of liquid storage tanks*. ASCE.

Veletsos, A. S., & Shivakumar, P. (1997). *Computer analysis and design of earthquake resistant structures, Chapter 15—Tanks containing liquids or solids*. London: Computational Mechanics Publications.

Veletsos, A. S., & Tang, Y. (1986). Dynamics of vertically excited liquid storage tanks. *Journal of Structural Engineering, 112*(6), 1228–1246.

Virella, J. C., Godoy, L. A., & Suárez, L. E. (2006). Dynamic buckling of anchored steel tanks subjected to horizontal earthquake excitation. *Journal of Constructional Steel Research, 62*(6), 521–531.

Virella, J. C., Suarez, L. E., & Godoy, L. A. (2005). Effect of pre-stress states on the impulsive modes of vibration of cylindrical tank-liquid systems under horizontal motions. *Journal of Vibration and Control, 11*(9), 1195–1220.

Wozniak, R. S., & Mitchel W.W. (1978). Basis of seismic design provisions for welded steel oil storage tanks. In *API Refining Dep. 43rd Midyear, Meeting* (pp. 1–34).

Zagórski, A., Matysiak, H., Tsyrulnyk, O., Zvirko, O., Nykyforchyn, H., & Kurzydłowski, K. (2004). Corrosion and stress-corrosion cracking of exploited storage tank steel. *Materials Science, 40*(3), 421–427.

Zhou, J., & Zhao, M. (2021). Shaking table test of liquid storage tank with finite element analysis considering uplift effect. *Structural Engineering and Mechanics, 77*(3), 369–381.

CHAPTER FOUR

# Hurricane performance and assessment models

**Sabarethinam Kameshwar**
Department of Civil and Environmental Engineering, Louisiana State University, Baton Rouge, Louisiana

## 4.1 Introduction

Above-ground storage tanks (AST) used to store hazardous substances in petrochemical facilities, ports, and other industries are predominantly located near navigable water bodies such as rivers, oceans, and bayous to facilitate easy and economical transport of products. This proximity to water bodies makes ASTs susceptible to various hazards such as flooding, hurricane-induced storm surge, waves, and rainfall. In addition to susceptibility, the consequences of AST failures can be catastrophic to the surrounding communities and environment, as evidenced by past failures such as the Murphy oil (EPA, 2006) and the Exxon-Valdez spills (Maki, 1991; Palinkas et al., 1993).

To address such vulnerabilities, there are few release prevention programs/guidelines such as CalARP in California, USA, the Seveso III directive in the European Union, industrial safety and disaster prevention laws in Japan, and the guiding principles for NaTech accident prevention in OECD counties. To help implement such requirements, the Center for Chemical Process Safety has released a monograph (AIChE, 2019) that provides a checklist that can help prevent NaTech accidents and the Canadian standard on risk management (CAN/CSA-IEC/ISO 31010, 2010) provides a list of potential methods used to assess risk in a variety of ways. However, there is a lack of technical directives from design guidelines such as the American Petroleum Institute (API) 650 (2013) and API 620 (2002) since they lack mandatory provisions that can prevent hurricane failures such as anchoring ASTs to the foundation. The lack of provisions in design guidelines and NaTech prevention programs can lead to an unmitigated risk of AST failures. Furthermore, climate change can

*Above Ground Storage Tank Oil Spills*.
DOI: https://doi.org/10.1016/B978-0-323-85728-4.00010-3

exacerbate future hurricane hazards and induce additional failures due to extreme precipitation and flood events, thereby increasing the risk of AST failures.

Prevention of AST failures is essential to avoid environmentally catastrophic spills, adverse societal impacts, and high cleanup costs. AST failure prevention requires an understanding of the different mechanisms that can cause damage. In this regard, since hurricanes are multihazard events that can impose wind, surge, wave, and rain loads on ASTs, they can induce a variety of failure modes in ASTs. Therefore, this chapter will primarily focus on understanding the different types of failure mechanisms caused by hurricanes and identifying methodologies that can be used to estimate the likelihood of failure for the different mechanisms. In this context, the following section first identifies the different failure mechanisms caused by hurricane-induced wind, surge, wave, and rainfall. Then the existing methods available for modeling the likelihood of AST failure are discussed in Section 4.3. The penultimate section discusses the potential for future work based on the discussion presented in this chapter. The final section will present conclusions along with key insights obtained from this chapter.

## 4.2 Hurricane failure modes

Hurricanes are multihazard events with different types of loads and hazards such as strong winds, storm surge, wave loads, extreme precipitation, wind-borne debris, and storm surge and wave-driven debris. The following subsections discuss the damage modes associated with the above-mentioned hazards. This discussion is also summarized in Table 4.1.

### 4.2.1 Wind-induced failures

Among the wind-related failures observed in ASTs, buckling of the tank shell is one of the most common failure modes. This failure has been reported in past events since at least 1967 when Kundurpi et al. (1975) identified wind buckling failure of ASTs in England. Similarly, in the Caribbean, in 1989, hurricane Hugo caused wind buckling failure of

**Table 4.1** Hurricane-induced failure modes and performance models.

| Hazard | Failure mode(s) | Availability of performance assessment models | Select references |
|---|---|---|---|
| Wind | Shell buckling (local and global) | ✓ | Ansourian (1992), Godoy (2016), Teng (1996), Kameshwar and Padgett (2018a) |
| | Floating roof failure | ✓ L | Kuroda et al. (2012), Yoshida et al. (2012) |
| | Overturning | ✓ L | Olivar et al. (2020) |
| | Debris impact | ✓ L | Ramirez et al. (2019) |
| | Insulation failure | × | |
| Storm surge | Flotation | ✓ | Landucci et al. (2012), Antonioni et al. (2015), Kameshwar and Padgett (2015) |
| | Sliding | ✓ | Khakzad and Van Gelder (2017, 2018), Qin et al. (2020), Mayorga et al. (2019) |
| | Buckling | ✓ | Kameshwar and Padgett (2018b), Yang et al. (2020), Zuluaga Mayorga et al. (2019) |
| | Pipe and shell failure | × | |
| | Base plate failure | × | |
| | Debris impact | ✓ L | Bernier and Padgett (2020a) |
| Wave | Flotation | ✓ L | Bernier and Padgett (2019a) |
| | Sliding | ✓ L | |
| | Buckling | ✓ L | Bernier and Padgett (2019a, 2019b) |
| | Debris impact | × | |
| | Foundation erosion | × | |
| Rainfall | Floating roof failure | ✓ L | Sun et al. (2008), Bernier and Padgett (2020b) |
| | Flotation | ✓ (Flotation models for storm surge can be used) | |

*L*, Limited studies exist.

ASTs in St Croix, and in 1995 hurricane Marilyn resulted in wind buckling failures in St. Thomas (Flores & Godoy, 1998). More recently, in Hurricanes Rita and Katrina, in 2005, several ASTs suffered wind-induced shell buckling (Godoy, 2007). Typically, shell buckling is observed in ASTs with a very small aspect ratio (ratio of height to diameter), for example, shown in Fig. 4.1. However, buckling failures were also observed during hurricane Laura in 2020 in smaller ASTs with large aspect ratios, as shown in Fig. 4.2.

Wind buckling failures described here are generally classified as local or global buckling. Usually, the thickness of the shell courses is tapered, that is, their thickness is gradually reduced from the base of the tank to the top of

**Figure 4.1** Wind buckling failure (US EPA).

**Figure 4.2** Wind bucking failure in short tanks with a large aspect ratio (Roueche et al., 2021).

the tank, because ASTs are primarily designed to sustain the internal liquid pressure. The tapered shell thickness can lead to local buckling due to wind loads, which occur in the upper shell courses of ASTs, where the thickness is minimal. Fig. 4.3 shows an example of local wind buckling in an AST observed after hurricane Katrina. The amplitude of buckling deflection is about three to four times the thickness of the shell. Therefore, local buckling, in general, does not lead to structural failure, but it may limit the functionality of ASTs, especially if floating roofs are used. In contrast, global buckling of the tank shell causes deformations in a large part of AST's shell, which can lead to fractures in the shells and render the AST inoperable until repairs are completed (e.g., Figs. 4.1 and 4.2).

Roof failure is another commonly observed wind-induced failure mode for ASTs, especially for floating roofs (Godoy, 2007). Floating AST roofs typically consist of steel plates and pontoons that have sufficient buoyancy to float atop the stored contents. Wind loads can cause sloshing of the liquid inside ASTs (Yoshida et al., 2012), which can lead to out-of-plane deformations in the roof. Similarly, the wind loads can have significant variation within the floating roof, which can lead to wave-like, out-of-plane, vibrations in the roof's steel plates (Yoshida et al., 2013). Such out-of-plane deflections can cause stress concentrations, and fatigue failures, and render the floating roof inoperable.

In rare cases, unanchored ASTs with a large aspect ratio can fail due to overturning when they are empty (Olivar et al., 2020). This failure

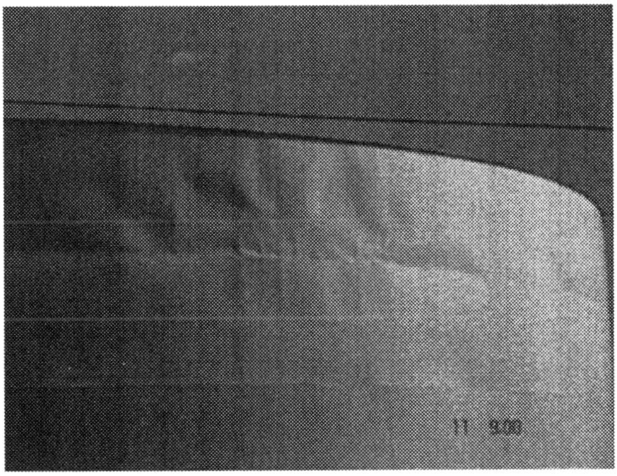

**Figure 4.3** Local wind buckling (Godoy, 2007).

specifically occurs when the overturning moments caused by wind loads exceed the overturning resistance provided by the self-weight of the tanks. Anchoring ASTs can prevent such failures. Wind-borne debris can act as projectiles and impact AST shells. Such impacts could cause extremely localized buckling in the tank shell and in extreme cases, the projectile impact can potentially rupture the shell. These two failure modes can lead to complete damage and hazardous spills. However, these two failures have not been commonly observed during hurricanes. In addition to these structural failure modes, failure of insulation has also been observed and reported by Godoy (2007) during hurricanes Katrina and Rita, shown in Fig. 4.4. Although insulation failure may not cause significant structural damage, it can still render ASTs inoperable. All the wind-induced failure modes are summarized in Table 4.1.

### 4.2.2 Storm surge failures

Flotation of ASTs is one of the most common failures modes during a hurricane storm surge. Flotation failures of ASTs have been widely reported during past hurricanes such as Katrina and Rita in 2005 (Godoy, 2007), Ike and Gustav in 2008 (Hyder, 2008; Sengul et al., 2012), and most recently during hurricane Laura in 2020 (Roueche et al., 2021). Fig. 4.5 shows flotation/sliding failure mode for the AST on the right after hurricane Katrina. Flotation failure happens when the buoyancy forces exceed the resistance against flotation, that is, the sum of self-weight, including the weight of the stored contents and the resistance provided by anchors (if any). ASTs are thin-walled structures; therefore, the weight of the tank shell is relatively low compared to the buoyancy forces. Furthermore, most ASTs are unanchored; consequently, unanchored empty

**Figure 4.4** Insulation damage (Godoy, 2007).

**Figure 4.5** Floatation/sliding failure (Krausmann et al., 2016).

tanks can float at inundation depths as low as three feet (~0.91 m). Even if ASTs are not completely empty, due to the low specific gravity of stored contents, ASTs can still float if the storm surge inundation depth is within a few feet of the fill level in the ASTs. Therefore, a fill level of 3–6 feet higher than the inundation depth is recommended (EPA, 2016).

Akin to flotation, the presence of flow currents during storm surge events can lead to sliding failure and dislocation of ASTs, shown in Fig. 4.5. Cozzani et al. (2010) have reported such failures based on the analysis of past failures. ASTs may slide if the resistance against sliding is exceeded by the hydrodynamic flow-induced forces. Typically, this resistance is provided by friction and anchors, if any. Buoyancy force reduces the frictional resistance against sliding, thereby further increasing the likelihood of AST dislocation due to sliding. Additionally, large debris, such as shipping containers, can act as projectiles and impact ASTs, which can further increase the likelihood of sliding failure. In such cases, even small currents (~ 1.5 m/s) can dislocate ASTs that are empty or have low fill levels (Bernier & Padgett, 2020a).

Dislocation of ASTs, either due to flotation or sliding, can initiate a wide variety of additional failures. For example, dislocation of ASTs can rupture AST shells where they are connected to the pipe and conversely, the pipes could also break (Necci et al., 2018). Fig. 4.6 shows an AST with broken pipelines. In addition to economic losses due to repair costs, pipe damage could also lead to spills (Krausmann et al., 2011); however, valves are typically closed before hurricanes. Therefore, the potential for

**Figure 4.6** Broken pipeline due to flotation/sliding (Necci et al., 2018).

hazardous spills due to pipe damage is reduced. In contrast, shell rupture can lead to hazardous material spills, which can be catastrophic for the surrounding communities and the environment. Such spills can also occur if dislocated buoyant ASTs settle back on uneven ground, which can result in failures like rupture of the base plate and the welds connecting the shell and the base plate. This failure led to the Murphy oil spill (EPA, 2006) in the aftermath of Hurricane Katrina (Godoy, 2007). Fig. 4.7 shows the damaged AST that caused the spill.

In cases where ASTs are restricted from dislocating, either due to anchors or due to the surrounding environment, storm surges can cause buckling of the tank shell. Such failures were observed during hurricane Katrina when topography prevented an AST from moving (Godoy, 2007; Necci et al., 2018). Fig. 4.8 shows storm surge buckling damage to an AST caused by hurricane Katrina. In contrast to wind-induced buckling, which primarily occurs in the upper shell courses, storm surge-induced buckling occurs in the lower shell courses and is more likely to be a global bucking. Since surge causes buckling in the lower shell courses, potential shell ruptures can lead to spills.

Furthermore, in addition to dislocating ASTs, debris projectiles driven by surges and waves can damage the tank shells. In comparison to windborne debris, surge and wave-driven debris can be significantly larger, which can buckle and puncture tank shells (Necci et al., 2018). All storm surge-induced failures, summarized in Table 4.1, can adversely affect the functionality of ASTs. Dislocated ASTs cannot be used immediately after the hurricane event and in case of pipe failure or shell rupture, additional time for repairs and cleanup will be required. Similarly, for buckling failure of ASTs, rupture may require extended repair times and even with

**Figure 4.7** Damaged AST in Murphy oil spill (EPA, 2006).

**Figure 4.8** Storm surge buckling (Necci et al., 2018).

rupture, buckled shell courses will disrupt the functionality of ASTs with floating roofs.

### 4.2.3 Wave-induced failures

Hurricane waves primarily impart lateral loads on ASTs and may exacerbate failures caused by storm surges. The presence of waves can increase the likelihood of AST dislocation and consequent failures of pipes and tank shells. For ASTs that are restricted from moving due to anchors or topography, the lateral loads due to wave impact can increase the probability of buckling. Similarly, waves can increase the likelihood of debris impact and the corresponding loads due to additional velocity imparted by the waves to debris. As discussed above, in Section 4.2.2, the consequences of wave-induced dislocation, buckling, and pipeline failure can range from hazardous spills to loss of functionality in the aftermath of hurricane events. In addition to exacerbating the storm surge-induced failures, large variations in wave conditions, such as breaking/nonbreaking waves, wave period, and wind speed significantly increase the uncertainty in the structural loads and AST damage. Consequently, these uncertainties may lead to a reduction in the accuracy of damage predictions models.

The widespread erosion and wave-induced damage and scour at structures in Bolivar Peninsula, TX, during hurricane Ike (Necci et al., 2018) highlights the potential for wave damage to ASTs and associated infrastructure. During past hurricanes such as Ike and Ivan, AST damage has been observed due to a combination of hurricane waves and surges leading to the severe crushing of dislocated of ASTs as shown in Fig. 4.9, leading to hazardous spills. Wave action can also cause scour and erosion at the base of ASTs, which can lead to differential settlement and cause base plate damage (Godoy & Sosa, 2003). In addition to inflicting structural damage on ASTs, the scouring potential of waves can also cause damage to other AST infrastructure. For example, waves can damage earthen secondary containment dikes and potentially damage onshore and offshore pipelines (Fredsøe, 2016). While such damage may not compromise the structural integrity of ASTs, it is likely to hinder the posthurricane functionality of ASTs and the facilities that use ASTs.

**Figure 4.9** Severely crushed and dislocated AST ("2013 | StateImpact Texas | Page 7," 2013).

Each of the wave-induced failure modes summarized in Table 4.1 has high damage potential and the consequences of damage can be severe. At present, only a small fraction of coastal ASTs may be vulnerable to them because most ASTs are situated farther inland, where the inundation depths are not sufficient for waves to form. Furthermore, other factors such as wind and topography also affect wave formation, increase wave energy dissipation, and reduce wave-induced lateral loads. Nevertheless, wave-induced failures should be investigated as more ASTs may become vulnerable to them as sea levels rise and the frequency and intensity of hurricanes increase.

### 4.2.4 Extreme precipitation induced failures

Floating roofs are often installed in ASTs that store volatile substances to reduce vapor pressure inside ASTs and thereby minimize the chances of ignition and explosions. Typically, floating roofs are designed to have sufficient buoyancy, enabling them to float atop the stored contents under atmospheric pressure. Typically, floating roofs consist of pontoons or radial bulkheads that provide buoyancy to the floating roofs. Furthermore, to avoid any stagnation of water on the floating roofs during rainfall events drains that can handle several inches (1 inch = 2.54 cm) of rainfall per hour are installed to divert the water away from the roofs. Therefore, floating roofs were not considered to be vulnerable to failure during rainfall events.

Until 2017, the failure of floating roofs was only documented due to earthquakes, lightning strikes, floods, and hurricane winds (Cozzani et al., 2010; Krausmann et al., 2011). However, the failure of 16 floating roofs during hurricane Harvey in 2017 exposed the vulnerability of floating roofs to extreme rainfall events (Bernier & Padgett, 2018; Necci et al., 2018). Fig. 4.10 shows floating roof failures observed during hurricane Harvey. Based on postevent investigations, three failure modes were identified (Bernier & Padgett, 2018, 2020b): (1) sinking of the roof due to excessive rainwater weight, (2) tiling of the roof due to asymmetric accumulation of water due to punctured roof pontoons or bulkheads, and (3) rupture or buckling of the roof due to excessive strains and/or stresses. All three failure modes were attributed to excessive accumulation of water on floating roofs due to clogged/insufficient drains. In addition to floating roof failures, excessive rainfall can cause containment dikes to flood. The ponding of rainwater within these dikes can also result in flotation failures discussed above. Both the failure modes are summarized in Table 4.1.

**Figure 4.10** Rainfall induced floating roof failures during hurricane Harvey (Bernier, 2019).

Typically, failure of floating roofs causes the roof to sink and the contents to spill on top of the floating roof, which may release hazardous volatile substances into the atmosphere and adversely affect the air quality. Furthermore, the drains on the floating roofs may also lead to spillage of the stored hazardous substances into the surrounding environment. However, the severity of these spills can be expected to be lower than spills caused by ruptures in the tank shell. Nevertheless, failure of floating roofs needs to be avoided since roof failures may render ASTs inoperable and incur cleanup and repair costs.

## 4.3 Hurricane performance assessment models

For each of the hurricane hazards and the corresponding failure modes in Table 4.1, the following subsections discuss available models (if any) that can be used for the performance assessment of ASTs. The availability of performance assessment models for the listed failure modes is summarized in Table 4.1, along with additional comments.

### 4.3.1 Wind load

#### *4.3.1.1 Buckling*

Numerous studies have been conducted on wind buckling of ASTs. Several studies have focused on estimating the wind loads and pressure distribution on shells and roofs of individual ASTs (MacDonald et al., 1988; Macdonald et al., 1990a; MacDonald et al., 1990b; Portela & Godoy, 2005a, 2005b; Uematsu et al., 2014). For closely-spaced groups of ASTs, several studies have experimentally assessed group effects, such as shielding and channeling (Macdonald et al., 1990a; MacDonald et al., 1990b; Portela & Godoy, 2007; Uematsu et al., 2015). Experimental studies have also focused on quantifying the prebuckling deflections for unstiffened (Uematsu & Uchiyama, 1985) and stiffened ASTs (Uchiyama et al., 1987) and static and dynamic buckling loads (Malik et al., 1979; Yasunaga & Uematsu, 2020).

Several studies have focused on developing analytical methods for predicting the buckling loads and pressures for ASTs (Bickell & Ruiz, 1967; Jerath & Ghosh, 1987; Jerath & Sadid, 1985; Malik et al., 1980). However, most studies have used finite element modeling to assess (1) the suitability of static and dynamic buckling analysis methods (Flores & Godoy, 1998; Sosa & Godoy, 2005; Yasunaga & Uematsu, 2020), (2) assessing buckling loads and pressures for a wide variety of AST configurations (Burgos et al., 2015; Flores & Godoy, 1999; Uematsu et al., 2014), (3) quantifying the effects of stiffening rings (Azzuni & Guzey, 2017; Bu & Qian, 2015; Lewandowski et al., 2015), (4) understating the effects of imperfections on the buckling behavior of ASTs (Godoy & Flores, 2002; Greiner & Derler, 1995), and understanding group effects (Burgos et al., 2014). This brief discussion highlights some of the existing studies that have assessed the wind performance of ASTs, for a comprehensive discussion of such studies, review papers may be referred to (Ansourian, 1992; Godoy, 2016; Teng, 1996).

These existing studies have significantly advanced the understanding of wind load performance of ASTs by developing methods to model ASTs, providing experimental data for verification and validation, highlighting the sensitivity to imperfections, and identifying methods to mitigate wind load failures. Predominantly, these studies are deterministic, that is, they lack systematic treatment of uncertainties emanating from loading, material, design, and geometric nonlinearities. Even though the need for probabilistic analysis for ASTs was identified several decades ago by highlighting the need for probabilistic quantification of buckling wind speed (Holroyd, 1985), limited studies exist. Probabilistic approaches are needed for future updates to design guidelines using more recent design philosophies like the load and resistance factor design approach, for risk assessment, and hazard mitigation for individual ASTs and at the regional level.

In this context, limited studies have probabilistically assessed the wind load performance of ASTs. The mean and standard deviation of wind buckling pressure was reported for four ASTs (Kameshwar & Padgett, 2016), more research is necessary to obtain such statistics for a large number of ASTs. Recent studies have also developed fragility models to quantify the hurricane performance of ASTs. Fragility (Porter, 2015) is a mathematical model which describes the probability of exceeding a performance threshold (e.g., deflection or buckling) given hazard intensity measures, structural characteristics, and material properties. Fragility models have been widely used for different structures, including ASTs, for seismic performance assessment (Fabbrocino et al., 2005; Salzano et al., 2003). Kameshwar and Padgett (2018a) developed logistic regression-based fragility models for wind-induced buckling of ASTs with fixed and floating roofs that were parameterized on geometric and material properties of ASTs for regional application. Fig. 4.11 shows the fragility curve for wind buckling failure of an AST with diameter = 15 m and height = 10 m. This figure shows the probability of failure, that is, buckling, on the $y$-axis for corresponding wind speeds shown on the $x$-axis.

While Kameshwar and Padgett used finite element modeling to develop the fragility models, Ramirez et al. (2019) and Olivar et al. (2020) used analytical equations that determine the buckling loads for ASTs to develop fragility models for wind-induced buckling. A drawback of these analytical equation-based fragility models is the inability to

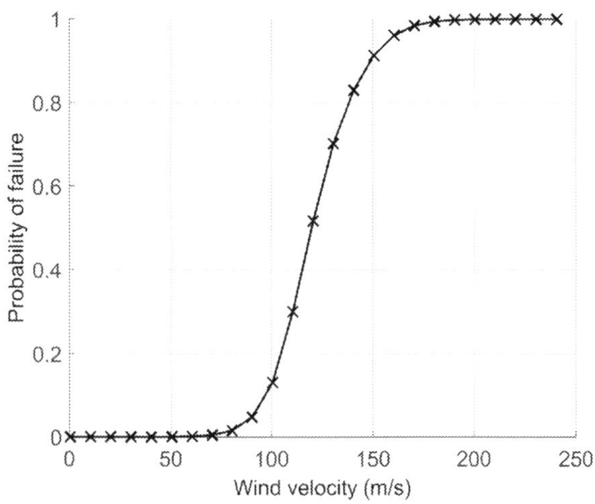


**Figure 4.11** Wind buckling fragility curve of an AST (Kameshwar, 2017).

incorporate the effects of imperfections and tapering shell thickness along with the height of the tanks.

#### 4.3.1.2 Floating roof failure

Wind loads can excite liquids stored in ASTs and trigger a sloshing response, which can lead to the failure of floating roofs. There is a large body of research on sloshing and roof failures during earthquakes (Matsui, 2007, 2009). However, limited studies have addressed wind-induced sloshing. Matsui, Uematsu, et al. (2009) performed wind tunnel tests to measure the wind pressure on floating roofs. Subsequently, they used it within the framework of linear potential theory to analytically determine the dynamic response of a floating roof. Matsui et al. observed that while the displacements caused by wind are low, the stresses are comparable to earthquake-induced stresses in floating roofs. Kuroda et al. (2012) performed simulations to first obtain the unsteady pressure and loads on the floating roofs, neglecting the movement of the roof. Yoshida et al. (2012) used these forces in computational fluid dynamics-based finite element simulations to assess the performance of floating roof decks. They observed wave-like patterns in the deck plate due to cyclic bending caused by wind loads and suggest the potential for fatigue failure of welded joints in the roof deck. Both these studies on floating roof failures are deterministic so they are unable to incorporate the effects of randomness in the roof wind loads and the observations are further studies are needed to generalize the observations for ASTs of different sizes.

#### 4.3.1.3 Other failures

Olivar et al. (2020) used an analytically modeled overturning moment and resisting moment due to self-weight to develop fragility curves for overturning failure of ASTs exposed to hurricane winds. Using a similar approach, Ramirez et al. (2019) also developed fragility models for ASTs considering wind-borne debris. Other failure modes such as failure of insulation have not been studied yet.

### 4.3.2 Storm surge loads

In contrast to wind loads, only recently, have storm surge loads been addressed. The following discusses the studies that have developed models to assess the storm surge performance of ASTs. Since these studies are recent, most of them are probabilistic and consider one or more sources of uncertainties: liquid level, material properties, and geometric imperfections.

#### 4.3.2.1 Dislocation failures (flotation and sliding)

Flotation failure can be modeled using the Archimedes principle. Similarly, sliding failure can be modeled by estimating loads using the well-established equations for hydrodynamic drag force on cylinders and the resistance can be modeled based on friction. This approach has been predominantly used in the existing studies that have developed flotation and sliding performance models for ASTs. In this context, Landucci et al. (2012) developed one of the earliest models to assess the probability of flotation and dislocation resistance of ASTs while considering uncertainties infill level. Although Landucci et al. developed these models for flood events, they can be applied for hurricane performance assessment as well, since the mechanisms of loading and resistance remain the same for the two hazards. This approach was also used by Antonioni et al. (2015) to assess the failure probability of ASTs exposed to floods. Khakzad and Van Gelder (2017, 2018) and Qin et al. (2020) developed fragility models for ASTs subjected to flood events for flotation and sliding. They used Bayesian networks to obtain the joint fragility for all three failure modes.

Kameshwar and Padgett (2015, 2018b) developed fragility models parameterized on AST's geometry, fill level, and inundation depth. A key difference between Kameshwar and Padgett and the studies mentioned earlier is the use of logistic regression, which enables the application of the fragility models for regional application. Furthermore, Kameshwar and Padgett also developed flotation fragility models for anchored ASTs considering uncertainties in anchor strength. Similarly, Mayorga et al. (2019) developed logistic regression-based fragility models for flotation and sliding failures. To understand the effects of variation in parameters on the ASTs' fragility, Yang et al. (2020) also used logistic regression. They identified fill level, floodwater depth, and flow velocity as key parameters and proposed mitigation strategies based on their findings such as anchoring ASTs and increasing the height of protective embankments around ASTs.

#### 4.3.2.2 Buckling failure

Kameshwar and Padgett (2015, 2018b) developed some of the earliest physics-based models to assess the storm surge-induced buckling performance of ASTs. They performed finite element simulations to develop logistic regression-based fragility models for buckling. The logistic regression fragility models were parameterized on tank geometry, material properties, the density of contents, and storm surge inundation depth to facilitate regional level application. In addition to the finite element analysis-based methods, studies have used simplified load-based analytical equations to develop fragility models for buckling failures (Khakzad & Van Gelder, 2017, 2018; Qin et al., 2020; Yang et al., 2020; Zuluaga Mayorga et al., 2019). However, the analytical equations used to estimate buckling loads have several drawbacks, such as they consider a perfect cylindrical shape for ASTs neglecting imperfections, which significantly affects the buckling capacity of ASTs. Moreover, these equations are applicable for ASTs with uniform shell thickness; however, ASTs have a tapering shell thickness profile. These drawbacks also propagate into the fragility models and limit their applicability to real ASTs.

#### 4.3.2.3 Other failure modes

Other potential failure modes such as debris impact, pipeline failure, and base plate failure due to storm surges have received little attention. Recently, Bernier and Padgett (2020a) developed fragility models for debris impact considering sliding failure and shell rupture. Their results show that damage to tank shells is more likely than sliding failure and filling ASTs with liquids a priori may not reduce the failure probability. Furthermore, their results show that ASTs with larger diameters are less likely to suffer damage due to debris impact since their failure is dominated by flotation and buckling failures.

#### 4.3.2.4 System failure

Unlike wind load failure, which is dominated by buckling failure, storm surge induces several failure models with a high likelihood. Therefore, several studies have developed system fragility models that combine the likelihood of different failure modes and provide an overall failure probability. In this regard, Kameshwar and Padgett developed logistic regression-based system fragility models considering flotation and buckling failure modes using a series assumption, where ASTs are considered to fail if any of the damage modes occurs. The fragility models were parameterized on AST geometry, material properties, stored contents, and inundation depth. Results from Kameshwar and Padgett (2018a, 2018b) showed that during a storm surge, flotation may dominate failures for unanchored tanks. But for excessively anchored tanks, failure may be dominated by buckling. For example, Fig. 4.12 shows the system fragility along with the flotation and buckling fragility of an anchored AST.

A similar approach was used by other researchers (Khakzad & Van Gelder, 2017, 2018; Qin et al., 2020; Yang et al., 2020) who have used

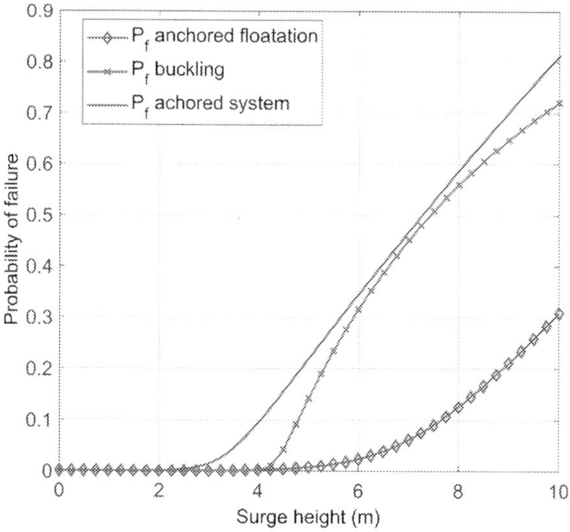


**Figure 4.12** System fragility of an anchored AST considering buckling and flotation.

Bayesian networks to develop system fragility models. The use of Bayesian networks facilitates the incorporation of correlation between different failure modes and facilitates improved system failure quantification when compared to the series system, assumption. Using this approach, Khakzad and Van Gelder (2017, 2018) evaluated system fragility by considering sliding, flotation, and buckling failure modes.

### 4.3.3 Wave loads

Bernier and Padgett further extended the storm surge fragility modeling of ASTs to include potential wave loads (Bernier & Padgett, 2019b) and concurrent surge, wind, and wave effects (Bernier & Padgett, 2019a) on flotation, buckling, and sliding failures of ASTs. For this purpose, they developed neural-network-based surge and wave load models, which were used during finite element simulations to develop logistic regression-based parameterized fragility models. Their results showed that large waves (e.g., 2 m wave height) can increase the buckling fragility of ASTs by 40%. Wave loads also increased the likelihood of sliding failure by 25%. The effects of current and wind were observed to be minimal with about a 5% increase in fragility. Although waves increased the fragilities significantly, their overall impact on a regional portfolio of ASTs may be limited by the number of ASTs exposed to inundations that are sufficiently deep and conducive for wave generation.

### 4.3.4 Rainfall loads

Rainfall-induced floating roof failures were only recently identified during hurricane Harvey in 2017. Therefore, at present, limited studies have addressed this issue. Sun et al. (2008) presented an approach to analyzing and designing floating roofs considering rainfall loads. However, their study did not incorporate uncertainties in material, geometry, or hazard characteristics. Bernier and Padgett (2020b) have developed probabilistic fragility models to predict the performance of floating roofs under extreme rainfall events. Bernier and Padgett first identified potential causes of roof failure mentioned in Section 4.2.4 and developed finite element models of pontoon roofs for each failure mode (sinking, tilting, and punctured pontoons). They varied AST geometry and propagated uncertainties in material and rainfall characteristics in these finite element simulations. The resulting data from the simulations were used to develop logistic regression-based fragility models that were parameterized on roof geometry, material characteristics, and accumulated rainfall. Their results show that a large accumulation of rainfall can lead to failures; however, well-designed drains can prevent failures. Therefore, factors that improve roof drainages, such as improved terrain drainage and higher fill levels, can reduce the likelihood of failure. Furthermore, they identified smaller diameter roofs to be more vulnerable to failure. While the study by Bernier and Padgett provides several key insights, future research is needed to study different types of roofs, designs, and improve rainfall accumulation modeling.

## 4.4 Discussion

The studies discussed above have significantly advanced the understanding of hurricane performance of ASTs. Table 4.1 summarizes the availabilities of studies and models for various failure modes, which highlight several knowledge gaps. ASTs are connected by a network of pipelines. However, the flood performance of pipelines, connections, and the tank shells around these connections have not been studied yet. As a result, there is a lack of knowledge on whether the pipeline, the connection, or the tank shell will break first during flood events. Furthermore, even for some of the most studied failure modes such as flotations and sliding, the effects of pipelines on the performance of ASTs are unknown, which is important to understand the potential for hazardous spills due to AST failure.

Floodwaters lead to uplift pressure on ASTs' bottom plate. However, the design of bottom plates, including the welds within the plate and the welds that connect it with the tank's walls, does not consider uplift pressure. Furthermore, none of the existing research studies have addressed the potential for bottom plate failure, which could lead to spills. The susceptibility of ASTs to debris impact during floods has been well established. However, the performance of ASTs subjected to debris impact is not well understood; at present, only shipping contained impact has been studied. The effects of different types of debris and their characteristics such as shape, size, mass, velocity, and stiffness are unknown.

Additionally, multihazard failure of ASTs also needs to be studied extensively since hurricane events may consist of multiple hazards, such as wind, surge, and wave, and ASTs may also be vulnerable to other hazards. Furthermore, ASTs' location near oceans and rivers and their use for storing potentially corrosive hazardous substances make them vulnerable to corrosion. Studies such as Bernier and Padgett (2019b) and Qin et al. (2020) have addressed the multihazard issue of hurricane-related hazards and floods and earthquakes were studied by Mayorga et al. (2019). Further research is necessary for a better understanding and design of ASTs for multihazard conditions and the effects of corrosion on the hazard performance of ASTs. In addition to further research, there is a need for a comprehensive comparative assessment of existing models to determine the limitations and applicability.

As discussed in Section 4.3, several studies have developed hurricane performance and fragility models, but there is a lack of understanding of when a failed AST leads to spills. So, there is a lack of models that can estimate the spill volume caused by failed ASTs. Moreover, none of the existing studies provide insights into the economic losses and restoration time associated with AST failures. Such models are necessary to estimate direct and indirect economic losses which can help inform risk analysis and mitigation strategies. Metrics such as spill volume, postevent functionality, and repair times are also necessary for resilience quantification. In this regard, models exist for earthquakes, which have been adapted for some hurricane failure modes (Kameshwar & Padgett, 2018a). However, extensive research is needed to develop physics-based models for estimating spills and other metrics.

At present, limited flood mitigation strategies for ASTs have been explored. Kameshwar and Padgett (Kameshwar & Padgett, 2018b, 2019) also developed mitigation strategies such as anchoring ASTs and the use of stiffening rings to prevent ASTs flotation and buckling. Based on the fragility models, Yang et al. (2020) also proposed mitigation strategies based on their findings, such as anchoring ASTs and increasing the height of protective embankments around ASTs. However, numerous potential mitigation options need to be evaluated for their effectiveness and practicality. Design guidelines lack hurricane mitigation measures for ASTs due to gaps in the understanding of the effectiveness of various measures and their practicality.

Among the damage modes which have been studied, only wind buckling and storm surge dislocation have been extensively studied. However, most of the existing studies on wind bucking are deterministic. For other failure modes, there are either a very limited number of models or there are no models at all. Therefore, even though the discussion in this section has highlighted several research gaps, all AST damage modes need to be studied to improve their safety during hurricanes.

## 4.5 Summary

This chapter has presented an overview of hurricane performance of ASTs highlighting the observed and anticipated failure modes for multiple hazards observed during hurricanes (wind, storm surge, waves, and extreme rainfall). For each failure mode, this chapter briefly discusses available performance assessment model(s), if any. The discussion highlights the lack of probabilistic studies for wind buckling assessment, which is necessary for probabilistic quantification of risk and selection of hazard mitigation. Furthermore, the discussion also highlights the limited availability of performance assessment models and highlights the need for additional research on all damage modes. Finally, several knowledge gaps were identified, which include: a lack of understanding of pipeline failures, base plate failure during a storm surge, debris impact, and multihazard performance. Additionally, the lack of repair cost, repair time, and spill volume estimation models was highlighted along with the lack of studies focusing on mitigating flood-induced AST failures.

## References

2013 | StateImpact Texas (2013). 7. <https://stateimpact.npr.org/texas/2013/page/7/> Accessed 6.15.21.

AIChE (2019). *CCPS monograph: Assessment of and planning for natural hazards*. American Institute of Chemical Engineer.

Ansourian, P. (1992). On the buckling analysis and design of silos and tanks. *Journal of Constructional Steel Research, 23*, 273–294. Available from https://doi.org/10.1016/0143-974X(92)90047-I, Special Issue on Structural Steel Research in Australia.

Antonioni, G., Landucci, G., Necci, A., Gheorghiu, D., & Cozzani, V. (2015). Quantitative assessment of risk due to NaTech scenarios caused by floods. *Reliability Engineering and System Safety, 142*, 334–345.

API (2002). *620: Design and construction of large, welded, low pressure storage tanks*. Washington, DC: API.

API (2013). *650: Welded steel tanks for oil storage*. Washington, DC: API.

Azzuni, E., & Guzey, S. (2017). Stability of open top cylindrical steel storage tanks: Design of top wind girder. *Journal of Pressure Vessel Technology*, 139. Available from https://doi.org/10.1115/1.4035507.

Bernier, C. (2019). Fragility and risk assessment of aboveground storage tanks during storm events (Doctoral dissertation, Rice University).

Bernier, C., & Padgett, J. E. (2018). Forensic investigation of aboveground storage tank failures during Hurricane Harvey using fragility models. In *Forensic Engineering 2018. Forging Forensic Frontiers*, (pp. 975–985). Reston, VA: American Society of Civil Engineers.

Bernier, C., & Padgett, J. E. (2019a). Buckling of aboveground storage tanks subjected to storm surge and wave loads. *Engineering Structures, 197*, 109388. Available from https://doi.org/10.1016/j.engstruct.2019.109388.

Bernier, C., & Padgett, J. E. (2019b). Fragility and risk assessment of aboveground storage tanks subjected to concurrent surge, wave, and wind loads. *Reliability Engineering and System Safety, 191*, 106571. Available from https://doi.org/10.1016/j.ress.2019.106571.

Bernier, C., & Padgett, J. E. (2020a). Probabilistic assessment of storage tanks subjected to waterborne debris impacts during storm events. *Journal of Waterway, Port, Coastal, and Ocean Engineering, 146*, 04020003. Available from https://doi.org/10.1061/(ASCE)WW.1943-5460.0000559.

Bernier, C., & Padgett, J. E. (2020b). Fragility assessment of floating roof storage tanks during severe rainfall events. *Journal of Performance of Constructed Facilities, 34*, 04020101.

Bickell, M. B., & Ruiz, C. (1967). Buckling. In M. B. Bickell, & C. Ruiz (Eds.), *Pressure vessel design and analysis* (pp. 516–528). London: Macmillan Education UK. Available from https://doi.org/10.1007/978-1-349-00129-3_15.

Bu, F., & Qian, C. (2015). A rational design approach of intermediate wind girders on large storage tanks. *Thin-Walled Structures, 92*, 76–81.

Burgos, C. A., Batista-Abreu, J. C., Calabró, H. D., Jaca, R. C., & Godoy, L. A. (2015). Buckling estimates for oil storage tanks: Effect of simplified modeling of the roof and wind girder. *Thin-Walled Structures, 91*, 29–37. Available from https://doi.org/10.1016/j.tws.2015.02.006.

Burgos, C. A., Jaca, R. C., Lassig, J. L., & Godoy, L. A. (2014). Wind buckling of tanks with conical roof considering shielding by another tank. *Thin-Walled Structures, 84*, 226–240. Available from https://doi.org/10.1016/j.tws.2014.06.007.

CAN/CSA-IEC/ISO 31010 (2010). Risk management—risk assessment techniques.

Cozzani, V., Campedel, M., Renni, E., & Krausmann, E. (2010). Industrial accidents triggered by flood events: Analysis of past accidents. *Journal of Hazardous Materials, 175*, 501–509.

EPA (2006). *Response to 2005 Hurricanes*. <https://archive.epa.gov/katrina/web/html/>. Accessed 10.18.20.

EPA (2016). Flood preparedness: Recommended best practices. Regional Response Team 6, Fact Sheet 103a.

Fabbrocino, G., Iervolino, I., Orlando, F., & Salzano, E. (2005). Quantitative risk analysis of oil storage facilities in seismic areas. *Journal of Hazardous Materials, 123*, 61–69. Available from https://doi.org/10.1016/j.jhazmat.2005.04.015.

Flores, F. G., & Godoy, L. A. (1998). Buckling of short tanks due to hurricanes. *Engineering Structures, 20*, 752–760. Available from https://doi.org/10.1016/S0141-0296(97)00109-0.

Flores, F. G., & Godoy, L. A. (1999). Forced vibrations of silos leading to buckling. *Journal of Sound and Vibration, 224*, 431–454. Available from https://doi.org/10.1006/jsvi.1999.2188.

Fredsoe, J. (2016). Pipeline–seabed interaction. *Journal of Waterway, Port, Coastal, and Ocean Engineering, 142*, 03116002. Available from https://doi.org/10.1061/(ASCE)WW.1943-5460.0000352.

Godoy, L. (2007). Performance of storage tanks in oil facilities damaged by Hurricanes Katrina and Rita. *Journal of Performance of Constructed Facilities, 21*, 441–449. Available from https://doi.org/10.1061/(ASCE)0887-3828(2007)21:6(441).

Godoy, L. A. (2016). Buckling of vertical oil storage steel tanks: Review of static buckling studies. *Thin-Walled Structures, 103*, 1 21. Available from https://doi.org/10.1016/j.tws.2016.01.026.

Godoy, L. A., & Flores, F. G. (2002). Imperfection sensitivity to elastic buckling of wind loaded open cylindrical tanks. *Structural Engineering and Mechanics, 13*, 533–542.

Godoy, L. A., & Sosa, E. M. (2003). Localized support settlements of thin-walled storage tanks. *Thin-Walled Structures, 41*, 941–955.

Greiner, R., & Derler, P. (1995). Effect of imperfections on wind-loaded cylindrical shells. *Thin-Walled Structures, 23*, 271–281. Available from https://doi.org/10.1016/0263-8231(95)00016-7.

Holroyd, R. J. (1985). On the behaviour of open-topped oil storage tanks in high winds. part II. structural aspects. *Journal of Wind Engineering & Industrial Aerodynamics, 18*, 53–73. Available from https://doi.org/10.1016/0167-6105(85)90074-1.

Hyder, M. (Ed.), (2008). *Oil spill intelligence report, assessment of hurricane ike damage continues*. Aspen Publishers.

Jerath, S., & Ghosh, A. K. (1987). Buckling of cylindrical shells under the action of non-uniform external pressure. *Computers & structures, 25*, 607–614. Available from https://doi.org/10.1016/0045-7949(87)90268-9.

Jerath, S., & Sadid, H. (1985). Buckling of orthotropic cylinders due to wind load. *Journal of Engineering Mechanics, 111*, 610–622.

Kameshwar, S. (2017). Multi-hazard fragility, risk, and resilience assessment of select coastal infrastructure.

Kameshwar, S., & Padgett, J. (2018a). Fragility and resilience indicators for portfolio of oil storage tanks subjected to hurricanes. *Journal of Infrastructure Systems, 24*, 04018003. Available from https://doi.org/10.1061/(ASCE)IS.1943-555X.0000418.

Kameshwar, S., & Padgett, J. E. (2015). Fragility assessment of above ground petroleum storage tanks under storm surge. In *12th international conference on applications of statistics and probability in civil engineering* (pp. 12–15). ICASP.

Kameshwar, S., & Padgett, J. E. (2016). Stochastic modeling of geometric imperfections in aboveground storage tanks for probabilistic buckling capacity estimation. *ASCE-ASME Journal of Risk and Uncertainty in Engineering Systems, Part A: Civil Engineering*, 2. Available from https://doi.org/10.1061/ajrua6.0000846.

Kameshwar, S., & Padgett, J. E. (2018b). Storm surge fragility assessment of above ground storage tanks. *Structural Safety, 70*, 48–58. Available from https://doi.org/10.1016/j.strusafe.2017.10.002.

Kameshwar, S., & Padgett, J. E. (2019). Stiffening ring design for prevention of storm-surge buckling in aboveground storage tanks. *Journal of Structural Engineering, 145*, 04019002. Available from https://doi.org/10.1061/(ASCE)ST.1943-541X.0002275.

Khakzad, N., & Van Gelder, P. (2017). Fragility assessment of chemical storage tanks subject to floods. *Process Safety and Environmental Protection, 111*, 75–84.

Khakzad, N., & Van Gelder, P. (2018). Vulnerability of industrial plants to flood-induced natechs: A Bayesian network approach. *Reliability Engineering and System Safety, 169*, 403–411. Available from https://doi.org/10.1016/j.ress.2017.09.016.

Krausmann, E., Cruz, A. M., & Salzano, E. (2016). *Natech risk assessment and management: Reducing the risk of natural-hazard impact on hazardous installations*. Elsevier.

Krausmann, E., Renni, E., Campedel, M., & Cozzani, V. (2011). Industrial accidents triggered by earthquakes, floods and lightning: Lessons learned from a database analysis. *Natural Hazards, 59*, 285–300.

Kundurpi, P. S., Johns, D. J., & Samavedam, G. (1975). Stability of cantilever shells under wind loads. *Journal of the Engineering Mechanics Division, 101*, 517–530. Available from https://doi.org/10.1061/JMCEA3.0002048.

Kuroda, S., Uejima, H., Ishida, K., Yoshida, S., Shiratori, M., Sekine, K., Tsuchida, T., & Iwata, K. (2012). Simulation for a floating roof behavior of cylindrical storage tank due to wind load: Part 1—CFD analysis. In *Pressure vessels and piping conference* (pp. 125–137). American Society of Mechanical Engineers.

Landucci, G., Antonioni, G., Tugnoli, A., & Cozzani, V. (2012). Release of hazardous substances in flood events: damage model for atmospheric storage tanks. *Reliability Engineering and System Safety, 106*, 200–216. Available from https://doi.org/10.1016/j.ress.2012.05.010.

Lewandowski, M. J., Gajewski, M., & Gizejowski, M. (2015). Numerical analysis of influence of intermediate stiffeners setting on the stability behaviour of thin-walled steel tank shell. *Thin-Walled Structures, 90*, 119–127.

Macdonald, P. A., Holmes, J. D., & Kwok, K. C. S. (1990a). Wind loads on circular storage bins, silos and tanks III. Fluctuating and peak pressure distributions. *Journal of Wind Engineering & Industrial Aerodynamics, 34*, 319–337. Available from https://doi.org/10.1016/0167-6105(90)90160-E.

MacDonald, P. A., Holmes, J. D., & Kwok, K. C. S. (1990b). Wind loads on circular storage bins, silos and tanks. II. Effect of grouping. *Journal of Wind Engineering & Industrial Aerodynamics, 34*, 77–95. Available from https://doi.org/10.1016/0167-6105(90)90150-B.

MacDonald, P. A., Kwok, K. C. S., & Holmes, J. D. (1988). Wind loads on circular storage bins, silos and tanks: I. Point pressure measurements on isolated structures. *Journal of Wind Engineering & Industrial Aerodynamics, 31*, 165–187. Available from https://doi.org/10.1016/0167-6105(88)90003-7.

Maki, A. W. (1991). The Exxon Valdez oil spill: Initial environmental impact assessment. Part 2. *Environmental Science & Technology, 25*, 24–29.

Malik, Z., Morton, J., & Ruiz, C. (1979). An experimental investigation into the buckling of cylindrical shells of variable-wall thickness under radial external pressure. *Experimental Mechanics, 19*, 87–92. Available from https://doi.org/10.1007/BF02324196.

Malik, Z., Morton, J., & Ruiz, C. (1980). Buckling under normal pressure of cylindrical shells with various end conditions. *Journal of Pressure Vessel Technology, 102*, 107–111. Available from https://doi.org/10.1115/1.3263288.

Matsui, T. (2007). Sloshing in a cylindrical liquid storage tank with a floating roof under seismic excitation. *Journal of Pressure Vessel Technology, 129*, 557–566. Available from https://doi.org/10.1115/1.2767333.

Matsui, T. (2009). Sloshing in a cylindrical liquid storage tank with a single-deck type floating roof under seismic excitation. *Journal of Pressure Vessel Technology*, 131.

Matsui, T., Uematsu, Y., Kondo, K., Wakasa, T., & Nagaya, T. (2009). Wind effects on dynamic response of a floating roof in a cylindrical liquid storage tank. *Journal of Pressure Vessel Technology*, 131. Available from https://doi.org/10.1115/1.3121538.

Necci, A., Girgin, S., & Krausmann, E. (2018). Understanding natech risk due to storms—Analysis, lessons learned and recommendations. JRC Technical Reports.

Olivar, O. J. R., Mayorga, S. Z., Giraldo, F. M., Sánchez-Silva, M., Pinelli, J.-P., & Salzano, E. (2020). The effects of extreme winds on atmospheric storage tanks. *Reliability Engineering and System Safety*, *195*, 106686. Available from https://doi.org/10.1016/j.ress.2019.106686.

Palinkas, L., Downs, M., Petterson, J., & Russell, J. (1993). Social, cultural, and psychological impacts of the exxon valdez oil spill. *Human Organization*, *52*, 1–13.

Portela, G., & Godoy, L. A. (2005a). Wind pressures and buckling of cylindrical steel tanks with a conical roof. *Journal of Constructional Steel Research*, *61*, 786–807. Available from https://doi.org/10.1016/j.jcsr.2004.11.002.

Portela, G., & Godoy, L. A. (2005b). Wind pressures and buckling of cylindrical steel tanks with a dome roof. *Journal of Constructional Steel Research*, *61*, 808–824. Available from https://doi.org/10.1016/j.jcsr.2004.11.001.

Portela, G., & Godoy, L. A. (2007). Wind pressure and buckling of grouped steel tanks. *Wind and Structures*, *10*, 23–44.

Porter, K. (2015). A beginner's guide to fragility, vulnerability, and risk. *Encyclopedia of Earthquake Engineering*, *2015*, 235–260.

Qin, R., Zhu, J., & Khakzad, N. (2020). Multi-hazard failure assessment of atmospheric storage tanks during hurricanes. *Journal of Loss Prevention in the Process Industries*, 104325. Available from https://doi.org/10.1016/j.jlp.2020.104325.

Ramirez, O., Mesa, A., Zuluaga, S., Munoz, F., Sanchez-Silva, M., & Salzano, E. (2019). Fragility curves of storage tanks impacted by strong winds. *Chemical Engineering Transactions*, 77, 91–96. Available from https://doi.org/10.3303/CET1977016.

Roueche, D., Kameshwar, S., Marshall, J., Mashrur, N., Kijewski-Correa, T., Gurley, K., Afanasyeva, I., Brasic, G., Cleary, J., Golovichev, D., Lafontaine, O., Lombardo, F., Micheli, L., Phillips, B., Prevatt, D., Robertson, I., Schroeder, J., Smith, D., Strader, S., ... Rodriguez, L. (2021). Hybrid preliminary virtual reconnaissance report-early access reconnaissance report (PVRR-EARR). *Designsafe-CI*. Available from https://doi.org/10.17603/DS2-NG93-SE16.

Salzano, E., Iervolino, I., & Fabbrocino, G. (2003). Seismic risk of atmospheric storage tanks in the framework of quantitative risk analysis. *Journal of Loss Prevention in the Process Industries*, *16*, 403–409. Available from https://doi.org/10.1016/S0950-4230(03)00052-4.

Sengul, H., Santella, N., Steinberg, L. J., & Cruz, A. M. (2012). Analysis of hazardous material releases due to natural hazards in the United States. *Disasters*, *36*, 723–743. Available from https://doi.org/10.1111/j.1467-7717.2012.01272.x.

Sosa, E. M., & Godoy, L. A. (2005). Nonlinear dynamics of above-ground thin-walled tanks under fluctuating pressures. *Journal of Sound and Vibration*, *283*, 201–215.

Sun, X., Liu, Y., Wang, J., & Cen, Z. (2008). Stress and deflection analyses of floating roofs based on a load-modifying method. *International Journal of Pressure Vessels and Piping*, *85*, 728–738. Available from https://doi.org/10.1016/j.ijpvp.2008.03.003.

Teng, J. G. (1996). Buckling of thin shells: Recent advances and trends. *Applied Mechanics Reviews*, *49*, 263–274. Available from https://doi.org/10.1115/1.3101927.

Uchiyama, K., Uematsu, Y., & Orimo, T. (1987). Experiments on the deflection and buckling behavior of ring-stiffened cylindrical shells under wind pressure. *Journal of Wind Engineering & Industrial Aerodynamics*, *26*, 195–211. Available from https://doi.org/10.1016/0167-6105(87)90044-4.

Uematsu, Y., Koo, C., & Yasunaga, J. (2014). Design wind force coefficients for open-topped oil storage tanks focusing on the wind-induced buckling. *Journal of Wind Engineering & Industrial Aerodynamics*, *130*, 16–29. Available from https://doi.org/10.1016/j.jweia.2014.03.015.

Uematsu, Y., & Uchiyama, K. (1985). Deflection and buckling behavior of thin, circular cylindrical shells under wind loads. *Journal of Wind Engineering & Industrial Aerodynamics*, *18*, 245–261. Available from https://doi.org/10.1016/0167-6105(85)90084-4.

Uematsu, Y., Yasunaga, J., & Koo, C. (2015). Design wind loads for open-topped storage tanks in various arrangements. *Journal of Wind Engineering & Industrial Aerodynamics*, *138*, 77–86. Available from https://doi.org/10.1016/j.jweia.2014.12.013.

Yang, Y., Chen, G., & Reniers, G. (2020). Vulnerability assessment of atmospheric storage tanks to floods based on logistic regression. *Reliability Engineering and System Safety*, *196*, 106721. Available from https://doi.org/10.1016/j.ress.2019.106721.

Yasunaga, J., & Uematsu, Y. (2020). Dynamic buckling of cylindrical storage tanks under fluctuating wind loading. *Thin-Walled Structures*, *150*, 106677. Available from https://doi.org/10.1016/j.tws.2020.106677.

Yoshida, S., Kuroda, S., Uejima, H., Ishida, K., Shiratori, M., Sekine, K., Tsuchida, T., & Iwata, K. (2012). Simulation for a floating roof behavior of cylindrical storage tank due to wind load: Part 2—sloshing response analysis. In *Pressure vessels and piping conference* (pp. 103–114). American Society of Mechanical Engineers.

Yoshida, S., Shiratori, M., Sekine, K., Kuroda, S., Uejima, H., Ishida, K., Tsuchida, T., & Serizawa, Y. (2013). Vibration of a floating roof in aboveground storage tank due to wind. *20th International Congress on Sound and Vibration*, *2013*(4), 3505–3512, 2013 ICSV.

Zuluaga Mayorga, S., Sánchez-Silva, M., Ramírez Olivar, O. J., & Muñoz Giraldo, F. (2019). Development of parametric fragility curves for storage tanks: A Natech approach. *Reliability Engineering and System Safety*, *189*, 1–10. Available from https://doi.org/10.1016/j.ress.2019.04.008.

CHAPTER FIVE

# Tank design

**Zhan Liu**
School of Mechanics and Civil Engineering, China University of Mining and Technology, Xuzhou, China

Tank design is greatly important in chemical engineering, energy engineering, and other engineering applications. Detailed design has obvious influences on the safety, durability, and reliability of different fuel storage tanks. The present workmainly focuses on the tank design in the room temperature scope. For the tank applied for cryogens, including liquid hydrogen, liquid oxygen, liquid nitrogen, and other low-temperature fluids, both the tank design and material selection should be given more considerations, and the related content is not involved in this chapter. The basic content on the tank design (Du, 2016; Jawad & Farr, 1989, 2018; Li & Li, 2010) is introduced in this section.

As we all know, the pressure vessels are usually welded with rotating shells whose middle surfaces are curved by rotation. Here, the rotation surface refers to the surface formed by a plane curve as the generatrix rotates around its axis in the plane. The middle surface of a rotating shell is a curved surface equidistant from the inner and outer surfaces of the shell. When the ratio of the outer diameter $D_o$ to the inner diameter $D_i$ of the rotating shell is $\leq 1.2$, the shell is called the rotating thin shell. The bending moment of the rotating thin shell is very small when it is loaded. If this moment is ignored, the stress analysis of the shell can be greatly simplified. The shell theory which ignores the bending moment, is called the torque-free theory or the thin-film theory. The solution to the shell problem solved by the torque-free theory is called the thin film solution, which is the basis of the pressure vessel design. Therefore, this section first introduces the torque-free theory of rotating thin shells.

*Above Ground Storage Tank Oil Spills.*
DOI: https://doi.org/10.1016/B978-0-323-85728-4.00007-3

# 5.1 Torque-free theory of rotating thin shells

## 5.1.1 Geometrical characteristics of general rotating thin shells

The geometrical features of the general rotating shell can be represented by the geometrical features of the middle plane, as shown in Fig. 5.1.

1. The longitude line and longitude line section cross-section. The line of the intersection between the plane and the middle plane over the axis of rotation refers to the longitude line, as shown in APC in Fig. 5.1A. The plane formed by the longitude line and the axis of rotation refers to the longitude line cross-section, as shown in APCO in Fig. 5.1A.
2. Normals. The line that passes through any point B on the longitude line and is perpendicular to the center plane is called the normal of the center plane at that point, as shown in $PK_2$ in Fig. 5.1B.
3. Latitude line and latitude line cross-section. The intersecting line between the cone section and the center plane formed by taking the normal as the generatrix around the axis of rotation is called the latitude line. The conical section (rotation section) refers to the latitude line cross-section.

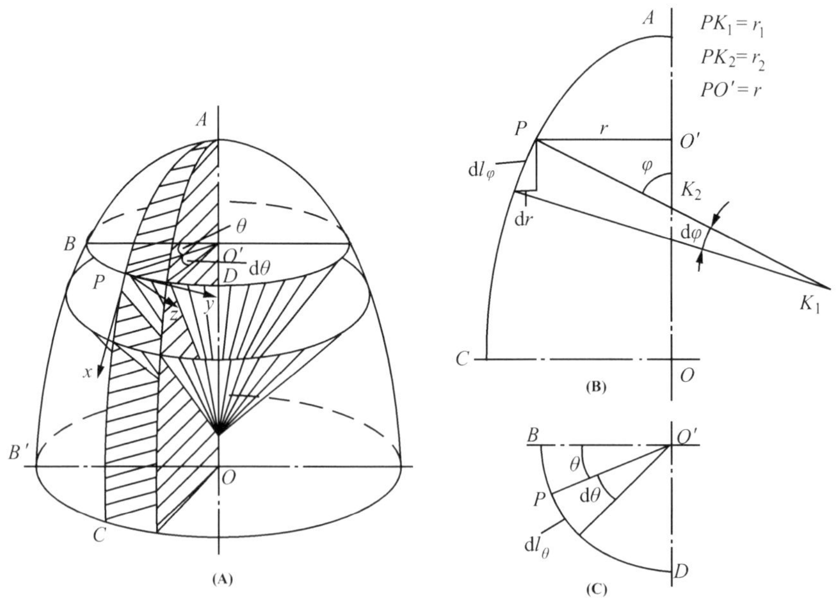


**Figure 5.1** Diagram of general rotating thin shells. (A) Global view, (B) Side view, and (C) Top view.

4. Parallel circles and radius of parallel circles. The intersection line between the plane and the center plane perpendicular to the axis of rotation is called the parallel circle. The radius of the parallel circle is shown in Fig. 5.1B, described by $r$.
5. Curvature radius. The curvature radius of the longitude line is called the first curvature radius, described as $K_1P$ in Fig. 5.1B. The curvature radius of the latitude line is called the second curvature radius, which can also be defined as the curvature radius of the intersection line at the point through which a plane is orthogonal to the latitude line of the point, described as $K_2P$ in Fig. 5.1B. The second curvature radius is selected with the length of the generatrix of the cone.
6. Longitudinal coordinates and circumferential coordinates. The position of the longitudinal line is determined by the angle $\theta$ measured from the plane of the generatrix. The position of the parallel circle is determined by the angle $\varphi$, and the position of any point on the middle plane can be determined by coordinates $\theta$ and $\varphi$. Specially, $\theta$ is called the meridional coordinate and $\theta$ the circumferential coordinate.

## 5.1.2 Geometric characteristics of several common shells

### 5.1.2.1 Cylindrical shell

Supposed that the radius of the middle plane is set as $R$ since the longitude line is a straight line, $r_1 = \infty$. As the plane which is perpendicular to the longitude line, coincides with the parallel circle, its radius meets the following equation $r_2 = r = R$.

### 5.1.2.2 Spherical shell

The curvature radius of all normal sections at any point of the surface in the spherical shell, is equal to the radius $R$ of the sphere, so $r_1 = r_2 = R$.

### 5.1.2.3 Ellipsoid shell

As Fig. 5.2 shows, the ellipse's major half axis is $a$ and its short half axis is $b$. For any point $B(x, y)$ in the ellipse, the related longitude equation is given as $\frac{x^2}{a^2} + \frac{y^2}{b^2} = 1$.

Driven from the differential calculus, the related curvature radius is given

$$r_1 = \left| \frac{\left[1+\left(\frac{dy}{dx}\right)^2\right]^{1.5}}{\frac{d^2y}{dx^2}} \right|, \frac{dy}{dx} = -\frac{b^2x}{a^2y}, \frac{d^2y}{dx^2} = -\frac{b^2(b^2x^2 + a^2y^2)}{a^4y^3} = -\frac{b^4}{a^2y^3}.$$

$$\text{Then } r_1 = \frac{1}{a^4b^4}(b^4x^2 + a^4y^2)^{1.5} = \frac{\left[a^4 - x^2(a^2 - b^2)\right]^{1.5}}{a^4b} \tag{5.1a}$$

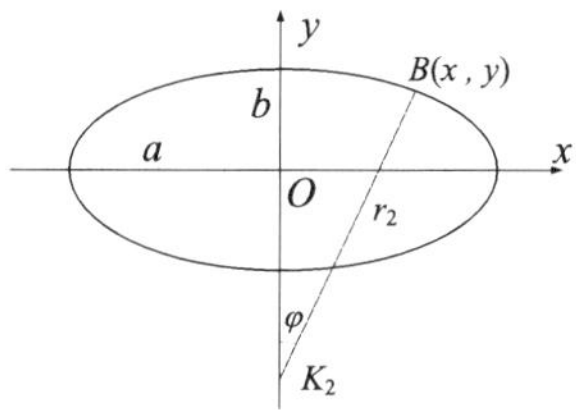


**Figure 5.2** Diagram of the ellipsoid shell.

$$r_2 = \frac{1}{b^2}(b^4x^2 + a^4y^2)^{0.5} = \frac{\left[a^4 - x^2(a^2 - b^2)\right]^{0.5}}{b} \tag{5.1b}$$

It can be seen from Fig. 5.2 that $x = r_2 \sin\varphi$. Assuming that $m = a/b$, substitute it into Eqs. (5.1a) and (5.1b), thus

$$r_1 = \frac{a^2b^2}{\left(a^2\sin^2\varphi + b^2\cos^2\varphi\right)^{1.5}} = ma\psi^3 \tag{5.2a}$$

$$r_2 = \frac{a^2}{\left(a^2\sin^2\varphi + b^2\cos^2\varphi\right)^{0.5}} = ma\psi \tag{5.2b}$$

Here, $\psi = \frac{1}{\sqrt{(m^2-1)\sin^2\varphi + 1}}$

Based on above expressions, there are some special cases. For example, while it is located at the vertices of an ellipsoid, $\varphi = 0$, $\psi = 1$, so $r_1 = r_2 = ma = a^2/b$. While it is located at the ellipsoidal equator, $\varphi = \pi/2$, $\psi = 1/m$, so $r_1 = a/m^2 = b^2/a$ and $r_2 = a$.

### 5.1.3 General equations of the torque-free theory

Under the axisymmetric condition, the equilibrium equation (Du, 2016; Li & Li, 2010) established based on the torque-free theory is given.

$$\frac{\sigma_\varphi}{r_1} + \frac{\sigma_\theta}{r_2} = \frac{p_z}{\delta} \tag{5.3}$$

This equation is called the Laplace equation.

$$N_\theta = r_2 p_z - \frac{J(\varphi)}{r_1 \sin^2\varphi} \tag{5.4}$$

This equation is the regional equilibrium equation.

As Fig. 5.3 shows, the physical meaning $J(\varphi)$ refers to the axial external load on the unit radian of the shell parallel circle.

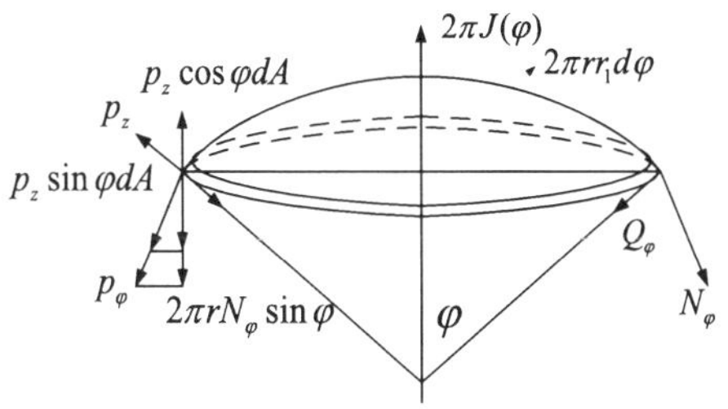


**Figure 5.3** Regional equilibrium of the rotating thin shells.

Take the area of the infinitesimal ring along the parallel circle $dA = 2\pi rr_1 d\varphi$, then the total axial load is

$$\int_0^{\varphi} (p_z \cos\varphi - p_\varphi \sin\varphi)\, dA = 2\pi \int_0^{\varphi} (p_z \cos\varphi - p_\varphi \sin\varphi)\, r \cdot r_1\, d\varphi = 2\pi \cdot J(\varphi)$$

The physical meaning of the regional equilibrium equation can be shown in Fig. 5.3, and the related expression is given as $2\pi r N_\varphi \sin\varphi = 2\pi J(\varphi)$. Here, $N_\varphi = \frac{J(\varphi)}{r\sin\varphi}$, and $J(\varphi) = \int_0^{\varphi} (p_z \cos\varphi - p_\varphi \sin\varphi)\, rr_1\, d\varphi$

The total axial load is denoted by Q, and it meets the following expression $Q = 2\pi J(\varphi) = 2\pi r N_\varphi \sin\varphi$.

Equations (5.3) and (5.4) are the basic formulas for the torque-free theory of rotating thin shells. While the bending moment is disregarded, it is assumed that the stress is uniformly distributed along with the wall thickness, hence the thin film stress of the shell can be determined. The longitudinal stress is expressed as $\sigma_\varphi = \frac{N_\varphi}{\delta}$, and the circumferential stress is expressed as $\sigma_\theta = \frac{N_\theta}{\delta}$.

## 5.1.4 Application conditions for torque-free theory

### *5.1.4.1 Geometric continuity*

It means the shell should have a continuous surface. There are some abrupt changes in the shell, such as the curvature change, wall thickness change, material change, etc. When the torque-free theory is adopted to analyze the shell stress in these special positions, obvious deformation disharmony will form. Moreover, the deformation disharmony usually leads to direct local bending, so the torque-free theory is not proper.

### *5.1.4.2 Continuous external load*

The external load on the shell should be continuous. If a concentrated force perpendicular to the wall surface of the shell, an obvious temperature difference, a torque action, and a stiffening ring are discontinuous, the shell will be in a torque state.

#### *5.1.4.3 Continuous constraint*

It mainly includes the following two points

1. The fixed form on the shell boundary should be freely supported. When the normal displacement and rotation angle on the boundary are constrained, the shell is bound to bend under the action of load, and cannot maintain the torque-free state.
2. The boundary force of the shell should be in the tangent plane of the shell surface. It is required that no transverse shear forces and bending moments form at the boundary. For example, the internal force should not generate and form at the boundary of the flange spherical head and the cylinder.

Hence, the existence of a torques-free state of a thin shell must satisfy the continuity of shell geometry, material, and stress load, and ensure that the shell has a free boundary. If these conditions are not satisfied, the torque-free theory cannot be adopted for the stress analysis. However, the solution of the torques-free theory is still valid for the places where there is no local bending, such as the places far away from the connection edge of the shell, the interface of the load, and the support of the container.

### 5.1.5 Application of torque-free theory

#### *5.1.5.1 Effect of gas pressure*

Gas pressure is an axisymmetric load, which is perpendicular to the surface of the shell and has the same value in different directions, as shown in Fig. 5.4. Under the constant internal pressure of the gas, $p_z = p = \text{constant}$ and $p_\varphi = 0$. As $r = r_2\sin\varphi$ and $r_1 d\varphi\cos\varphi = dr = d(r_2\sin\varphi)$, for the sealed shell at the top, the following equation is given,

$$\begin{aligned} J(\varphi) &= \int_0^\varphi rr_1\left(p_z\cos\varphi - p_\varphi\sin\varphi\right)d\varphi = \int_0^\varphi pr_1r_2\cos\varphi\sin\varphi d\varphi \\ &= p\int_0^\varphi (r_2\sin\varphi)d(r_2\sin\varphi) = p(r_2\sin\varphi)^2/2 \end{aligned} \tag{5.5}$$

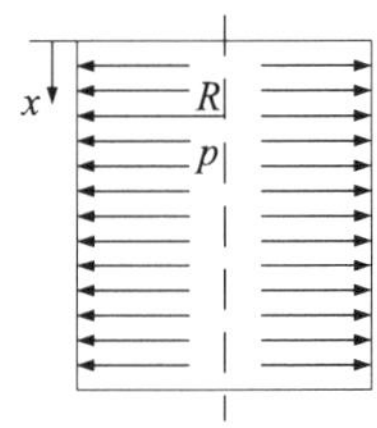


**Figure 5.4** Diagram of gas pressure distribution.

$$N_\varphi = \frac{J(\varphi)}{r_2\sin^2\varphi} = \frac{pr_2}{2} \text{ or } \sigma_\varphi = \frac{pr_2}{2\delta} \tag{5.6a}$$

$$N_\theta = r_2\left(p - \frac{N_\varphi}{r_1}\right) = N_\varphi\left(2 - \frac{r_2}{r_1}\right) \text{ or } \sigma_\theta = \sigma_\varphi\left(2 - \frac{r_2}{r_1}\right) \tag{5.6b}$$

Thus, for the rotating shell subjected to the gas pressure $p$, once the radius values of the shell $r_1$ and $r_2$ are determined, both the thin-film stress and the structure deformation can be calculated.

#### 5.1.5.2 Effect of liquid pressure

The liquid pressure is a static pressure and the pressure at each point varies with the depth of the liquid. The pressure is zero at the liquid surface, and the liquid column pressure is expressed as $\rho gh$, at the depth from the liquid surface $h$. Here, $\rho$ is the liquid density, as shown in Fig. 5.5.

The liquid pressure is an axisymmetric load only when the shell locates in the vertical direction and its axis is perpendicular to the ground, and the hydraulic pressure is a nonaxisymmetry load in other cases. The liquid pressure, which is perpendicular to the surface of the shell, is expressed as $p_\varphi = 0$, and $p_z = \rho gh$.

Under the axisymmetric condition, the expression of $J(\varphi)$ can be transformed into

$$\begin{aligned} J(\varphi) &= \int_0^\varphi rr_1\left(p_z\cos\varphi - p_\varphi\sin\varphi\right)d\varphi + C \\ &= \rho g\int_0^\varphi hr_1r_2\cos\varphi\sin\varphi d\varphi + C \end{aligned} \tag{5.7}$$

Once the generalized equation of the shell is determined, the relation between $h$ and $\varphi$ can be calculated, so Eq. (5.7) can be solved. The constant $C$ is calculated and determined based on the boundary conditions, so $J(\varphi)$ could be solved. Therefore, the internal forces can be calculated.

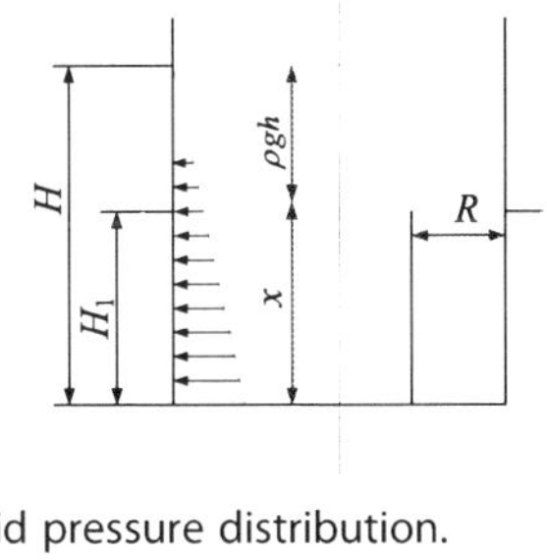


**Figure 5.5** Diagram of liquid pressure distribution.

$$N_\varphi = \frac{J(\varphi)}{r_2 \sin^2 \varphi} \tag{5.8a}$$

$$N_\theta = r_2\left(p_z - \frac{N_\varphi}{r_1}\right) = \rho g h r_2 - \frac{r_2}{r_1} N_\varphi \tag{5.8b}$$

## 5.2 The edge problem

### 5.2.1 Reason for the formation of discontinuous stress

In actual engineering applications, most of the shell structures are connected by several simple shell combinations. The parallel circle at the junction of two components is called the connection edge. At the junction of two shells, if the two shells are regarded as free bodies, there is free deformation under the action of internal pressure. At the present condition, the film displacement and angle of rotation at the junction are generally not equal. However, as the two shells are connected, the displacement and angle of rotation of the two shells at the junction must be equal. In this case, a constraint is formed near the connection of the two shells, so the local bending deformation of the shell at the connection is generated and evolved. Subjected to the stress deformation, the local stress is formed at the connection edge as well, so that the total stress in this area increases correspondingly.

As the whole structure is discontinuous, the stress increases rapidly in the local area near the edge of the connection, which is called the discontinuity effect or edge effect. The local stress caused by this effect is called the discontinuous stress or marginal stress.

### 5.2.2 Calculation method for discontinuous stress

The discontinuous stress can be calculated according to the general theory of the shell, but the related calculation method is greatly complicated. To solve the actual issues in engineering applications, a simple solution is adopted. Usually, the shell stress solution is divided into two parts. The first one is the thin film solution, that is, the shell solution based on the torque-free theory. It is caused by external load and increases with external load, so it is one kind of force, which must meet the equilibrium relation between the internal and external force and moment. While its value is larger than the material yield point, the destruction of the material or large deformation may be generated and formed. The second one is the moment solution. It is the solution of the moment theory for the free boundary under the effects of edge force and edge moment when the connecting edge of two shells is cut. The stress is caused by the constraints of adjacent parts materials or structural stress constraint itself, so it has the characteristic of the self-limited. Therefore, the local yield or small deformation generates once the stress exceeds the material yield point. To get better stress distribution results, different shell deformations at the connection edge can be coordinated. By superimposing the above two solutions, the final solution can be obtained, which could keep the structure of the composite shell in continuity.

Here, a composite shell connected by a hemisphere shell and a cylindrical shell is taken as an example.

While the connecting edges of the hemispheric shell and the cylindrical shell under the internal pressure are cut along the parallel circle, the film deformation of the two shells is shown in Fig. 5.6B. The parallel

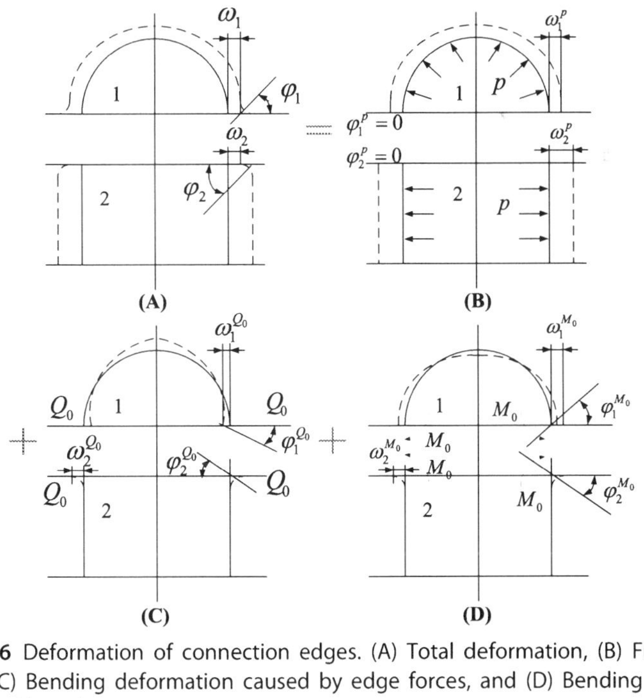


**Figure 5.6** Deformation of connection edges. (A) Total deformation, (B) Film deformation, (C) Bending deformation caused by edge forces, and (D) Bending deformation caused by edge moments.

circular longitudinal displacements of the two shells are not equal, that is, $\omega_1^p \neq \omega_2^p$. As the two shells are connected and belong to a continuous structure, the edge forces $Q_0$ and edge moments $M_0$ are generated at the junction of the two shells, which causes the bending deformation, as shown in Fig. 5.6C and D. According to the continuity condition in the bending deformation, the following expressions are given, $\omega_1 = \omega_2$ and $\varphi_1 = \varphi_2$. While the bending deformation and the film deformation are in superposition, the total deformation of the two shells at the junction must be equal, thus the continuity equation of the edge deformation can be written as

$$\omega_1^p + \omega_1^{Q_0} + \omega_1^{M_0} = \omega_2^p + \omega_2^{Q_0} + \omega_2^{M_0}$$

$$\varphi_1^p + \varphi_1^{Q_0} + \varphi_1^{M_0} = \varphi_2^p + \varphi_2^{Q_0} + \varphi_2^{M_0}$$

where, $\omega^p, \omega^{Q_0}, \omega^{M_0}$ and $\varphi^p, \varphi^{Q_0}, \varphi^{M_0}$ denote the radial displacement of parallel circles and the longitude angle generated at the connection of the shell related to $p$, $Q_0$ and $M_0$ respectively. Here, subscript 1 refers to the hemispherical shell, and subscript 2 refers to the cylindrical shell. The relations between $p$, $Q_0$, $M_0$ and displacement are obtained by the torque-free theory, while the relations between $p$, $Q_0$, $M_0$ and rotation angle are obtained by the torque theory. As Fig. 5.6C and D show, the left half of the cylinder is taken as the object. The radial displacement $\omega$ is negative outward, and the rotation angle $\varphi$ is positive counterclockwise.

By substituting $p$, $Q_0$, $M_0$ and the deformation relations into the above two equations, two edge loads $Q_0$ $M_0$ can be obtained, and then the edge bending solution can be solved as well. Superimposed with the thin film solution, the full solution can be obtained.

## 5.2.3 Characteristics and treatments of discontinuous stress

### *5.2.3.1 Characteristics of discontinuous stress*

1. Locality. The composite shells with different structures have different edge stresses at the connecting edges. Some have significant edge effects, and the corresponding stress can reach a large value. For different conditions, a common feature is that the related influence range is very small, and these stresses only exist in the local area near the joint. This property refers to the locality of the discontinuous stress.
2. Self-limited. As the adjacent shell has different deformations of thin films in the joint, the edge stress is generated correspondingly. Here, the shell deformation on the edge of the connection is usually caused

by elastic constraints. For the shell made of plastic material, when the plastic deformation is produced in the local area of the edge connection, the elastic restraint begins to ease, the deformation is not developed continuously, and the edge stress is generated correspondingly. This nature is called the self-limited discontinuous stress.

#### *5.2.3.2 Treatment of discontinuous stress in engineering problems*

As the discontinuous stress is usually distributed locally and self-limited, the detailed calculation is generally not conducted in the design process for the plastic material shell subjected to static load. To solve this issue, the method of local treatment on the structure is adopted to reduce the stress level.

However, for the shell made of brittle material, when it is subjected to fatigue load or low temperature, high discontinuous stress may lead to fatigue failure or brittle failure of the shell. Hence, the discontinuous stress should be calculated according to the relevant provisions during the design process, and the maximum discontinuous stress should be limited within the proper range.

## 5.3 Design of inner pressure cylinder

The cylindrical tank is the most common type of pressure vessel. Due to its simple structure, being easy to manufacture and install accessories, the cylindrical tank is widely used as reactors, heat exchangers, separators, and medium and small volume storage tanks.

### 5.3.1 Strength calculation of internal pressure cylinder

#### *5.3.1.1 Tank design*

According to the stress analysis, when a cylindrical shell with a middle diameter of $D$ and a wall thickness $\delta$, is subjected to a uniform medium internal pressure $p$, the following radial and circumferential thin-film stresses are generated in the tank wall:

$$\sigma_\varphi = \frac{pD}{4\delta}, \sigma_\theta = \frac{pD}{2\delta} \tag{5.9}$$

where $\delta$ refers to the calculated thickness with the unit of mm; $D$ refers to the diameter of the middle surface of the cylinder with the unit of mm.

Obviously, according to the first strength theory and the third strength theory, the equivalent stress at each point on the cylinder wall is $\frac{pD}{2\delta}$, that is $\sigma_1 = \sigma_\theta$. Based on the thin film stress intensity condition, the following equation is given.

$$\sigma_1 = \sigma_\theta = \frac{pD}{2\delta} \leq [\sigma]^t \tag{5.10}$$

where $[\sigma]^t$ is the allowable stress or permissible stress of steel at the design temperature, with the unit of MPa.

The cylinder of the container is generally welded by a steel plate. Considering that the welding seam may have negative effects on the strength of the cylinder, the allowable stress of the cylinder should be expressed by the allowable stress of the steel plate multiplied by the welding joint coefficient $\phi$. Therefore, Eq. (5.10) is changed to

$$\sigma_1 = \sigma_\theta = \frac{pD}{2\delta} \leq [\sigma]^t\phi \tag{5.11}$$

In addition, the nominal diameter of the cylinder is generally determined by technological conditions. For the rolled cylinder, its internal diameter $D_i$ is usually adopted with the expression of $D = D_i + \delta$. Hence, Eq. (5.11) is changed into

$$\frac{p(D_i + \delta)}{2\delta} \leq [\sigma]^t\phi \tag{5.12}$$

So, $\delta$ can be solved and expressed as

$$\delta = \frac{pD_i}{2[\sigma]^t\phi - p} \tag{5.13}$$

During the actual engineering design, the calculated pressure $p_c$ is usually selected as $p$ in Eq. (5.13), so

$$\delta = \frac{p_c D_i}{2[\sigma]^t\phi - p_c} \tag{5.14}$$

where $\delta$ means the calculated thickness with the unit of mm, $p_c$ means the calculation pressure with the unit of MPa, $D_i$ means the inner diameter of a cylinder with the unit of mm, $\phi$ means the weld joint coefficient, and $[\sigma]^t$ means the allowable stress of steel at the design temperature with the unit of MPa.

When a cylinder is made of seamless steel pipe, the nominal diameter is its outer diameter $D_o$, expressed by $D_i = D_o - 2\delta$. By substituting into Eq. (5.14), it can be obtained

$$\delta = \frac{p_c D_o}{2[\sigma]^t \phi + p_c} \tag{5.15}$$

### 5.3.1.2 Tank check

When the cylinder size is determined, the cylinder strength needs to be checked according to Eq. (5.16).

$$\sigma_t = \frac{p_c(D_i + \delta_e)}{2\delta_e} \leq [\sigma]^t \phi \tag{5.16}$$

where $\delta_e$ refers to the effective thickness with the unit of mm, and $\sigma_t$ refers to the calculated stress of the cylinder at the design temperature and its unit is MPa.

The maximum permissible working pressure $[p_w]$ of the cylinder is expressed as,

$$[p_w] = \frac{2\delta_e[\sigma]^t \phi}{D_i + \delta_e} \tag{5.17}$$

Equations (5.14) and (5.15) are derived from the thin film stress of the cylinder according to the maximum tensile stress criterion, so the above equations can only be used for a certain thickness range. If the thickness is too large, the actual stress situation is greatly different from the assumption that the stress is uniformly distributed along with the thickness, therefore, the above equations cannot be used. According to the thin-film theory, it is only applicable in the immediate range $\delta/D \leq 0.1$, $K \leq 1.2$. However, for an engineering design, the maximum tensile stress criterion is adopted and the material design coefficient is introduced into the determination of allowable stress, so the applicable thickness range can be expanded. That is to say, the stress intensity of the inner wall of the cylinder is within the yield point of the material when the maximum pressure is extended.

## 5.3.2 Determination of design technical parameters

The design technical parameters of the pressure vessels mainly include the design pressure, design temperature, welding joint coefficient, allowable stress, thickness and additional amount, etc.

#### 5.3.2.1 The inner diameter of the container $D_i$

The inner diameter of the container should conform to the nominal diameter of the container. Because the cylinder of the container should be matched with the flange and support, the cylinder system has a series of nominal diameters, and the standards of flange and support are formulated according to the series of a nominal diameter of the container.

For the cylinder welded with steel coil, it is stipulated that the inner diameter of the cylinder is taken as the nominal diameter in China, and its series dimensions are as follows: 300, 400, 500, 600, 700, 800, 900, 1000, 1200, 1400, 1600, 1800, 2000, 2200, 2400, 2600, 2800, 3000, 3200, 3400, 3600, 3800, and 4000 mm.

If the tube is made of seamless steel pipe, the outer diameter of the tube is specified as the nominal diameter of the tube, and the series dimensions are as follows: 159, 219, 273, 325, 377, and 426 mm.

#### 5.3.2.2 Working pressure $p_w$ and design pressure $p$

The working pressure $p_w$ refers to the maximum pressure that can be reached at the top of the vessel under the normal working conditions. Design pressure $p$ refers to the maximum pressure set at the top of the vessel, and its value is not lower than the working pressure. Both the design pressure and the corresponding design temperature are usually used as the condition of the design load.

The design pressure of the internal pressure vessels shall be determined in accordance with the following regulations.

- When the safety relief device is installed on the container, the design pressure shall not be lower than the opening pressure of the safety valve and the bursting pressure of the bursting disc device.

  Generally, the opening pressure $p_z$ of the safety valve is determined according to the working pressure $p_w$ of the container, it satisfies the equation of $p_z \leq (1.05 - 1.1)p_w$. For a container with a bursting disc, the opening pressure meets $p_z \leq (1.15 - 1.75)p_w$; When $p_z < 0.18\text{MPa}$, the value $p_z$ can have appropriate increases. The design pressure of the container should be equal to or slightly higher than $p_z$, that is, $p \geq p_z$.
- For vessels containing liquefied gases, the design pressure shall be determined according to the possible attained metal temperature under the working conditions within the specified range of filling coefficient.

The designed storage capacity of the pressure vessel containing liquefied natural gas shall not exceed the calculated value of Eq. (5.18).

$$W = \phi_V d_t V \tag{5.18}$$

where $W$ is the storage capacity with the unit of kg, $\phi_V$ is the loading coefficient, generally with the value of 0.9, the coefficient can be greater than 0.9 after actual measurement, but not greater than 0.95. $d_t$ is the density of the saturated liquid at the design temperature, with the unit of kg/m$^3$, and $V$ is the volume of pressure vessel with the unit of m$^3$.

#### 5.3.2.3 Calculated pressure $p_c$

The calculated pressure refers to the pressure used to determine the element thickness at the appropriate design temperature, including the hydrostatic pressure. The hydrostatic pressure subjected by the element can be negligible when it is less than 5% of the design pressure.

#### 5.3.2.4 Design temperature

Design temperature refers to the metal temperature of the set component of the container under the normal operating conditions (the average temperature along the component metal section). Both the design temperature and design pressure are treated as the design load conditions.

The following points should be noted for determining the design temperature of the container.

- The design temperature shall not be lower than the highest possible temperature of the component metal in working conditions. For metal temperatures below 0°C, the design temperature shall not be higher than the lowest possible temperature for the component metal.
- When the metal temperature of each part of the container is different in the working state, the design temperature of each part can be set separately.
- The metal temperature of the element may be calculated by heat transfer equation or determined on a similar vessel already in use, or by the temperature of the internal medium.

There is a corresponding relationship between the design temperature and design pressure. When the pressure vessel has different operating conditions, the design conditions of the vessel should be adopted according to the combination of the most severe pressure and temperature, instead of the design temperature and design pressure according to the single most severe conditions under different operating conditions.

#### 5.3.2.5 Allowable stress

The allowable stress is the allowable strength of the material of the container shell, and the ratio of the limit value of the material strength failure criterion to the corresponding material design factor (also known as the safety factor) is taken. The allowable stress of the material must be selected reasonably during the design process. If the allowable stress is too small, it will make the designed parts too bulky, and results in the waste of the material. Otherwise, it will make the elements too thin and easy to be damaged.

The limit value of the material strength failure criterion can be expressed in various ways, such as yield point $\sigma_s$, tensile strength $\sigma_b$, endurance strength $\sigma_D$, creep limit $\sigma_n$, etc. The limit value should be determined according to the failure type.

Below the creep temperature, the minimum tensile strength $\sigma_b$ of the material at normal temperature, the stress at the yield point temperature $\sigma_s$ or normal temperature, or design temperature $\sigma_s^t$ are divided by their respective material design coefficients, the minimum value of the above three parameters is usually taken as the allowable stress in the design of the pressure vessel.

$$[\sigma] = \min\left\{\frac{\sigma_b}{n_b}, \frac{\sigma_s}{n_s}, \frac{\sigma_s^t}{n_s}\right\} \tag{5.19}$$

That is to say, the tensile strength and yield point are used to control the allowable stress in the design of the compression element. For vessels made of ductile materials, according to the elastic failure design criterion, the maximum stress intensity of the overall vessel should be lower than the yield point of the material, so the allowable stress should be determined based on the yield point. Now, in the design of pressure vessels, the tensile strength is selected as the calculation standard of allowable stress in many standards, and its purpose is mainly to prevent fracture failure to a certain extent.

When the design temperature of carbon steel or low alloy steel exceeds 420°C, the design temperature of chromium-molybdenum alloy exceeds 450°C, and the design temperature of austenitic stainless steel exceeds 550°C, the creep may occur. Therefore, the allowable stress based on the high-temperature creep limit or rupture strength must be considered simultaneously.

$$[\sigma]^t = \frac{\sigma_D^t}{n} \quad or \quad [\sigma]^t = \frac{\sigma_n^t}{n} \tag{5.20}$$

Material design coefficient is a coefficient of the strength insurance, which is mainly used to ensure the safety of the intensity of pressure elements have enough safety reserve, its size is closely related to the accuracy

**Table 5.1** Determination of allowable stress for materials used in steel pressure vessels.

| **Material** | **Allowable stress/MPa** |
|---|---|
| Carbon steel, low alloy steel, carbon body high alloy steel | $\min\left\{\frac{\sigma_b}{3.0}, \frac{\sigma_s}{1.6}, \frac{\sigma_s^t}{1.6}, \frac{\sigma_D^t}{1.5}, \frac{\sigma_n^t}{1.0}\right\}$ |
| Austenitic high alloy steel | $\min\left\{\frac{\sigma_b}{3.0}, \frac{\sigma_s(\sigma_{0.2})}{1.5}, \frac{\sigma_s^t(\sigma_{0.2}^t)^a}{1.5}, \frac{\sigma_D^t}{1.5}, \frac{\sigma_n^t}{1.0}\right\}$ |

[a]For pressure parts made of austenitic high alloy steel, when the design temperature is below the creep range and a small amount of permanent deformation is allowed, the allowable stress can be appropriately increased to the compression element made of austenitic high alloy steel $0.9\sigma_s^t(\sigma_{0.2}^t)$, but not more than $\frac{\sigma_s^t(\sigma_{0.2}^t)}{1.5}$.

of stress calculation, the uniformity of material performance, the exact degree of load, manufacturing process and factors such as the advanced level of management as well as the testing level. The determination of the material design coefficient requires not only certain theoretical analysis but also the accumulation of long-term practical experience.

Table 5.1 gives the basis for the determination of the allowable stress for steel materials except for bolt materials in GB150−1998 (2003) of China. The required stress in design calculation can be obtained directly from the allowable stress table or Table 5.1. However, it must be noted that the allowable stress of steel plates often decreases with the increase in thickness or temperature of steel plates. The allowable stress of the bolt shall be determined based on different states and diameters of the materials. To ensure the tightness of the bolted flange connection structure, the elastic deformation of bolts must be strictly controlled. In general, the allowable stress value of bolt material is lower than that of other compression element materials. Meanwhile, to prevent the small-diameter bolt from breaking during installation, the allowable stress of the small-diameter straight-down bolt is lower than that of a large-diameter bolt.

#### *5.3.2.6 Weld joint coefficient $\phi$*

For the welded vessels, there may be slag inclusion, non-penetration, cracks, pores, and other welding defects in the welding seam. Coarse grains are easily formed in the heat-affected zone of the welding seam, which will reduce the strength or plasticity of the base metal. Therefore, the welding seam often becomes a weak link in the strength of the container. To compensate for the weakening of the overall strength of the vessel, the weld joint coefficient is introduced into the strength calculation. The weld joint coefficient represents the ratio of the strength of the

**Table 5.2** Weld joint coefficient value of steel pressure vessel.

| Type of joint form | The ratio of nondestructive testing | Value of $\phi$ |
|---|---|---|
| Double welded butt joint and full penetration butt joint | 100% | 1.0 |
| | Local detection | 0.85 |
| Single side welded butt joint (with backing plate close to base metal along the whole length of the weld root) | 100% | 0.90 |
| | Local detection | 0.80 |

weld metal to the base metal and reflects the degree to which the strength of the vessel is weakened. It is denoted by $\phi$.

There are many factors affecting the coefficient of welded joints. Among different influence factors, the weld joint coefficient is mainly related to the type of welded joints, the requirements of nondestructive testing of welds, and the length ratio.

The weld joint coefficient of the steel vessel can be selected according to Table 5.2.

#### *5.3.2.7 Thickness and additional thickness*

The calculated thickness is considered and given in Eqs. (5.14) and (5.15), but the additional thickness quantity $C$ is not considered in the above formulas. $C$ is composed of the negative thickness deviation of steel $C_1$ and the corrosion allowance $C_2$, namely $C = C_1 + C_2$, and the thinning amount $C_3$ is not included. The processing thinning amount is generally determined by the manufacturer rather than the designer according to the specific manufacturing process and the actual thickness of the plate. Therefore, the actual thickness at the factory may not be completely consistent with the drawing thickness.

The relationship between different thicknesses is shown in Fig. 5.7.

1. The calculated thickness $\delta$ refers to the thickness calculated according to Eqs. (5.14) and (5.15), but the additional thickness is not included. If necessary, the required thickness of other loads shall be included.
2. The design thickness $\delta_d$ refers to the sum of the calculated thickness and the corrosion allowance, that is, $\delta_d = \delta + C_2$.
3. The nominal thickness $\delta_n$ refers to the thickness rounded upward to the standard specification of steel after adding the design thickness with the negative deviation of steel thickness, which is marked on the drawing. That is to say $\delta_n = \delta_d + C_1 + \Delta$. Here, $\Delta$ is the rounded value.

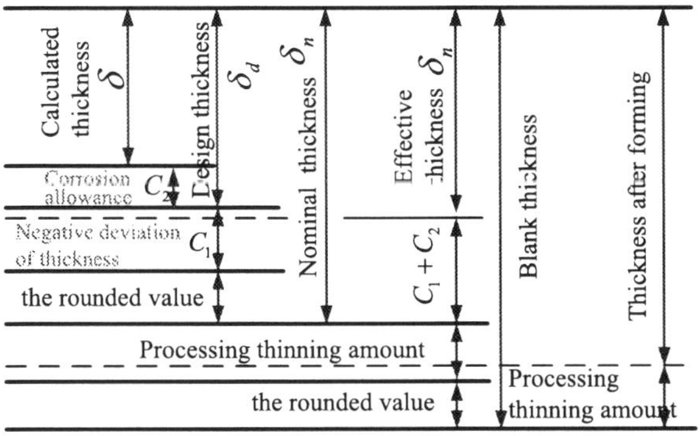


**Figure 5.7** Relationship among different thicknesses.

4. The effective thickness $\delta_e$ refers to the nominal thickness minus the corrosion allowance and the negative deviation of steel thickness, namely $\delta_e = \delta_n - C_1 - C_2$.
5. The additional thickness $C$ of steel is composed of the negative thickness deviation $C_1$ and the corrosion allowance $C_2$.

   The negative thickness deviation $C_1$ of steel plate or steel pipe shall be selected according to the nominal thickness of the corresponding steel standard. When the negative thickness deviation of the steel is not higher than 0.25 mm and does not exceed 6% of the nominal thickness, it is advisable to treat $C_1 = 0$.

   To prevent the thickness of the container from weakening and thinning due to corrosion and mechanical wear, the corrosion allowance shall be considered. For the components with corrosion or wear, the corrosion allowance shall be determined based on the expected life of the container and the corrosion rate of the medium to the metal materials.

$$C_2 = K_a B \tag{5.21}$$

where $K_a$ is corrosion rate with the unit of mm/a, and $B$ is the design life for the container, usually 10–15 years.

Moreover, the value $C_2$ can be directly selected according to the corrosion rate. When the corrosion rate of the material is $0.05 - 0.1$ mm/a, $C_2 = 1 - 2$ mm for the single-side corrosion, and $C_2 = 2 - 4$ mm for the double-side corrosion. When the corrosion rate of the material is $\leq 0.05$ mm/a, $C_2 = 1$ mm for the single-side corrosion, and $C_2 = 2$ mm for the double-side corrosion.

Different corrosion allowances can be used when the corrosion degree of each component of the container is different.

While the carbon steel or low alloy steel container is used for compressed air, water vapor, or water, the corrosion margin should not be less than 1 mm.

For stainless steel, when the corrosion of the medium is very small, the corrosion margin can be ignored.

**6.** Minimum thickness $\delta_{\min}$. For vessels with low pressure, the thickness calculated according to the strength formula is very thin, which often brings difficulties to mechanical manufacture, transportation, and hoisting. Therefore, to meet the requirements of manufacture technology and the stiffness requirements in the process of transportation and installation, the minimum thickness excluding the corrosion margin is stipulated for the shell after processing and forming.

- For carbon steel or low alloy steel containers, $\delta_{\min}$ is not less than 3 mm.
- For high alloy steel vessels, $\delta_{\min}$ is not less than 2 mm.

## 5.4 Design of internal pressure spherical shell

The calculated thickness of the spherical shell is calculated based on Eq. (5.22), which applies to the design pressure $p_c \leq 0.6[\sigma]^t\phi$.

$$\delta = \frac{p_c D_i}{4[\sigma]^t\phi - p_c} \tag{5.22}$$

where, $\delta$ is the calculated thickness of the spherical shell with the unit of mm, $D_i$ is the inner diameter of the spherical shell with the unit of mm, $p_c$ is the calculation pressure with the unit of MPa, $[\sigma]^t$ is the allowable stress of spherical shell material at the design temperature, its unit is MPa and $\phi$ is the weld coefficient.

The film stress of the spherical shell is checked according to Eq. (5.23).

$$\sigma^t = \frac{p_c(D_i + \delta_e)}{4\delta_e} \leq [\sigma]^t\phi \tag{5.23}$$

where $\delta_e$ is the effective thickness of the spherical shell, the related unit is mm.

## 5.5 Design of internal pressure dished head

The dished head is an important part of the container. The common dished head includes a convex head, cone head, and flathead. These heads have their advantages in strength and manufacture. The selection of the dished head depends on the requirements of process conditions, the difficulty in processing, and the consumption of materials.

For the strength calculation of the head under uniform internal pressure, as the head is connected with the cylinder, it is necessary to consider both the film stress caused by the internal pressure of the head and the discontinuous stress at the connection with the cylinder.

The total stress at the joint is related to the geometry and the ratio of the head size to the thickness of the cylinder. However, when the design formula for the thickness of the head is derived, the internal pressure film stress is mainly used as the basis, and the stress enhancement effect caused by the discontinuity effect is introduced into the calculation formula of the thickness calculation in the form of stress enhancement coefficient. The stress enhancement coefficient is derived from the moment theory and modified by experiment.

When the dished head is designed, the type and parameter recommended in the head standard should be selected firstly, and then conduct the strength or stability calculation according to the compression condition to determine the appropriate thickness.

### 5.5.1 Internal pressure convex dished head

The convex head includes the hemispherical head, ellipsoidal head, dished head, and spherical crown head, as shown in Fig. 5.8.

#### *5.5.1.1 Hemispherical head*

The hemispherical head consists of a half-spherical shell, as shown in Fig. 5.8A. Therefore, this form meets the requirement of the thin film

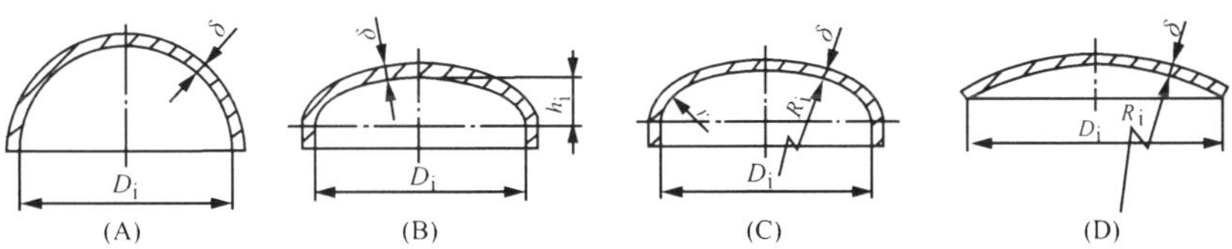


**Figure 5.8** Different kinds of convex heads. (A) Hemispherical head, (B) Ellipsoid head, (C) Dished head, and (D) Spherical crown head.

stress theory, and the thickness calculation formula is the same as that of the spherical container.

$$\delta = \frac{p_c D_i}{4[\sigma]^t \phi - p_c} \tag{5.24}$$

Meanwhile, in order to meet the elasticity requirements, the applicable scope of Eq. (5.24) is limited to $p_c \leq 0.6[\delta]^t \phi$, which is equal to $K \leq 1.33$.

After taking into account the corrosion margin and the negative deviation of steel thickness, it is rounded up to the standard steel plate specification, which is the nominal thickness of the head $\delta_n = \delta + C_1 + C_2 + \Delta$.

The processing thinning amount of the head stamping shall be determined by the manufacturer according to personal experience, but the actual thickness of the head shall not be less than its nominal thickness.

The hemispherical head has the same advantage as the spherical vessel, and it has the minimum surface area for the same vessel. Under the same pressure conditions, the required wall thickness is the thinnest, so it saves material and has good strength. However, due to the large depth of the hemispherical head, the overall stamping molding is difficult, especially when the diameter is relatively small. For the hemispherical head with a large diameter (diameter is >2.5 mm), several steel plates can be formed by tail-welding, but the size is not too accurate, and the workload of splicing is large. When the head is made of top circle and flap splicing, the weld direction is only allowed to be radial and circumferential, and the minimum distance between nonintersecting welds should not be less than 3 times of the nominal thickness of the head, and not less than 100 mm.

#### *5.5.1.2 Ellipsoid head*

The ellipsoidal head is composed of a half ellipsoidal shell and a short cylinder, as shown in Fig. 5.8B. The function of the straight edge is to make the welding seam avoid the connection edge of the semiellipsoid shell and the cylinder shell, and avoid the adverse situation that the welding thermal stress and the edge stress overlap. The straight edge height of the head usually varies in the range of 25–50 mm. The straight edge section can avoid the sudden change of the radial curvature radius at the joint weld of the head and the cylinder and improve the stress of the weld. As the curvature of the ellipsoid part of the head changes smoothly and continuously, the stress distribution is relatively uniform. Moreover, the depth of the ellipsoid head is small, which makes it easy to press and shape.

Therefore, it is one of the heads that is widely used in medium and low pressure vessels now.

The stress in the ellipsoidal head subjected to internal pressure includes the film stress caused by internal pressure and the discontinuous stress at the joint of the head and cylinder. The results show that, under certain conditions, the ratio of the maximum stress in the ellipsoidal head to the stress in the circular film is related to the ratio of the long axis $a$ to the short axis $b$ of the ellipsoidal head. The details are shown by the dotted line in Fig. 5.9. The experimental results show that the position and magnitude of the maximum stress vary with the change $a/b$.

The shape coefficient $K$ is introduced to consider the influence of ellipsoidal head stress on head strength.

$$K = \frac{1}{6}\left[2 + \left(\frac{D_i}{2h_i}\right)^2\right] \tag{5.25}$$

where $D_i$ is the inner diameter of the head with the unit of mm, and $h_i$ is the head surface depth, with the unit of mm.

The calculation formula of the thickness of the ellipsoidal head in China's container standard is given as follows.

$$\delta = \frac{Kp_cD_i}{2[\sigma]^t\phi - 0.5p_c} \tag{5.26}$$

Based on the ellipsoidal theory, this formula is derived according to the theory of the maximum principal stress, which takes into account the bending stress at the flange of the head, and adjusts it with the stress

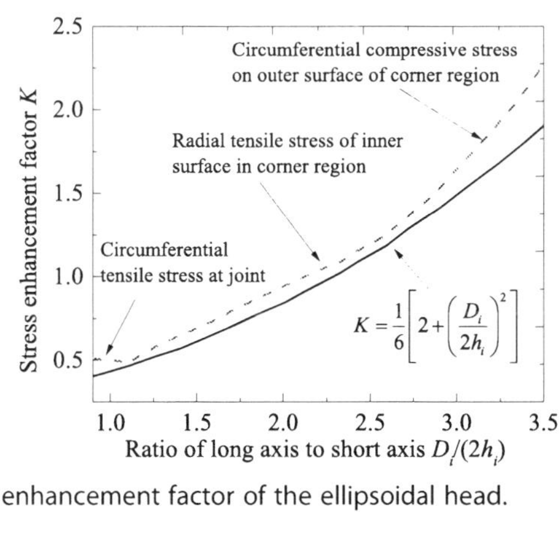


**Figure 5.9** Stress enhancement factor of the ellipsoidal head.

enhancement coefficient $K$. The enhancement coefficient is rounded by the proposed curve calculated by Coates and modified by experiments. This formula is firstly adopted by ASME (2004a, 2004b, 2007a, 2007b, 2010, 2013), and it is still used now.

For standard ellipsoidal head $a/b = 2$, $K = 1$.

$$\delta = \frac{p_c D_i}{2[\sigma]^t \phi - 0.5 p_c} \tag{5.27}$$

The maximum allowable working pressure of the ellipsoidal head is determined by Eq. (5.28).

$$[p_w] = \frac{2[\sigma]^t \phi \delta_e}{K D_i + 0.5 \delta_e} \tag{5.28}$$

The above formula avoids the yield of the head from the aspect of strength. However, according to the stress analysis, the standard ellipsoidal head under internal pressure has a high circumferential compressive stress at the transition angle, so although the ellipsoidal head under internal pressure can meet the strength requirements, circumferential folds may still occur and lead to local buckling failure. In particular, the large diameter, and the thin-walled oval head is easy to lose stability in the elastic range and suffer from damage. So far, this problem has been investigated deeply, and several design methods have been proposed, but the calculation process is quite complicated. The method of limiting the minimum thickness of the ellipsoidal head is generally adopted in engineering. As Chinese GB150−1998 (2003) stipulates, the effective thickness of a standard oval head should not be less than 0.15% of the inner diameter of the head, and the effective thickness of a nonstandard oval head should not be less than 0.30%.

#### *5.5.1.3 Dished head*

A dished head is also known as a ball head with a flanged edge, and consists of 3 parts, as shown in Fig. 5.8C. The first part is the central spherical part with radius $R_i$, the second part is the straight side part with height $h_i$, and the third part is the transition zone connecting the two parts, namely the surface with a radius of curvature $r$.

The dished head is different from the ellipsoidal head. From the perspective of geometry, it is a discontinuous surface. At the junction of two surfaces with different curvature radii, there is large bending stress due to an obvious change of curvature. As a result of the superposition of the bending stress and the film stress, the stress in this part is much higher than that in other parts. Therefore, the dished head differs from the ellipsoidal head, its stress distributions are not uniform, so it is not an ideal selection in engineering applications. The main advantage of the dished head is that it is easy to process and shape manually. For the spherical and folded parts, it can be shaped by manual forging, and processed and manufactured at the installation site. As large edge stress exists in the dished head, the stress condition of the dished head is not as good as the ellipsoidal head. This is also the main weakness of the dished heand. In addition, because of long processing time of manual forging, the oxide skin falls off seriously when heated, and after many forgings, the processing of thinning is relatively large. Therefore, the dished head is not used in most factories and is replaced by the ellipsoidal head now. Dished heads are generally used only when large atmospheric or low-pressure cylindrical tanks are manufactured in the installation site.

Because of the large edge stress, the torque theory should be used strictly to analyze and calculate the stress of the dished head under internal pressure condition, with complicated solution process. The failure study of the dished head shows that the total stress of the transition annular shell, including the discontinuous stress, is always higher than that of the central sphere. The ratio of the maximum total stress of the transition annular shell to the maximum total stress of the central sphere can be expressed by the formula $M = \frac{20(r/R_i)+3}{20(r/R_i)+1}$, shown as the dotted line in Fig. 5.10. Moreover, Marker derived an approximate correction coefficient based on the maximum total stress of the spherical part, expressed by Eq. (5.29).

$$M = \frac{1}{4}\left(3 + \sqrt{\frac{R_i}{r}}\right) \tag{5.29}$$

where $M$ is the stress enhancement coefficient of the dished head, also known as the shape coefficient, which means the total stress in the transition zone of the dished head is $M$ times larger than the stress of the spherical section. The shape coefficient is introduced to consider the influence of the edge stress caused by the abrupt curvature change of the warp at the junction of the transition circular arc and the sphere, and its value is shown as the solid line in Fig. 5.10. Accordingly, the thickness calculation formula of the dished head can be obtained by multiplying the formula of the hemisphere shell thickness $M$.

$$\delta = \frac{Mp_cR_i}{2[\sigma]^t\phi - 0.5p_c} \tag{5.30}$$

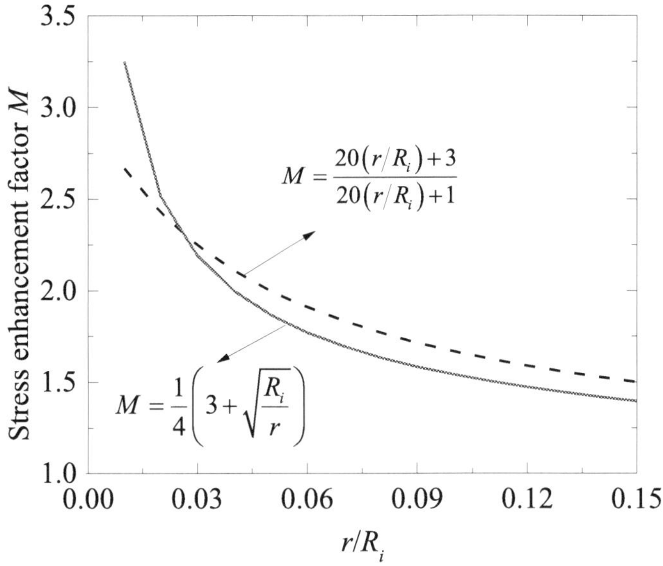


**Figure 5.10** Stress enhancement factor of the dished head.

The maximum allowable working pressure of the dished head under internal pressure is calculated by Eq. (5.31).

$$[p_w] = \frac{2[\sigma]^t \phi \delta_e}{MR_i + 0.5\delta_e} \tag{5.31}$$

The above formula is based on the basic formula of a spherical shell, which takes into account the bending stress and tensile stress at the connecting edge of the head, and adjusts by the stress enhancement factor $M$. Its value is calculated from the test and has been used in the ASME (2010, 2013) code for more than 30 years

It can be seen from Fig. 5.10 that the strength of the dished head is related to the radius of the transition zone $r$. If $r$ is too small, the heat stress will be too large. Therefore, the shape of the head is limited to $r \geq 0.01D_i$, $r \geq 3\delta$ and $R_i \leq D_i$. For the standard dished head, $R_i = 0.9D_i$, and $r = 0.17D_i$.

Similar to the ellipsoidal head, the transition zone of the dished head under the action of internal pressure also has the circumferential bending problem. For this reason, GB150—1998 (2003) stipulates that the effective thickness of the dished head $M \leq 1.34$, should not be less than 0.15% of the internal diameter. The effective thickness of the dished head shall not be less than 0.30%

#### 5.5.1.4 *Spherical crown head*

If the radius of the butterfly head is set as $r = 0$, it becomes the spherical crown head, so the spherical crown head is also called the unflanged ball head. There is a direct connection between the sphere and the cylinder, as shown in Fig. 5.8D. It is easy to see the related structure is simple and the manufacture is convenient. It is commonly used as the middle head of two independent pressure chambers in the container. As there is no corner transition at the joint between the sphere and the cylinder, considerable discontinuous stresses form and generate on the head and the cylinder near the joint, and the stress distribution is not reasonable. The T-joint of the head to the cylinder must be fully welded.

The calculated thickness of the spherical crown head under internal pressure (pressed on a concave surface) is calculated by Eq. (5.32).

$$\delta = \frac{Qp_cD_i}{2[\sigma]^t\phi - p_c} \tag{5.32}$$

where $Q$ is the coefficient, which can be obtained from the related standards.

In any case, the thickness of the cylinder connected to the spherical crown head shall not be less than the thickness of the head. Otherwise, a strengthening section transition connection shall be set between the head and the cylinder. The thickness of the strengthening section of the cylinder shall be the same as that of the sealing head.

The inner radius of the spherical crown head is generally not larger than the diameter of the barrel body, which is usually taken with $R_i = (0.9\text{-}1.0)D_i$. In the connection between the spherical part and the cylinder, there is no common tangent line between the two shells, and the radius of curvature has a sudden change, so the edge stress is quite large. Generally, it is only used in low pressure situations.

### 5.5.2 Internal pressure cone head thickness calculation

The axisymmetric conical shells can be divided into unflanged conical shells and flanged conical shells, as shown in Fig. 5.11.

For the large end of the conical shell, the structure without flaps can be adopted if the conical shell is at the half vertex angle $\alpha \leq 30^{\circ}$, as shown in Fig. 5.11A. If $\alpha > 30^{\circ}$, the folding structure with a transition section shall be adopted, otherwise the design should be carried out according to the stress analysis method. The corner radius $r$ of the transition section of

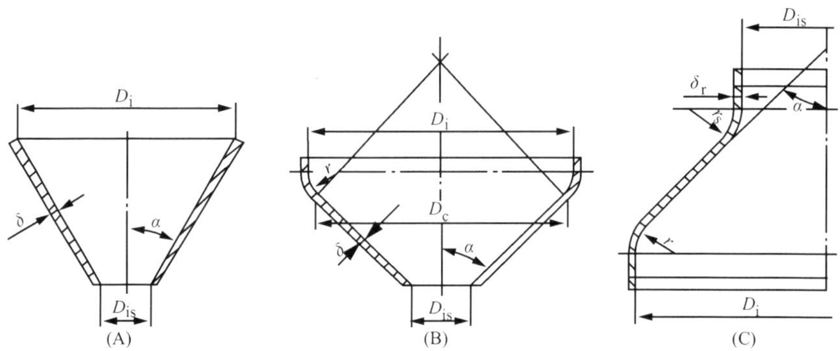


**Figure 5.11** Structure type of conical shell. (A) Unflanged conical shells, (B) Conical shell with big end folded edge, and (C) Flanged conical shell.

the large end folded conical shell shall be less than 10% of the inner diameter $D_i$ of the large end of the head, and not less than 3 times of the thickness of the transition section, as shown in Fig. 5.11B.

For the small end of the conical shell, the flangeless structure can be adopted if the conical shell is at the half vertex angle $\alpha \leq 45^\circ$, as shown in Fig. 5.11A. If $\alpha > 45^\circ$, the flanged structure with a transition segment should be adopted. The corner radius $r_s$ of the transition section of the small end flanged conical shell should not be less than 5% of the inner diameter $D_{is}$ of the small end of the head, and not less than three times of the thickness of the transition section, as shown in Fig. 5.11C.

When the conical shell is at the half vertex angle $\alpha > 60^\circ$, the thickness of the conical shell should be calculated according to the flat cover and can also be determined by the stress analysis method.

If necessary, the conical shell can also be composed of several conical shell segments with different thicknesses at the same half vertex angle. The connection between the conical shell and the cylinder shall adopt a full penetration welding structure.

The strength of the conical shell is determined by the thin film stress caused by the internal pressure of the conical shell and the edge stress at the connection between the two ends of the conical shell and the cylinder. To design the conical shell, the thickness of the conical shell, and the thicknesses of the large end strengthening section and small end strengthening section of the conical shell should be calculated respectively. If only one thickness component is considered, the maximum thickness of the above sections is taken.

If the conical shell is used as the head, to reduce the edge stress of the joint, the following two methods can be adopted. The first one is to increase the thickness of the head and cylinder near the joint, and this method is called the local strengthening. The second one is to increase the transition arc between the head and the cylinder body. This kind of head is called a conical head with a folded edge.

#### 5.5.2.1 Conical shell without folding under internal pressure

1. The thickness of the conical shell.

   According to the torque-free theory, the maximum thin-film stress is the circumferential stress $\sigma_\theta$ at the large end of the conical shell.

$$\sigma_\theta = \frac{pD}{2\delta\cos\alpha} \tag{5.33}$$

   As $D = D_c + \delta\cos\alpha$, the thickness calculation formula can be obtained based on the maximum tensile stress criterion.

$$\delta_c = \frac{p_c D_c}{\left[2[\sigma]^t\phi - p_c\right]\cos\alpha} \tag{5.34}$$

where $D_c$ is the inner diameter of the conical shell with the unit of mm, $\delta_c$ is the thickness of the conical shell with the unit of mm, and $\alpha$ and is the half-apex angle of the conical shell.

   When the conical shell is composed of several conical shell segments with different thicknesses at the same half vertex angle, the value of $D_c$ in above formula is selected with the inner diameter of the large end of each conical shell segment.

2. The large end of the conical shell.

At the junction between the large end of the conical shell and the cylinder, the curvature radius changes, and the radial internal forces of the two shells cannot be completely balanced, so the conical shell will cause a transverse thrust to the edge of the cylindrical shell. Because of the geometric discontinuity at the joint and the existence of transverse thrust, significant edge stresses generate at the joint edges of two shells. Since the edge stress is self-limited, the maximum stress can be limited within $3[\sigma]^t$. The relationship between $p/\left([\sigma]^t\phi\right)$ and $\alpha$ can be obtained according to the present condition, as shown in Fig. 5.12.

According to Fig. 5.12, if the coordinate point $\left[p/\left([\sigma]^t\phi\right), \alpha\right]$ is located above the curve in the figure, there is no need to strengthen, and the thickness is still calculated according to Eq. (5.32). If the coordinate

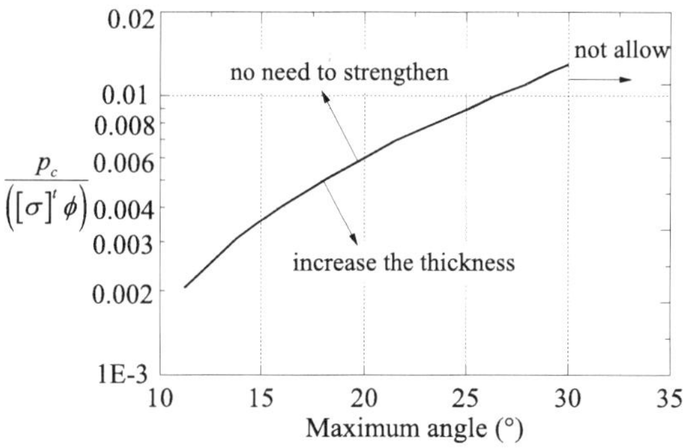


**Figure 5.12** Reinforcement diagram of the big end connection of conical shell.

point $\left[p/([\sigma]^t\phi), \alpha\right]$ is below the curve in the figure, it needs to increase the thickness to strengthen. A strengthening section shall be set between the conical shell and the cylinder. The strengthening section of the conical shell and the cylinder shall have the same thickness, and the thickness is calculated according to Eq. (5.35).

$$\delta_r = \frac{Qp_cD_i}{2[\sigma]^t\phi - p_c} \tag{5.35}$$

where $D_i$ is the inner diameter of the large end of the conical shell, with the unit of mm, $Q$ is the stress increment coefficient, and $\delta_r$ is the calculated thickness of the reinforcement section of the conical shell and its adjacent cylinder with the unit of mm.

In any case, the thickness of the reinforcing section shall not be less than that of the connected conical shell. The length of the reinforcement section $L_1$ of the conical shell shall not be less than $2\sqrt{\frac{0.5D_i\delta_r}{\cos\alpha}}$. The length of the reinforcement section $L$ of the cylinder shall not be less than $2\sqrt{0.5D_i\delta_r}$.

The calculation method of the thickness at the small end of the conical shell is similar to that at the large end. Please refer to the corresponding specifications for details.

### *5.5.2.2 Flanged conical shell under internal pressure*

**1.** The large end of the conical shell.

Its thickness is calculated according to Eqs. (5.36) and (5.37), and the larger value is taken.

The thickness of the transition section at the big end of the conical shell is similar to that of the dished head.

$$\delta = \frac{Kp_cD_i}{2[\sigma]^t\phi - 0.5p_c} \tag{5.36}$$

where $K$ is the coefficient, which can be obtained from Table 5.3.

The thickness of the conical shell connected with the transition section is calculated according to Eq. (5.37).

$$\delta = \frac{fp_cD_i}{[\sigma]^t\phi - 0.5p_c} \tag{5.37}$$

where $f$ is the coefficient and expressed as $f = \frac{1 - 2r/D_i(1 - \cos\alpha)}{2\cos\alpha}$, which can be obtained from Table 5.4. $r$ is the corner radius of the transition section at the big end of the folded conical shell with the unit of mm.

**Table 5.3** Values of coefficient $K$.

| $\alpha/(^\circ)$ | Value of $r/D_i$ | | | | | |
|---|---|---|---|---|---|---|
| | **0.10** | **0.15** | **0.20** | **0.30** | **0.40** | **0.50** |
| 10 | 0.6644 | 0.6111 | 0.5789 | 0.5403 | 0.5168 | 0.5000 |
| 20 | 0.6956 | 0.6357 | 0.5986 | 0.5522 | 0.5223 | 0.5000 |
| 30 | 0.7544 | 0.6819 | 0.6357 | 0.5749 | 0.5329 | 0.5000 |
| 35 | 0.7980 | 0.7161 | 0.6629 | 0.5914 | 0.5407 | 0.5000 |
| 40 | 0.8547 | 0.7604 | 0.6891 | 0.6127 | 0.5506 | 0.5000 |
| 45 | 0.9253 | 0.8181 | 0.7440 | 0.6402 | 0.5635 | 0.5000 |
| 50 | 1.0270 | 0.8944 | 0.8045 | 0.6765 | 0.5804 | 0.5000 |
| 55 | 1.1608 | 0.9980 | 0.8859 | 0.7249 | 0.6028 | 0.5000 |
| 60 | 1.3500 | 1.1433 | 1.0000 | 0.7923 | 0.6337 | 0.5000 |

**Table 5.4** Value of coefficient $f$.

| $\alpha/(^\circ)$ | Value of $r/D_i$ | | | | | |
|---|---|---|---|---|---|---|
| | **0.10** | **0.15** | **0.20** | **0.30** | **0.40** | **0.50** |
| 10 | 0.5062 | 0.5055 | 0.5047 | 0.5032 | 0.5017 | 0.5000 |
| 20 | 0.5257 | 0.5225 | 0.5193 | 0.5128 | 0.5064 | 0.5000 |
| 30 | 0.5619 | 0.5542 | 0.5465 | 0.5310 | 0.5155 | 0.5000 |
| 35 | 0.5883 | 0.5573 | 0.5663 | 0.5442 | 0.5221 | 0.5000 |
| 40 | 0.6222 | 0.6069 | 0.5916 | 0.5611 | 0.5305 | 0.5000 |
| 45 | 0.6657 | 0.6450 | 0.6243 | 0.5828 | 0.5414 | 0.5000 |
| 50 | 0.7223 | 0.6945 | 0.6668 | 0.6112 | 0.5556 | 0.5000 |
| 55 | 0.7973 | 0.7602 | 0.7230 | 0.6486 | 0.5743 | 0.5000 |
| 60 | 0.9000 | 0.8500 | 0.8000 | 0.7000 | 0.6000 | 0.5000 |

2. The small end of the conical shell.

Two cases should be considered for the small end of the conical shell. With the half vertex angle of the conical shell $\alpha \leq 45^{\circ}$, if the small end is not folded, the thickness of the small end should be calculated according to the calculation method of the thickness of the small end of the conical shell without folded edge; if the small end is folded, the thickness of the transition section of the small end should be calculated separately.

From the aspect of the stress strength, the conical head is not as good as the convex head, but better than the flat cover. The conical head with a larger diameter and thinner wall thickness is easy to manufacture, but the conical head with a smaller diameter and thicker wall thickness is difficult to manufacture.

Because of the discontinuity of the structure, the stress distribution of the conical shell is not ideal, but its special structure is conducive to the discharge of solid particles and suspended or viscous liquid. Hence, it can also be used as the intermediate transition section of cylinders with different diameters. It is widely used in medium and low pressure vessels. A conical head is generally used for discharge of high viscosity or suspension in atmospheric or low pressure vessels. For example, the conical head with the half vertex angle $\alpha > 30^{\circ}$ is often used as the bottom of the low pressure reaction tank. On the one hand, it is easy to drain the viscous or suspended materials; on the other hand, its edge stress is not large, so it is not necessary to adopt strengthening measures. Moreover, it is easy to manufacture. The outlet section of the reactor and the separation tower body with different diameters are usually connected by conical heads with folded edges at both ends, which is also called the variable diameter section. That is to say, it can be used to connect cylinders with different diameters into a unity.

#### 5.5.2.3 Flathead

Flathead is a kind of flat component. Many parts of the pressure vessel are composed of flat plates or ring plates. The common shape includes the circular plate or the circular plate with a hole in the center. The calculation of the plate thickness is based on the stress analysis of a circular plate. According to the theory of the flat plate, there are two bending stresses in the wall of a flat plate under uniform load. One is the radial bending stress, and the other is the tangential bending stress. For the circular plate with a uniformly distributed load, the maximum stress is generated in the center of the circular plate, and the radial bending stress is equal to the tangential bending stress.

In the theoretical analysis, the peripheral support of a flat plate is regarded as the fixed support or simple support. When the flat cover is connected with the cylinder, the real support is neither fixed support nor simple support, but between the fixed support and simple support. The location of the maximum stress is related to the specific connection structure and the size of the cylinder. Therefore, the empirical formula based on the circular plate theory, is often used in engineering calculations. The supporting condition around the flat cover is reflected by the coefficient $K$. The smaller the value $K$ is, the closer the edge of the flat cover is to the fixed support; otherwise, it is closer to the simple support.

Due to the difference in the connection structure between the circular plate head and the cylinder body and the size parameters of the cylinder body, the maximum stress of the flat cover may occur in the central part of the connection part of the cylinder and the flat cover, both can be expressed as follows.

$$\sigma_{\max} = \pm Kp\left(\frac{D}{\delta}\right)^2 \tag{5.38}$$

Considering that the flathead may be made of welded steel plate, the welding joint coefficient is introduced into the allowable stress. Based on the criterion of the maximum tensile stress, the thickness calculation formula of circular flat cover is obtained.

$$\delta_p = D_c\sqrt{\frac{Kp_c}{[\sigma]^t\phi}} \tag{5.39}$$

where $\delta_p$ is the calculated thickness of the flat cover, with the unit of mm, $K$ is the structural characteristic coefficient, and $D_c$ is the calculated diameter of the flat cover, with the unit of mm.

For different structural forms, the value of $K$ is different, which has a great relationship with the connection type of the cylinder.

Because the flathead is in the disadvantageous state of bending subjected to external pressure, the wall thickness of the flathead is much larger than that of the cylinder with the same diameter, and the flathead will cause greater boundary stress on the cylinder. Therefore, although it is simple in structure and easy to manufacture, it is not widely used in pressure equipments.

#### *5.5.2.4 Selection of head*

The selection of the head should be based on the requirements of working conditions. The influence factors of the shape of the head, the related stress distribution law, and the difficulty of stamping, welding, and assembly, should be considered, and a comprehensive technical and economic analysis should be conducted as well.

1. Geometry. Under the same volume, the surface area of the hemispherical head is the smallest, and the ellipsoidal head and dished head are almost the same.
2. Mechanics. Under the condition of the same diameter, wall thickness, and working pressure, the stress of the hemispherical head is the smallest, and the two-way thin-film stress is equal, and it is evenly distributed along the radial line. If the cylinder is connected with the same wall thickness, the maximum stress near the edge is not significantly different from the stress of the film.

   The stress of the ellipsoidal head is not as uniform as that of the hemispherical head, but better than that of the dished head. The stress at the top is the largest, and there is circumferential compressive stress at the equator. When $D_i/(2h) = 2$, it is connected with the cylinder with the same wall thickness, the ellipsoidal head can reach the same strength as the cylinder.

   The biggest disadvantage of the dished head in mechanics is that it has a small flanged radius $r$. The existence of this flanging zone makes the radial line of the head discontinuous, which results in a larger radial bending stress and circumferential compressive stress. However, the smaller the size $r/R$, the higher these stresses in the flaps. Therefore, the circumferential cracks and axial wrinkling may occur.

   When $r = 0$, the dished head becomes a spherical cover without folding, the mechanical properties of the head are not good, and the peak stress appears in the local area of the folding point, and the weld seam at the folding point will become a potential hazard source. The fillet weld of the head and the cylinder is a full penetration structure.

   The conical head is widely used in chemical containers. The main reason is that the conical shell is conducive to the uniform distribution of fluid and discharge. According to the mechanical characteristics of the conical head, the strength of the cone top is very large, and the hole in the cone top generally does not need reinforcement.
3. Manufacturing and material consumption. All kinds of heads are generally made by beating, stamping, rolling, or explosive molding. Hemispherical and ellipsoidal heads are usually made by stamping. Large hemispherical heads can be made by stamping ball petals firstly, and then assembled by welding. Dished heads are usually made by

beating, stamping, or explosive molding, and the folded part can be made by rolling or knocking.

If analyzed from the manufacturing process, sizes of the diameter and wall thickness increase with the depth of the head, so it is more difficult to make. It is more obvious when the high-strength steel is used. The whole stamping hemispherical head is not as easy to manufacture as the ellipsoidal head. The ellipsoidal head must have the correct geometry of the ellipsoidal surface mold, and manual percussion manufacturing. The manufacturing flexibility of the ellipsoidal head is large, and the cone top part of the cone head is difficult to roll. When the cone top angle is small, the combined head is sometimes used for reducing the manufacturing difficulties and reducing the height of the vertebral body.

From the aspect of material saving, the hemispherical head consumes the least amount of material, the dished head consumes 30% more material than the ellipsoidal head, and the conical head does not save material.

## 5.6 Pressure test

In addition to the defects of the material itself, other defects in the manufacture and the use of the vessel also exist. Therefore, to assess the impact of defects on the safety of the pressure vessel, the pressure test should be carried out after the manufacture of the pressure vessel or during the regular inspection. Pressure test includes the pressure bearing test and airtightness test. Pressure bearing test refers to the hydraulic or pneumatic test under the condition of exceeding the design pressure. The airtightness test refers to the pneumatic test under the condition of equal to or lower than the design pressure.

For internal pressure vessels, the purpose of the pressure test is to examine whether defects will cause damage or crack due to rapid expansion and leakage under excess design pressure and check the sealing performance of the bearing structure. For the external pressure vessel, under the action of external pressure, the defects in the vessel can't crack under the action of compressive stress. The critical buckling stress under external pressure is mainly related to the geometric size and manufacturing accuracy of the vessel and has nothing to do with the defects. Generally, the external pressure test is not used to assess its stability, while the internal pressure test is used to check whether there are penetration defects.

## 5.6.1 Pressure bearing test

There are two kinds of pressure bearing tests, hydraulic test, and air pressure test. It is the first pressure test of the container before it is used, and the test pressure is higher than the highest working pressure of the container, so the possibility of explosion of the container is greater than that when it is used. Under the same pressure and container, the higher the compression coefficient of the test medium, the more the energy stored in the container, and the more dangerous the explosion. Therefore, the fluid with small compressibility should be selected as the test medium. At room temperature, the compressibility of water is much smaller than that of gas, and the source is abundant, so it is commonly used as a test medium. The air pressure test is used only when water or other liquids cannot be filled into the container due to structural or supporting reasons, and operating conditions do not allow residual liquids.

### *5.6.1.1 Test medium*

1. Usually, the medium of the hydraulic test is clean water. If other liquids are adopted, it should make sure the liquid must not cause the risk. Moreover, the temperature of the liquid during the test should be lower than its flash point or boiling point.

   When a hydraulic test is carried out with water for an austenitic stainless steel pressure vessel, the chloride ion content in water should be strictly controlled below 25 mg/L. After passing the test, the water stains shall be removed immediately.
2. The gas used in the pneumatic test shall be dry, crystallized air, nitrogen, or other inert gases.

### *5.6.1.2 Test pressure*

1. Hydraulic test

   The experimental pressure should meet the following equation.

$$p_T = 1.25p\frac{[\sigma]}{[\sigma]^t} \tag{5.40}$$

where $p_T$ is the test pressure with the unit of MPa, $p$ is the design pressure with the unit of MPa, $[\sigma]$ is the allowable stress of container component materials at test temperature with the unit of MPa, and $[\sigma]^t$ is the allowable stress of the container component material at design temperature, with the unit of MPa.

2. Air pressure test
   The experimental pressure should meet the following formula.

$$p_T = 1.15p\frac{[\sigma]}{[\sigma]^t} \tag{5.41}$$

where $p_T$ is the test pressure with the unit of MPa, $p$ is the design pressure with the unit of MPa, $[\sigma]$ is the allowable stress of container component materials at test temperature with the unit of MPa, and $[\sigma]^t$ is the allowable stress of the container component material at design temperature, with the unit of MPa.

Since the pneumatic test is more dangerous than the hydraulic test, the test pressure is lower than the hydraulic test pressure. Hence, 100% radiographic or ultrasonic testing should be carried out for the butt joint on the vessel.

#### *5.6.1.3 Stress check*

1. Before the hydraulic test, the cylinder stress should be checked according to the following formula.

$$\sigma_T = \frac{p_T(D_i + \delta_e)}{2\delta_e} \le 0.9\sigma_s\phi \tag{5.42}$$

2. Before the air pressure test, the cylinder stress should be checked according to the following formula.

$$\sigma_T = \frac{p_T(D_i + \delta_e)}{2\delta_e} \le 0.8\sigma_s\phi \tag{5.43}$$

where $\sigma_s$ is the yield point of cylinder material at test temperature with the unit of MPa, $\phi$ is the welded joint coefficient of the cylinder, and $p_T$ is the experimental pressure, with the unit of MPa.

#### *5.6.1.4 Test temperature*

1. During the hydraulic test of carbon steel, 16Mn, and normalized 15MnVR steel vessels, the liquid temperature shall not be lower than 5°C. For other low alloy steel vessels, the liquid temperature during the hydraulic test shall not be lower than 15°C. If the nonductile transition temperature of the material increases due to the plate thickness and other factors, the temperature of the test liquid should increase correspondingly.

2. For carbon steel and low alloy steel vessels, the medium temperature shall not be lower than 15 °C during the air pressure test.

#### *5.6.1.5 Test method*

1. Hydraulic test

   During the test, an exhaust valve should be set at the top of the container, and the air in the container should be exhausted absolutely. During the test, the observation surface of the container shall be kept dry.

   During the test, the pressure should rise slowly. After reaching the specified test pressure, the pressure holding time is generally not less than 30 minutes. The pressure is then reduced to 80% of the specified test pressure and maintained long enough to inspect all soldered joints and connections. If theleakage is observed, the repair and retest should be conducted.

   For the jacketed vessels, the hydraulic test of the inner cylinder is carried out firstly, and then the jacket is welded after passing the test. If there are no other issues after inspection, the hydraulic test in the jacket can be carried out. Before the pressure test, the compressed air in 0.4–0.5 MPa should be injected into the opening reinforcement ring of the vessel to check the weld quality, and the test hole on the reinforcement ring should be opened during the pressure test.

   After the completion of the hydraulic test, the liquid shall be drained and the interior shall be blown dry with compressed air.

2. Air pressure test

During the test, the pressure should rise slowly to 10% of the specified test pressure, and is no more than 0.05 MPa, the pressure should be maintained for 5 minutes, and then the initial leakage inspection should be carried out on all welded joints and connection parts. If there is any leakage, the test shall be conducted again after repair. Once the initial leak inspection is qualified, the pressure continues to increase slowly to 50% of the specified test pressure and then increases to the specified test pressure gradually according to the level difference of 10% of the specified test pressure for each stage. After the pressure holding for 10 minutes, reduce the pressure to 87% of the specified test pressure, and keep it for enough time before checking the leakage again. In case of leakage, the test shall be conducted again according to the above regulations after repair.

#### 5.6.1.6 Acceptable quality level

1. There shall be no leakage and visible deformation during the hydraulic test. There shall be no abnormal sound during the test, and no cracks formed on the surface of the material with the specified lower limit of tensile strength $\geq 540$ MPa by nondestructive testing.
2. During the air pressure test, the pressure vessel is qualified if there is no abnormal sound, no air leakage, and no visible deformation through the inspection of soap liquid or other leak detection liquids.

### 5.6.2 Airtightness test

If the medium is inflammable, toxic, and highly hazardous, or no trace leakage is allowed in the design, the airtightness test must be carried out after the pressure test is qualified. The pressure of the airtightness test depends on whether the vessel is equipped with a safety relief device. If there is no safety relief device on the vessel, its airtightness test pressure value is generally 1.0 time of the design pressure. However, if there is a safety relief device on the vessel, to ensure the normal operation of the safety relief device, its airtightness test pressure value should be lower than the opening pressure of the safety valve or the design bursting pressure of the fragments. It is recommended to take 1.0 time of the maximum working pressure of the vessel.

- Test medium. The test medium should be dry, crystalline air, nitrogen, or other inert gases.
- Test pressure. It is required in the regulations that the pressure of air-tightness test is the design pressure of the pressure vessel.
- Test temperature. The temperature of the test gas for pressure vessels made of carbon steel and low alloy steel shall not be lower than 5°C.
- Test method. During the test, the pressure should rise slowly and hold for 10 minutes after reaching the specified test pressure. Then, all welded joints and connecting parts should be checked for leakage. Small containers can also be immersed in water for inspection. In case of leakage, the hydraulic test and air tightness test shall be conducted again after repair.
- Qualified standard. If there is no leakage after inspection, it is qualified when the pressure is maintained for no less than 30 minutes.

The airtightness test is dangerous and should be carried out after the hydraulic test is qualified. Before the airtightness test, the safety accessories on the container shall be fully assembled.

## 5.7 Summary

The basic procedure of the tank design is introduced in this section. The design theory, the key design parameters, the tank type, and the tank test are described in detail. The related introduction could supply the technical reference for the structure design of different kinds of storage tanks. However, for some tanks used in extreme conditions, such as ultra-high pressure, extremely low temperature, and high corrosion conditions, the present introduction may not be acceptable, and the related tank should be specially designed.

## References

ASME (2004a). *Boilers & pressure vessel code, VIII division 1. Rules for construction of pressure vessels.*

ASME (2004b). *Boilers & pressure vessel code, VIII division 2. Rules for construction of pressure vessels.*

ASME (2007a). *Boilers & pressure vessel code, VIII division 1. Rules for construction of pressure vessels.* New York: The America Society of Mechanical Engineers.

ASME (2007b). *Boilers & pressure vessel code, VIII division 2. Rules for construction of pressure vessels.* New York: The America Society of Mechanical Engineers.

ASME (2010). *Boilers & pressure vessel code, VIII division 2. Alternative rules, rules for construction of pressure vessels.*

ASME (2013). *Boilers & pressure vessel code, VIII division 2. Alternative rules, rules for construction of pressure vessels.*

Du, Y. (2016). *Fundamental of pressure vessel design.* Beijing: China Petrochemical Press. (In Chinese).

GB150—1998 (2003). *Steel pressure vessels.* Beijing: Standards Press of China.(In Chinese).

Jawad, M. H., & Farr, J. R. (1989). *Structural analysis and design of process equipment* (2nd ed.). Wiley, American Institute of Chemical Engineers, Inc.

Jawad, M. H., & Farr, J. R. (2018). *Structural analysis and design of process equipment* (3rd ed.). Wiley, American Institute of Chemical Engineers, Inc.

Li, F., & Li, Q. (2010). *Pressure vessel and process equipment design.* Beijing: Chemical Industry Press. (In Chinese).

CHAPTER SIX

# On buckling of oil storage tanks under nearby explosions and fire

**Luis A. Godoy[1,2], Rossana C. Jaca[3] and Mariano P. Ameijeiras[1,4]**
[1]Institute for Advanced Studies in Engineering and Technology, CONICET/UNC, Córdoba, Argentina
[2]Mechanical and Aerospace Engineering Department, West Virginia University, Morgantown, WV, United States
[3]Faculty of Engineering, National University of Comahue, Neuquén, Argentina
[4]Faculty of Exact, Physical and Natural Sciences, National University of Córdoba, Córdoba, Argentina

## 6.1 Introduction

Typical tanks in the oil industry are designed as metal cylindrical shells with a fixed or floating roof. A fixed dome roof may be used in the smaller tanks, whereas fixed conical roofs are used for larger diameters. In a simplified model, a vertical tank is characterized by its diameter $D$, height $H$, and thickness $h$. For small to medium size tanks ($D < 40$ m), the most common configurations include a shallow dome or a conical roof supported by rafters, whereas large tanks ($D > 40$ m) are opened at the top and with an internal floating roof that moves with the fluid level. The dimensions depend on the fluid stored: Tanks that store chemical products have diameters in the order of $D = 12$ m; this may increase up to $D = 28$ m for light hydrocarbon fuels. Open-top tanks with $D = 70$ m are used to store heavy hydrocarbon fuels (Noret et al., 2012), and very large designs of open-top tanks with $D = 100$ m and $H = 25$ m have been reported in several countries (Lu et al., 2019). Most large tanks that are open at the top require a wind girder to stabilize the circular shape; such tanks may also include intermediate ring stiffeners.

Oil storage tanks are seldom found in isolation from other structures and, in most cases, they are part of what is known as a tank farm, in which there are dozens or hundreds of tanks to store flammable products (Sengupta et al., 2011).

Because buckling is a major concern in the design of shells such as those used in the oil industry, there are several references collecting the

*Above Ground Storage Tank Oil Spills.*
DOI: https://doi.org/10.1016/B978-0-323-85728-4.00004-8

state of the art from various perspectives (Godoy, 2016, Rotter & Schmidt, 2008, Zingoni, 2015).

The American Petroleum Institute regulations (API 650, 2021), state that oil and fuel storage tanks should be designed to withstand pressures from the fuel stored, wind, and earthquakes, but it does not specify design considering explosions and fire. API advice is the standard of design of tanks at present: A recent report on the optimization of an oil storage tank design based on API 650 illustrates that no attention is given to fire and explosion scenarios (Agboola et al., 2021).

This is in contrast with evidence collected for many years in this industry, in which it is found that the most frequent causes of accidents leading to damage or failure of tanks are associated with fire and explosions (see, e.g., Chang & Lin, 2006, Persson & Lönnermark, 2004, Planas-Cuchi et al., 1999).

According to Myers (1997), oil spills are the most influential factor driving changes in the design codes of oil storage tanks. Oil spills may cause damage to the environment, leading to air, soil, and subsurface water contamination (Riazi, 2021), and they are also one of the main sources of fire and explosions in tank farms. Such accidents have economic and social consequences for the various stakeholders, and during the last two decades, there have been changes in the public perception of the dangers involved (Nolan, 2014). Cleaning an oil spill is not an easy task, and may become both expensive and full of surprises (Fingas, 2013).

Given the frequency of accidents associated with oil spills leading to fire and explosions, the dramatic consequences of such accidents, affecting the infrastructure of a petrochemical plant and nearby buildings, and the possibility of causing environmental contamination, it is surprising to notice that limited attention has been given to this topic in national regulations and design codes.

Both fire and explosions may affect a single tank but there are cases in which fire propagates to other tanks in what is known as a domino effect. During the last decade, some attention has been given to damage produced by explosions and fire in tank farms, affecting nearby tanks and leading to domino effects (see, e.g., Reniers & Cozzani, 2013).

Accidents are typically initiated by the failure of a tank farm component, a sensor in equipment, or by human error. The cause of an explosion may also be associated with an intentional act of sabotage or terrorism, which may start with an explosive placed to destroy a tank. Taveau (2012a, 2012b) reports examples of explosions affecting tanks in France and other countries in an investigation to highlight the potential structural consequences and the possible sequence of events leading to structural failure. Further examples of buckling, global flexure, and knockover of tanks under explosions are given by Noret et al. (2012).

In most accidents in a tank farm, an explosion starts with the release of a hydrocarbon fuel most likely as a consequence of a system component failure, such as the failure of a sensor controlling the fuel level inside a tank. As stated by Burgan "both oxygen and a source of ignition are required for this fuel to ignite. An explosion will occur if the circumstances of the release permit the buildup of an explosive mixture of hydrocarbon gas, or vapor and air, followed by a delayed ignition." (Burgan, 1995, 157).

Fire and explosions are frequently part of the same accident. Explosions that occur in a tank farm are usually associated with fire and seldom are the cases in which only an explosion occurs. In many cases, domino effects follow an explosion or fire causing burning and destruction of neighboring tanks, but tanks located at larger distances that are not reached by fire frequently fail under high pressures due to a shock wave. From the point of view of the analysis, both effects (fire and explosions) are treated as separate events because their structural consequences are different. Further, fire produces a gradual increase in temperatures with time, whereas an explosion is a sudden increase in pressures arriving at the structure.

This chapter reviews the main features involved in the investigation of structural consequences due to fire and explosions affecting tanks in the oil industry. The sources of fire or explosions are considered to be located outside a target tank, whereas internal explosions in tanks are not emphasized here.

The outline of the chapter is as follows: Section 6.2 considers evidence from real accidents involving fire and/or explosions in the past, emphasizing the mechanisms of initiation of the event. The 2009 accident at Bayamon in Puerto Rico is first described and selected cases are reported next to illustrate differences and similarities. The analysis of explosions and their effects on vertical storage tanks is the subject of Sections 6.3–6.5. An introductory description of explosions and a summary of efforts made to obtain pressures via small-scale testing is the subject of Section 6.3. Section 6.4 deals with computational simulations used in the literature to represent the effects of explosions, including simplified and advanced approaches. This section aims to discuss features of an explicit expression representing time and space-dependent pressures due to blast loads or strategies to obtain them. The structural behavior of tanks due to explosions is discussed in Section 6.5. Both static and dynamic nonlinear behavior are presented for a case study, including elastic behavior and plasticity.

The analysis of fire is the subject of Sections 6.6–6.8. An introductory description of fire in tank farms is presented in Section 6.6, together with a brief mention of available tests. Section 6.7 focuses on modeling from the source (a flame) to the target tank; both simplified and advanced models are discussed. The structural behavior of tanks under heat from a nearby fire is considered in Section 6.8. A summary of computational results of critical temperatures is presented. Postcritical features of the response are shown in a case study. Special features of tanks under fire are also discussed in this section. Finally, the current needs for research and recommendations for further developments are discussed in Section 6.9.

## 6.2 A review of selected accidents involving explosions and fire in tank farms

Although explosions and fire occur in tank farms with some frequency, they are seldom reported in the technical literature. Some reasons prevent the open disclosure of information, including insurance and legal issues. And there are other reasons: Oil companies have pressure to respond to regulators and investors regarding pollution and worker safety so that each year they establish their own environmental goals. Such goals are established in terms of the number of incidents, not in terms of other indicators, such as oil spill volumes, so that a huge oil spill may be encapsulated in one single accident. Further, top executives of such companies get annual rewards for meeting these goals (MacMillan, 2021).

This section summarizes a small number of cases involving accidents of fire and/or explosions in fuel storage tanks. The cases have been selected to illustrate different aspects, including the source of an accident, the oil spill from the tank, and the ignition process, and to illustrate the recurrent nature of some causes. Fire and explosions occur frequently and in countries with different levels of economic development, in some cases in countries where enforcement of engineering codes of practice and safety codes is done rigorously.

### 6.2.1 Case study: The Bayamon Accident in Puerto Rico, 2009

An accident of large extent and consequences occurred on October 23, 2009, at the oil depot and refinery Caribbean Petroleum Corporation

(CAPECO) in Puerto Rico, located in Bayamon some 6 km west of the city of San Juan (CSB, 2015). Puerto Rico does not produce oil and crude oil is transported by cargo ships to a port and sent by an oil duct to plants like CAPECO for processing and distribution. The plant had some 40 tanks.

Interest, in this case, is due to the large dimensions of the accident; further, one of the authors of this chapter was the leader of a research team investigating the accident (Batista-Abreu & Godoy, 2011, 2013, Godoy & Batista-Abreu, 2012). A few years before the accident, the research team visited the CAPECO plant to document the tanks and their distribution. Typical vertical tanks in this plant had diameter $D = 30$ m, height $H = 12$ m, with a shallow conical roof. The total storage capacity of this plant was approximately three million barrels. Earth dikes around each tank were used as a secondary level of security to reduce contamination in case of oil spills.

In the early morning of October 23, 2009, a series of explosions occurred, followed by fire, and secondary explosions and fire of lower intensity were detected during the next hour. The last explosion took place eight hours after the first one. High temperatures were recorded on that day in Puerto Rico, with a maximum of 34°C and a minimum of 26°C, with a maximum wind speed of 29 km/h. Rain was not reported on that day.

Flames soon propagated and reached 30 m in height, as shown in Fig. 6.1. One hour after the first explosion there were five burning tanks;

**Figure 6.1** The Bayamon accident in Puerto Rico, October 23, 2009 (CSB, 2015).

this number increased to 11 burning tanks in two hours and five hours from the start of the accident there were 18 burning tanks. A total of 50% of the area dedicated to fuel storage together with 21 tanks were severely affected by the accident. A huge column of fire and smoke, reaching 6 km in height, could be observed from many places on the island of Puerto Rico.

The firefighters arrived 20 min after the first explosion and could control the main points of fire five hours later, but fire extinction was only completed two days later, with the consequence that some storage tanks were burning at various places for 60 h.

The initial scenarios considered by the federal agencies involved in the postevent investigations were being discarded as the research proceeded, and in the final explanation, it was believed that the initiation of the accident was caused by the failure of a sensor detecting the level of fuel in one tank (identified as Tank 409) with a fixed roof that was being filled at the time of the accident. An oil spill occurred as a consequence of this event and fuel was found outside one tank. This oil spill started a chain of events: Once outside the tank, the fuel produced a vapor cloud that easily expanded to an area of approximately one square km, until a source of ignition was found (CSB, 2015). Although it was not possible to identify with certainty the source of ignition, it is believed that the fire started due to sparks from some electric equipment and then propagated through the vapor cloud to reach several neighboring tanks.

A total of 32 tanks were involved in the accident, of which 20 were destroyed. It was found that some open-top tanks collapsed with steel melting at high temperatures; other open-top tanks remained standing but with large deformations at the top courses of the cylindrical shell, and some tanks with a fixed roof showed severe deformations at the top of the cylinder and on the roof. Photographs of each of these groups of failure modes are illustrated in Fig. 6.2.

The mechanism leading to the accident in Puerto Rico was very similar to that found to occur in the oil depot in Buncefield, United Kingdom, in 2005, in which failure of a sensor was the starting cause of the accident.

## 6.2.2 Brief description of other accidents

Understanding the causes of accidents and their consequences requires reviewing case studies, and this is reported in this section by several

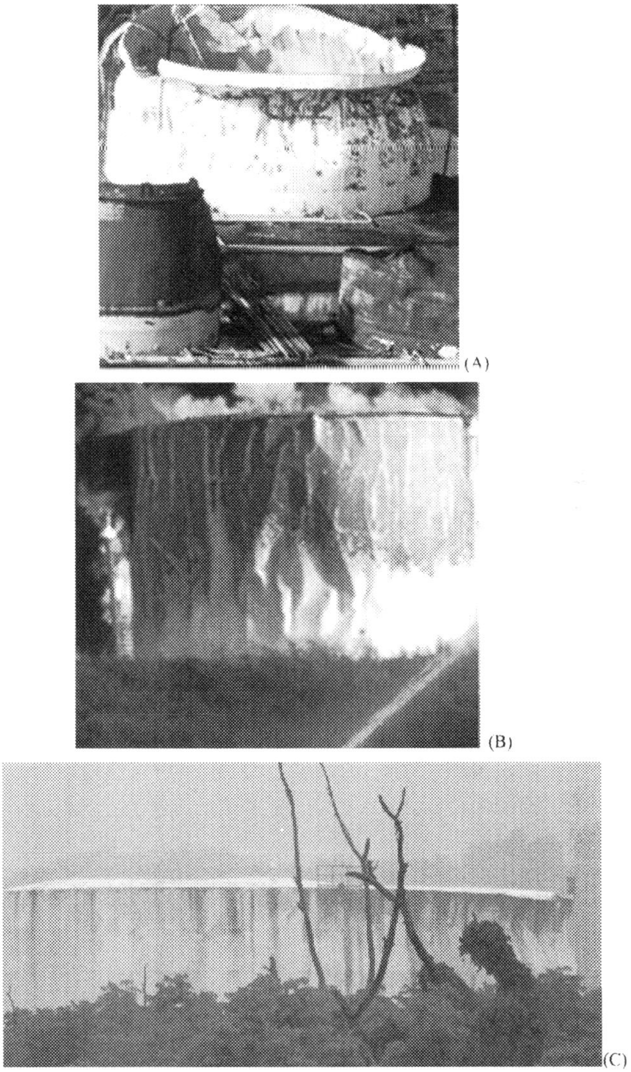


**Figure 6.2** Examples of tanks that failed during the accident in Bayamon, Puerto Rico in 2009. (A) Failure of an open-top tank; (B) tank with a fixed roof that was severely damaged but remained standing after the fire; (C) tank with a fixed roof that was severely damaged by an explosion. *Reproduced from (A) Batista-Abreu, J., & Godoy, L. A. (2011). Investigación de causas de explosiones en una planta de almacenamiento de combustible en Puerto Rico.* Revista Internacional de Desastres Naturales, Accidentes e Infraestructura Civil, 11*(2), 109–122 (in Spanish); (B) Godoy, L. A., & Batista-Abreu, J. C. (2012). Buckling of fixed roof aboveground oil storage tanks under heat induced by an external fire.* Thin-Walled Structures, 52, *90–101*

examples. Special attention is given here to accidents that occurred during the 21st century, namely, Guam, United States in 2002; Texas City, United States in 2005; Buncefield, United Kingdom in 2005; and Malargüe, Argentina in 2014. This is a short list compared with accidents that took place during the last two decades, and they are taken as illustrations of accident mechanisms and their consequences. The list would be considerably increased if cases in Asia were included (see, e.g., Lu et al., 2019, Mishra et al., 2014).

*The accident in the isle of Guam, United States, in 2002.* This accident should be considered in two steps: the first one occurred in July 2002 with typhoon Chata'an affecting a small fuel storage plant in the isle of Guam located in the Pacific Ocean. One of the tanks, having a fixed roof on top and a floating roof inside, suffered shell buckling due to wind, with radial deformations that were sufficient to block the vertical displacement of the floating roof. The tank was not repaired, with the consequence that the internal roof could easily block its functioning. A second typhoon, named Pongsona, reached the island in December 2002; in this case, there was sand transported by wind, and this produced friction on the steel walls of the tank, thus inducing static electricity inside the tank. It is believed that this effect started an ignition process and fire at this source tank finally propagated to other tanks in the facility.

Several tanks were destroyed as a consequence of the fire. In some tanks, damage affected the complete structure in elevation; but in other tanks, it was possible to observe a lower part not affected by fire, and only at the top there were large plastic deformations; this was clear evidence of the stabilizing incidence of the fuel stored in a tank, and leads to the observation that an empty tank is the worst scenario is cases of fire.

*The accident in Texas City, United States, in 2005.* There are several tank farms in Texas City, and the accident described next occurred at the British Petroleum refinery in March 2005. The source of the accident was the rapid overfilling of one tank, in which the sensors that should be detecting the level of fluid stored were not working properly, and there were also problems in the alarm system that should stop the fuel influx in the event of reaching a high level. An oil spill occurred reaching the outside of the tank, followed by the development of a vapor cloud that propagated at floor level throughout the plant. The mechanism of ignition was perhaps associated with the start of a truck engine in the plant.

As a consequence of this accident, 15 workers died, dozens were injured, and 50 tanks were destroyed or damaged.

*The accident in Buncefield, United Kingdom, in 2005.* On December 11, 2005, an explosion occurred at the Hertfordshire Oil Storage Terminal, a tank farm at Buncefield, Hertfordshire, United Kingdom, owned by two companies: Total United Kingdom and Texaco. This was one of the largest fuel storage terminals in the United Kingdom.

According to the postevent investigations, the first explosion, which was also the largest one, was initiated by the failure of a sensor controlling the fuel level in a tank while being filled with fuel, so that pressure accumulated on an internal roof and reached the ventilation ducts in the fixed roof, causing a loss of fuel and its contact with air. This formed a vapor cloud with an estimated depth of 5 m. Vapor reached a nearby station of pumps and engines where it must have ignited leading to a large-scale deflagration (Taveau, 2012c). According to Atkinson, "the blast almost invariably caused a rapid escalating fire in many tanks surrounded by the vapor cloud, even if they contained relatively high flashpoint materials such as diesel" (Atkinson, 2011, 382). It is believed that pressures exceeded 200 kPa within the vapor cloud.

The explosion was accompanied by fire which took two days to be extinguished, with the final destruction of some 20 tanks. Several buildings in the area were also affected by the explosions at the terminal.

The Buncefield incident, as it was known, was the subject of a government investigation to establish responsibilities and avoid the occurrence of similar incidents in the future. An extensive report was made public in 2008 (Buncefield, 2008), which is perhaps the main document available for a postevent investigation. The recommendations include topics such as taking measures to prevent fuel from escaping a tank and preventing the occurrence of a flammable vapor cloud in cases of fuel leaking.

The explosion was characterized as having unusually high strength. Research following the Buncefield incident indicated that peak overpressures were in the order of 200 kPa (Taveau, 2012c).

Tanks located relatively far from a source of fire were not affected by temperature and yet they buckled due to blast pressures. To illustrate the structural effects of the Buncefield explosion, Fig. 6.3 shows a tank that did not catch fire but had large plastic deformations caused by the blast loads. This indicates that the final effect of the explosion on this tank was not elastic buckling but involved permanent strains in the shell with displacements that are a fraction of the shell radius. Elastic buckling may have occurred at the onset of the process in some tanks, but it must have escalated to turn this into plastic buckling. The buckled configuration in Fig. 6.3 has a small number of waves around the circumference.

**Figure 6.3** Damage to a fuel storage tank due to explosion in Buncefield, United Kingdom, 2005. This tank was not affected by fire. *Reproduced from Atkinson, G. (2011). Blast damage to storage tanks and steel-clad buildings.* Process Safety and Environmental Protection, 89, *382–390.*

*The accident in Malargüe, Argentina, in March 2014.* A fire in a YPF (the national oil company) plant at Malargüe, Argentina, occurred in March 2014. A tank exploded and affected several others in the plant. The aereal pictures in Fig. 6.4 show the destruction and damage of tanks and the oil spills produced as a consequence of the accident. This is an isolated area and other buildings were not affected by the accident. This is in contrast with accidents in tanks located in populated areas, such as that reported in Santos, Brazil, in 2013.

*Recent accidents in China.* Accidents in China seldom appear in the press or journal papers. One should include here an explosion in a chemical plant in Zhangzhou, Fujian Province, China, on April 6, 2015; and an explosion at a petrochemical plant in Rizhou, Shangdon Province, China, on July 20, 2015. Fire followed a blast at the accident in a chemical plant in Fujian, whereas the reverse sequence occurred at the Shangdon accident, in which a fuel leak led to a fire and this was followed by the explosion of four spherical tanks. These tanks were filled with liquid hydrocarbon.

*Explosion as an act of terrorism in Bouches du Rhone, France, July 2015.* This is an explosion/fire case affecting tanks that started with an

(A)

(B)

**Figure 6.4** Tanks damaged during the explosion and fire in Malargüe, Mendoza, Argentina, 2014. (A) Overview of tanks after the accident; (B) Destruction of one tank at Malargüe.

intentional act. A couple of explosions reached two tanks in a petrochemical complex in Bouches du Rhone, not far from the city of Marseille, France. The explosions took place on the national day of France, July 14, 2015, and they were attributed to a criminal act, although the motive remained unclear.

The tanks had a floating roof and a wind girder at the top, together with a stiffening ring. The diameter of the tanks was in the order of 60 m, with a volume capacity of 40,000 $m^3$. One of the tanks contained oil and the other one was filled with a highly flammable product.

There were two simultaneous explosions of tanks separated by 500 m from each other. A third explosive was also found at the site but it did not ignite the fuel stored. The explosions were followed by fire, and the smoke column caused by the fire could be seen from a long distance from the plant.

It is estimated that 1100 plants in France could be a target of similar criminal acts.

### 6.2.3 Common features of accidents and lessons learned

Vilchez et al. (1995) reported operations of emptying and filling in storage tanks as one of the main situations leading to accidents. Within the context of chemical industries, these authors investigated the risk of starting fires and explosions in storage tanks (Planas-Cuchi et al., 1999). Several mechanisms were identified, including the release of fuel stored (this occurred in 44.6% of cases), nearby fire (in 22% of cases), explosions (in 18.8% of cases), and gas cloud (in 8.6% of cases). In most cases, the specific type of explosion or fire was not known. The causes of accidents were identified as mechanical failure in 22% of cases considered, human failure accounted for 20% of cases, and an external event occurred in 9.6% of cases. The main mechanisms leading to accidents were (1) air entrance during the operation to empty a tank or in an empty tank, causing deflagration; (2) the release of gas or vapor through ventilation openings, leading to an unconfined deflagration; and (3) the release of combustible liquid in cases of breaking off an infill pipe, leading to pooling fire.

In several cases, including accidents in Texas City, Buncefield, and Bayamón, a common feature was present: the failure of a sensor that should shut the fuel entrance to a tank based on the level of stored fuel. The consequence of overfilling a tank is often an oil spill through the ventilation ducts or the junction between the cylinder and the roof. Other ways in which loss-of-containment events may occur in atmospheric tanks have been discussed by Casal (2018). A vapor cloud has been reported to develop from the oil spill and rapidly expands in an area of a plant. Several possible sources of ignition have been identified, including sparks from engines or vehicles.

There are, of course, other possible mechanisms, as illustrated by the accident in Guam, in which fire started inside a tank and not from a vapor cloud. A comparison of the events in Bayamon and Guam is shown in Fig. 6.5. In the accidents in Bayamon and Buncefield, the source tank is different from the target tank; in the accident in Guam, the fire starts in the source tank.

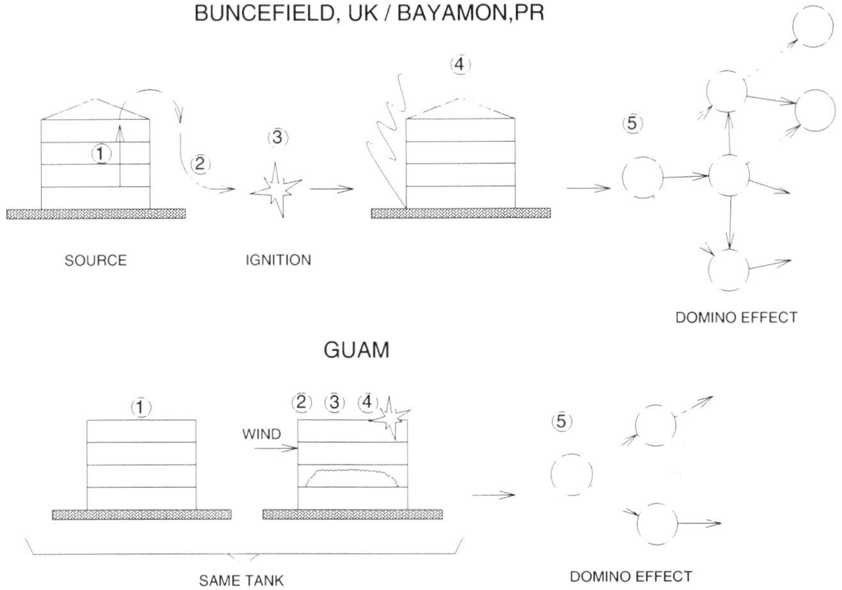


**Figure 6.5** Sequence of events leading to explosion and fire in a tank farm. (1) Source tank; (2) Oil spill; (3) Ignition; (4) Fire in a target tank; (5) Domino effect.

There are several reasons to pursue studies of fire in tanks: First, to reconstruct the sequence of events that took place during the accident. In forensic investigations, such as those that follow an accident in a tank farm, investigators attempt to postulate the mechanisms that occurred in a structure based on the deformation pattern identified during postevent inspections. From a structural engineering perspective, fire and explosions induce large deflections in the shell and nonlinear material behavior inducing plasticity and degradation. In cases of fire, the material behavior depends on the temperature levels reached before the fire is extinguished.

Second, to improve current design procedures. In the United States the National Institute of Standards and Technologies is in charge of investigating the consequences of natural disasters, such as earthquakes, hurricanes, and others, to understand if there are new modes of failure that were not considered by the current design provision and should be included in a revised version. Their role is to identify common features in accidents to make sure that they are accounted for in future designs.

Third, to assign responsibilities and liabilities in an accident. In cases of accidents, there are conflicting interests among stakeholders, including the owner of the plant, insurance companies, local government and authorities, environmental protection agencies, and citizens directly or indirectly affected by the accident. For example, three different companies stored fuel in the Buncefield facility, so it was highly relevant to identify which one was responsible for starting the accident.

Further discussions on this topic may be found in the text of Assael and Kakosimos (2010, 287–292), where causes of the destruction of tanks are identified.

## 6.3 Effects due to explosions

### 6.3.1 Basic features of explosions affecting nearby tanks

An explosion is defined as a sudden, large-scale liberation of energy that is violently dissipated by shock waves. In chemical explosions, the energy arises due to the rapid oxidation of combustible fuel, such as those that are stored in oil refineries and oil depots. The source of the explosion is a relatively small area with air at elevated pressures and high temperatures in the order of 3000°C. Hot gas expands and generates a volume change with a layer of compressed air that contains most of the energy liberated by the explosion. The waves expand with a spherical shape and move the layer of highly compressed air from the source to surrounding objects.

The accident that occurred in Buncefield was triggered by a deflagration, caused by an unconfined vapor cloud; on the other hand, confined explosions are known as detonations and are frequently associated with explosives. The speed of wave propagation is higher in detonations, with high pressures on surrounding objects lasting in the order of tens of milliseconds. Deflagrations, on the other hand, are associated with high pressures acting for a hundred milliseconds on nearby objects.

The main variables controlling the magnitude of an explosion include the fuel mass, type of hydrocarbon and its physical state, type of release, dispersion and mixing of the fuel with the surrounding air, the timing of ignition, ventilation, and confinement, and location of the ignition source (see, e.g., Burgan, 1995, 157). The height of the vapor cloud in a deflagration is also an important factor influencing effects on neighboring tanks.

Thus, a wave front expands in an explosion, with a zone of compressed air that induces an overpressure in the air on top of environmental pressure and a dynamic pressure caused by flow. The distance between a given point and the source of an explosion is known as the stand-off distance. Often the term is employed to establish the distance of an object with respect to the source.

The free-field pressures carried by the shock wave may be visualized in terms of distance and time, both considered with respect to the source. A schematic representation of the pressure wave in a free field as a function of the distance from the source is shown in Fig. 6.6 for different times (see, e.g., US Army Corps of Engineers, 2008).

The shock wave is characterized by an overpressure peak at the front (pressure on top of the atmospheric pressure), with a significant decrease toward the source of the explosion, as shown by curves at $t_1$ to $t_3$ in Fig. 6.6. As the wave moves away from the source, the peak pressure decreases (the peak for $t_3$ is lower than for $t_1$). However, at a more advanced time (curve $t_5$ in Fig. 6.6), the pressure variation toward the source falls to values lower than the atmospheric pressure, and this is called a negative phase.

Another classical way to represent the free-field effects of explosions is shown in Fig. 6.7, in a plot of pressures versus time; this shows a typical

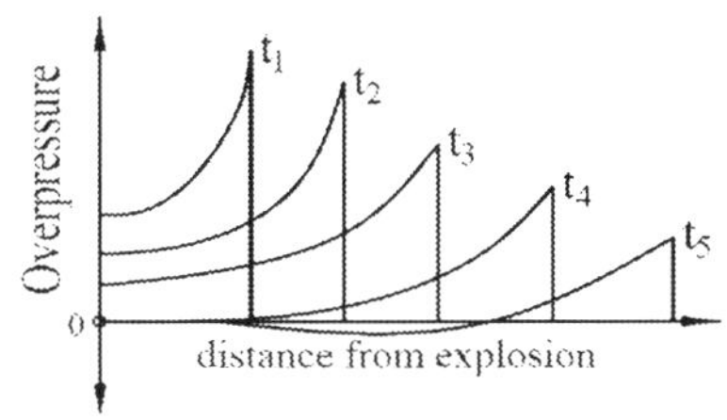


**Figure 6.6** Free-field overpressure as a function of distance to the source of explosion, for different times from initiation of the explosion.

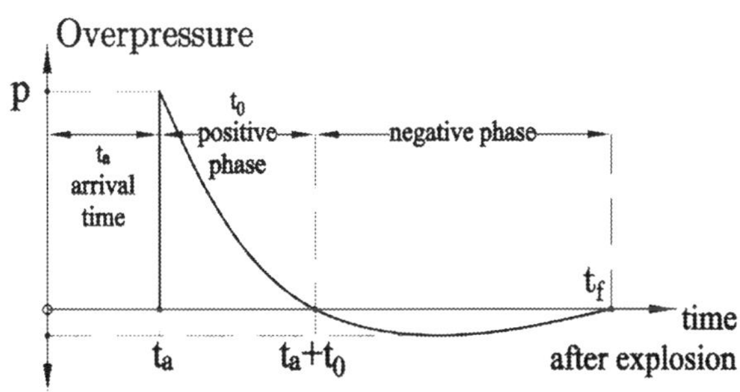


**Figure 6.7** Pressure as a function of time at a point located at distance *R* from the source (free-field.)

overpressure evolution at a specific location when the shock-wave impacts the structure.

At a given distance from the source of the explosion, the peak pressure of the wave arrives at time $t_a$ (known as time of arrival) and the effect at this location finishes at $t_f$. The positive part of the pressure takes place for $t_a < t \leq t_a + t_0$ and the negative part occurs at $t_a + t_0 < t < t_f$.

The overpressure carried by the shock wave suddenly increases when the wave impacts a structure; this is called the reflected pressure, and the maximum intensity will be identified as $p_0$. Time histories of reflected pressures have the same variation as that shown in Fig. 6.7 (with amplitudes depending on distance).

Test variables and results are frequently normalized using scale parameters so that they can be employed for other sources/target configurations. There are many possible explosive substances and to work with comparable magnitudes the mass of the actual explosive is normalized with respect to an equivalent TNT charge; this is identified as $W$. The explosion itself is usually characterized by a normalized variable $\overline{Z}$, given by

$$\overline{Z} = \frac{\overline{R}}{\sqrt[3]{W}} \left[\frac{m}{kg^{1/3}}\right] \tag{6.1}$$

where $\overline{R}$ is the stand-off distance and $W$ is the TNT equivalent mass of the explosive.

First, consider the effects of an explosion on a vertical flat surface. There are several models and methods to predict overpressures and their effects in terms of the explosive charge and run-off distance (see, e.g., Baker et al., 1977, Taveau, 2012b, 2012c, and others). If the shock wave finds a plane object perpendicular to the direction of propagation, the reflection increases overpressures to a maximum value $p_r$ (known as reflected pressure). Based on values of $\overline{Z}$, the US Army Corps of Engineers (2008) developed ways to estimate other variables in an explosion, such as the peak reflected pressure $p_0$, and the duration $t_0$ of the positive phase of the explosion.

The pressure duration of the explosion at the target is a very short time, measured in [ms]. Its significance largely depends on the relation between the pressure duration and the natural period of the structure. For the thin-walled structures considered in this work, typical natural periods are in the order of 150 ms; thus, pressure durations of approximately 30 ms should be considered impulsive loads. It is often assumed that only the positive part of the pressure is of significance and the effects due to the negative part tend to be neglected. This simplification causes a larger impulse to be considered and constitutes a safe assumption for the analysis.

To illustrate the use of such a procedure, consider a mass of 250 kg of TNT placed at 75 m from a structure; this causes an incident pressure in the order of 12 kPa, with a positive phase lasting for $t_0 = 32$ ms. As a reference value, Noret et al. (2012) state that overpressures in the range between 7 and 50 kPa may cause damage to tanks, depending on specific conditions. However, peak values are only part of the information required and pressure signatures are also of importance.

The shape of the pressure versus time curve (called the impulse signature) is an important feature to be considered. The simplest signature is a triangular impulse, which has often been considered in the literature; however, other pressure signatures have also been proposed to refine pressure estimates.

## 6.3.2 Evidence from small-scale testing of pressures reaching a tank

Most data available in the literature refer to reflected pressures caused by explosions on a flat vertical surface; however, these are not the same as pressures acting on a cylindrical surface, for which new space and time distribution should be obtained. Thus, such pressures need to be evaluated via physical tests or by using Computational Fluid Dynamics (CFD) simulations. A brief description of test results carried out at various laboratories is presented in the section.

A test may be performed on a thick-walled cylinder, from which time and pressure distributions can be obtained with negligible deflections. Such shells should be instrumented with pressure gauges to evaluate pressures around the circumference and in elevation. Pressures obtained from rigid cylinders are used as input to perform a structural analysis of flexible tanks. Alternatively, tests may be carried out using flexible wall cylinders, in which case the structural response is of direct interest: the buckling process is recorded using high-speed cameras and the final deformation may be inspected.

Fig. 6.8 shows a schematic representation of the test setup under controlled conditions. The intensity at the source is characterized by $W$; the distance between source and target is $\overline{R}$; the structural dimensions are the diameter $D$ and height $H$, with shell thickness $h$. The ground between source and target is assumed to be flat.

The range of interest to the oil industry includes pressures between 20 and 150 kPa, with blast duration between 20 and 100 ms.

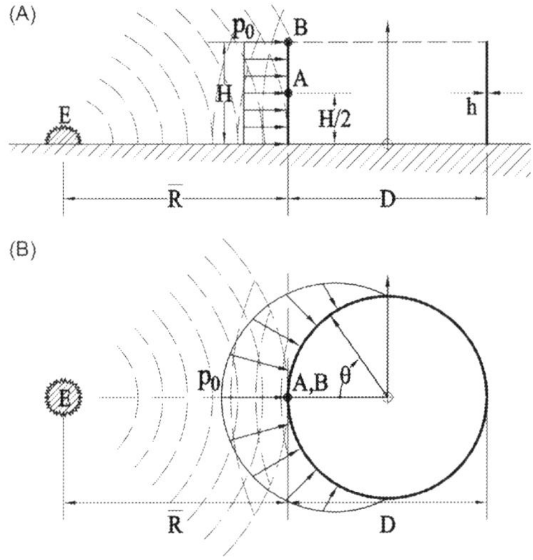


**Figure 6.8** Schematic representation of test of small-scale tank under controlled explosion. (A) Side view; (B) plan view.

Early experiments on thick cylindrical shells ($10 < R/h < 36$) reported by Lindberg (1964) identified dynamic buckling loads associated with the plastic flow. For thinner aluminum shells clamped at the ends ($H/D = 0.92$ and $R/h = 480$), Lindberg noticed a plastic buckling mode with short wrinkles; however, he stated that elastic buckling modes take place before the development of plastic buckling modes with a large number of wrinkles. Further studies were reported by Lindberg (1987).

Perhaps the largest effort performed up to date to understand the behavior of tanks under the effects due to explosions has been performed by a group of researchers in France (Duong et al., 2012a, 2012b, Noret et al., 2012). The main goal of the study was to establish vulnerability analysis, in which damage levels are related to explosion intensity. Tests were performed on an instrumented small-scale tank to record overpressure values, pressure duration, and their spatial distribution. The recorded information was subsequently employed using simplified structural models.

The tests were performed on a small standardized table with dimensions 2.5 m × 1.5 m. The detonation was materialized as a spherical charge of propane/oxygen. Rigid models were fabricated with PVC on a scale with respect to prototype tanks as found in industrial applications, and details of the scale rules may be found in Duong et al. (2012a). The results of peak reflected pressure, impulse, and duration of positive phase are given in the paper, with differences with respect to those found on flat surfaces. Pressure distributions around the circumference were performed by rotating the model at an angle of 15 degrees around the circumference for each test.

Another set of tests was conducted in 2012 at the University of North Carolina at Charlotte, to evaluate pressures due to explosions on a rigid cylindrical shell modeling an open-top tank (Weggel & Whelan, 2013). The dimensions of the model were $D = 0.914$ m, $H = 0.732$ m, $h = 9.5$ mm, with an explosive at a run-off distance $\overline{R} = 1.8$ m. Pressure sensors were placed around the circumference and in elevation to evaluate dynamic pressures.

The reflected pressures normalized with respect to the maximum pressure can thus be written in the form:

$$p_r(\theta) = \sum_{n=0}^{8} c^p{}_n \times \cos(n\omega\theta) \tag{6.2}$$

where $\omega = 0.0199$ and the $c^p{}_n$ coefficients are values resulting from the tests.

Tests carried out at Harbin University in China have been recently reported by Lu et al. (2019). Interest in this research was on the behavior of very large open-top vertical tanks. The specific model investigated had dimensions $D = 923$ mm, $H = 297$ mm, and $h = 1.2$ mm, with a scale from prototype to a model of 108. The model was filled with water up to a level of 260 mm, which is 87% of the height of the tank. Three pressure gauges were used to measure blast pressures on the outer part of the shell facing the explosion. The tests were accompanied by a finite element model carried out to validate the computational work. A description of the pressure variation around the circumference or in elevation was limited to the three gauges placed in the model.

## 6.4 Modeling pressures due to explosions reaching a target tank

### 6.4.1 Simplified models of pressure distribution around tanks due to a nearby explosion

Simplified blast overpressures have been proposed in the literature to model effects due to an explosion. The simplest equation is a triangular decay in time and is written as:

$$p = p_0\left(1 - \frac{t}{t_0}\right) \tag{6.3}$$

where the pressure is characterized by the two variables, $p_0$ and $t_0$. A more complex model is given by:

$$p = p_0\left(1 - \frac{t}{t_0}\right)\exp\left(-k_2 t/t_0\right) \tag{6.4}$$

where $k_2$ is a decay coefficient.

The space distribution of pressures is here taken as a distribution adapted from Putelat and Triantafyllidis (2014):

$$p = p_0\left(1 - \frac{t}{t_0}\right)\exp\left(-k_2 t/t_0\right) \quad \exp\left(-k_1{}^2\theta^2\right) \tag{6.5}$$

where the coordinate $\theta$ is measured from the point of maximum pressures; the scalar parameter $p_0$ is the maximum value of the reflected pressure; the term $\exp\left(-k_1{}^2\theta^2\right)$ is the normalized distribution around the circumference, and $k_1$ is a pressure decay coefficient around the circumference. This equation does not take into account the fact that the pressures do not arrive at all points around the shell at the same time; further, it does not account for different durations of the positive phase. These effects were investigated by the authors (Ameijeiras et al., 2014) and it was shown that they only have a minor incidence on the dynamic buckling of the shell.

### 6.4.2 Advanced models of the source of an explosion and its consequences on tanks

The pressures caused by an explosion have been the subject of much research during the past decades, but a review of such contributions is outside the scope of this chapter. Only the main conclusions and modeling idealizations are reported in this section.

There are two main ways to describe an explosion at the source using computational tools: One is utilizing physical parameters, that is, high temperature and pressure in a small spherical volume that liberates and expands, producing a shock wave. A more complex way is to model the chemical reactions involved in the explosion.

Complex numerical simulations based on LS-DYNA were reported by Duong et al. (2012a) considering a flexible model and a rigid model. This was done to better understand the incidence of fluid-structure coupling on the results. To reduce the required computer resources needed for the analysis, the wave front was simulated as an initial flow. To tackle the computational problem, the mesh was divided into four zones: blast wave generation, free-field propagation, the tank structures, and the cylinder inside. These authors found that the error due to neglecting fluid-structure interaction is in the order of 1%, except for meridians at $\theta < 45$ degrees, for which the errors were higher but in any case, smaller than 5% (Duong et al., 2012a). These results suggest that uncoupling both phenomena (fluid flow and structural deformations) has a small impact and can be accepted for engineering purposes to model tanks under blast loads.

A CFD simulation was used by Hu and Zhao (2016) to investigate the loads acting on tank walls due to an internal explosion. The explosion itself was simulated using a $k$-$\varepsilon$ turbulence model and a combustion model based on the eddy dissipation within the framework of CFD. In the knowledge of the authors, equivalent simulations for external explosions reaching a tank are not available at present.

Simulations based on multiphysics approaches are a promising avenue of research, but this has not been explored at present within the context of tank farms. The framework of contributing disciplines includes gas dynamics coupled with viscous, chemical, and turbulence effects (see, e.g., Gelfand et al., 2012).

## 6.5 Structural behavior of tanks under impulsive loads

### 6.5.1 Computational modeling

The finite element evaluation of deformations, buckling, and plasticity of tanks under blast loads requires first understanding the static behavior and then the dynamic response of the shell.

Static analyses using finite element general-purpose packages can evaluate Linear Bifurcation Analysis (LBA), Geometrically Nonlinear Analysis with Imperfections (GNIA), and Geometrically and Material Nonlinear Analysis with Imperfections (GMNIA); such analyses are described, for example, by Rotter and Schmidt (2008). Nonlinearity is usually followed with step-by-step algorithms, such as that proposed by Riks (1979). Typical dynamic analyses employ either implicit or explicit algorithms and, for a given time step, it is possible to compute variables along with the transient motion.

The dynamic response of cylindrical shells to blast and pulse loads was already investigated in the last century (see, e.g., Islam et al., 1992, Ruis et al., 1989). The influence of the liquid stored in a tank was considered by Ngo et al. (2007) using a simplified blast pressure acting on the shell, together with a coupled Eulerian-Lagrangian approach to represent the fluid and the solid. Sloshing was taken into account in this model. It was found that hoop stresses were maximum at the top of the shell and above the fluid level and that the stresses increased for higher constraints assumed at the bottom of the shell. Sloshing was more important in tanks with a large aspect ratio of $H/D$.

## 6.5.2 Dynamic buckling criteria

Two main effects occur in shells under blast pressures: buckling and plasticity. Plasticity thresholds are well defined in terms of yield criteria, but dynamic buckling under impulsive loads is a less explored area and will be briefly discussed in this section.

In the 1960s, Bernard Budiansky proposed a criterion of dynamic buckling based on the geometrically nonlinear dynamic motion of the structure (Budiansky & Roth, 1962, Budiansky, 1967). Budiansky evaluated the motion for several load levels and stated that a structure reached dynamic buckling if, for a small increment in load, there was a large (non-proportional) increment in the displacement reached during the motion. One way to visualize this process was by plotting the transient motion (displacement vs time) at various load levels. Consideration was given to what is now known as the pseudo-equilibrium path, that is, a relation between the applied load and the largest displacement of the structure during the transient motion. This is a qualitative criterion and requires computing the transient response at several load levels.

The Budiansky criterion has been successful in the evaluation of dynamic buckling under step loading and for a pulse with a duration not less than 3 s, but it has some problems for short duration blast loads. Under seismic effects, in which there is a reversing load, a change in slope rather than a jump in the pseudo-equilibrium path has been detected (Buratti & Tavano, 2014, Virella et al., 2006) and subsequently employed by Kubiak (2013) as a criterion for dynamic buckling. The methodology is also used for the identification of thermal buckling in shells, which is known as the "Inflection Point Method" (see, e.g., Bhagat & Jeyaraj, 2018, and references cited there).

Critics of the Budiansky criterion argued that it does not offer a precise identification of the load when the structure reaches instability, that it does not account for the possibility of bifurcation, and that it does not account for imperfection sensitivity (Kleiber et al., 1987). The last objection is not a limitation of the criterion but it has to do more with the cost of computations because one can compute the transient motion using various imperfections in the geometry of a shell.

Lee was the first to explicitly investigate the possibility of bifurcations for the transient motion, and this he called "quasi-bifurcation" (Lee, 1981). Simitses (1990) investigated dynamic buckling using energy principles based on information obtained from the static analysis under the same load. An extension of this procedure was recently used by (Ameijeiras & Godoy, 2021a) for horizontal tanks in the oil industry.

More efficient implementations of computational dynamic buckling were proposed in the 1980s by Kleiber et al. (1987) and Burmeister and Ramm, 1990, in which the transient motion is computed and the eigenvalues of the system are checked at each time step. Along the same lines, Kroplin and Dinkler (1986) investigated the eigenvalues of the tangent matrix along with the dynamic motion. Dynamic buckling is not a closed topic and new proposals have been under discussion for some years (see, e.g., Ameijeiras & Godoy, 2021b).

### 6.5.3 Structural behavior of open-topped tanks with a wind girder under an explosion

A tank opened at the top, with a wind girder, and simply supported at its base, is next investigated as a case study. The overall geometry is given by $D = 15$ m, $H = 12$ m, with a uniform thickness of $h = 7.5$ mm, leading to geometric relations $H/D = 0.8$ and $R/h = 1000$. The dimensions of the wind girder were evaluated as specified in API 650 (2021), and for a wind speed of 48 m/s. The resulting cross-section modulus of the wind girder was $Z = 132$ cm$^3$. This section modulus was modeled as an increased thickness of 57.5 mm at the top 240 mm of the cylindrical shell.

Properties of A-36 steel are assumed, with values of modulus of elasticity $E = 200$ gPa, Poisson's ratio $\nu = 0.3$, density $\rho = 7850$ kg/m$^3$, and yield stress $\sigma_y = 250$ mPa. An elastoplastic material behavior was assumed following the von Mises yield criterion and an associated flow rule.

Based on previous experience with this class of structures, convergence is reached with a mesh density having a maximum element size of $0.1875 \times 0.1875$ m$^2$. The number of shell elements with linear interpolation

(SR4 in the ABAQUS nomenclature) used was 16,315 for the cylindrical part of the tank.

First, the static response of the shell was investigated by means of LBA and then a model based on GMNIA was implemented, both within the ABAQUS environment. Riks' (1979) method was used to follow the nonlinear equilibrium path under a static pressure with the same spatial distribution as that assumed for the dynamic analysis (Eq. 6.5).

The static behavior was evaluated using LBA leading to a critical buckling pressure $p_0 = 5.62$ kPa. The eigenvector computed from this analysis is shown in Fig. 6.9, with about five waves in the circumferential direction and one-half wave in the vertical direction.

The equilibrium path, shown in Fig. 6.10A, was computed with GMNIA. The assumed geometric imperfection in the cylinder has a shape and an amplitude. Based on previous studies in this field, the shape is taken as the eigenmode associated with the lowest eigenvalue in the LBA. This is frequently assumed in shell buckling studies because it leads to the highest imperfection sensitivity in most cases. The amplitude $\xi$ is specified as equal to the shell thickness.

The path reaches a maximum at $p_0 = 3.8$ kPa and the radial displacement is $u_r/h = 4.75$ at that load level. The displacement is measured at the mid-height of the shell on the most loaded meridian.

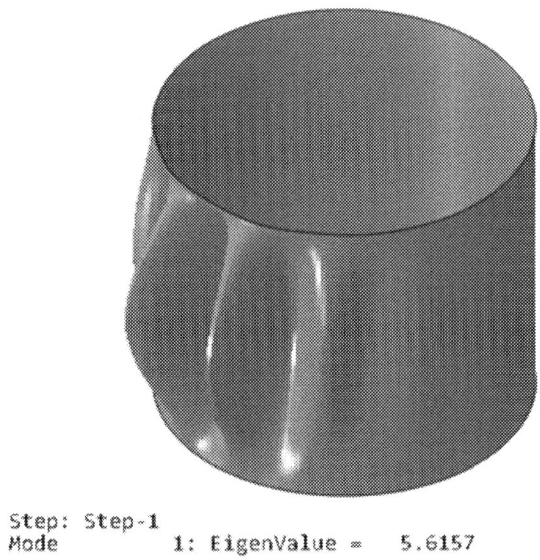


**Figure 6.9** Eigenmode for a tank with a wind girder, for the lowest critical pressure $p^C = 5.62$ kPa.

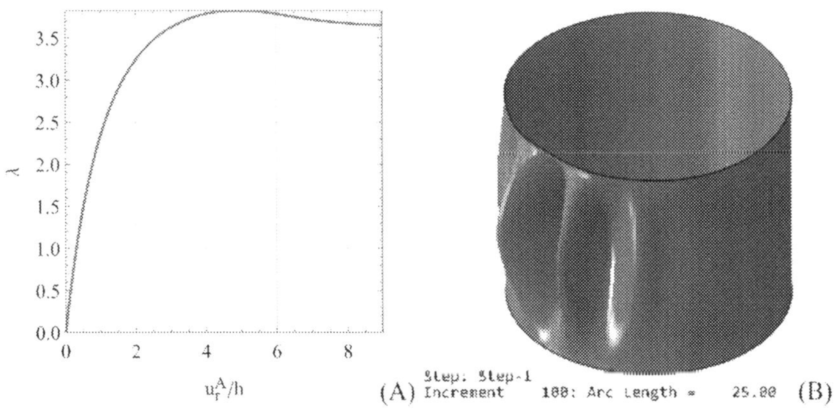


**Figure 6.10** Open-top tank with a wind girder: static results. (A) Equilibrium path; (B) deflected configuration at the state of maximum pressure.

Notice that there is a 32% decrease in pressure from the bifurcation to the maximum in the nonlinear analysis, with a reduction from 5.62 to 3.8 kPa. However, the deflected configuration in both cases is the same.

From this point onwards the path becomes stable and a slight stiffening of the shell occurs. The shape of the shell in Fig. 6.10B is substantially the same as the eigenmode, so the imperfection grows with increasing pressure until the maximum in the path is reached.

For the assumed dimensions of the girder, the eigenvector and the mode shape at the state of maximum load are very similar to those found for a fixed roof tank.

The dynamic response is next reported. The fundamental period of this tank is $t_n = 0.18$ s. Notice that the ratio $t_0/t_n = 0.16$ is smaller than 0.25 so the load can be considered impulsive. The associated eigenvector is plotted in Fig. 6.11.

The number of shell elements (SR4 in the ABAQUS nomenclature) used in the dynamic analysis was 16,315 based on convergence studies. Explicit time integration was carried out, with an approximate time increment of 0.03 ms.

For a pressure applied during 30 ms, the transient response was computed in the range 35 kPa $< p_0 <$ 175 kPa, and the results are shown in Fig. 6.12. The first peak in displacements is the largest one, and the peak occurs at approximately $t = 50$ ms, independently of the value of $p_0$. Notice that the pressures only act during 30 ms, thus the peak occurs under zero applied load. Following this first peak, the displacements rapidly decay with time.

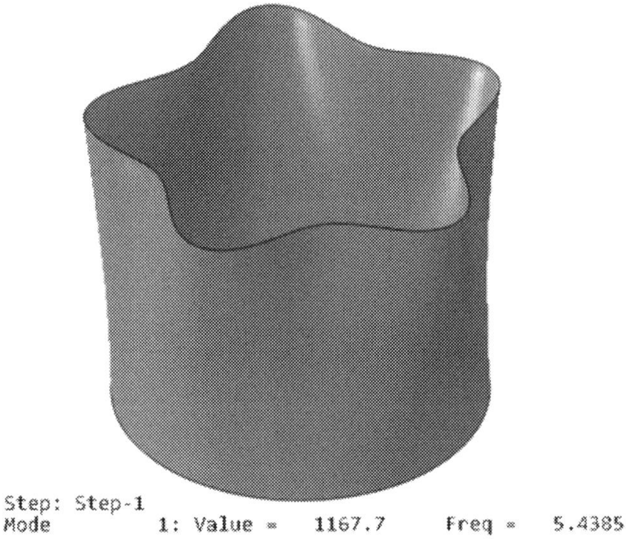


**Figure 6.11** Eigenmode associated with natural period $t_n = 0.18$ s.

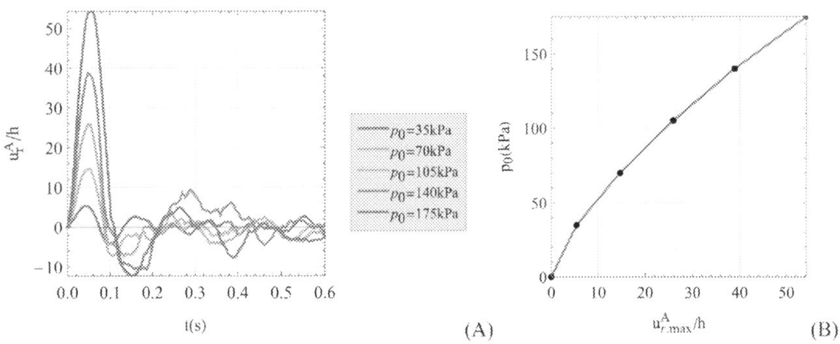


**Figure 6.12** Open top tank with a wind girder: dynamic results under blast pressures, for $t_0 = 30$ Ms, and 35 kPa $< p_0 <$ 175 kPa. (A) Transient response; (B) Pseudo-equilibrium path.

The peak displacements may be plotted versus values of $p_0$ to draw the pseudo-equilibrium path in Fig. 6.12B. This path is nonlinear, and the initial slope of the path decreases with increasing $p_0$ until a new slope is reached. A tangent can be drawn to the two mentioned segments, to illustrate the change in behavior that takes place at approximately $p_0 = 30$ kPa. This change in the pattern along the pseudo-equilibrium path has been identified as a criterion for dynamic buckling (Virella et al., 2006).

The values of the largest displacements for each pressure $p_0$ have been plotted in the pseudo-equilibrium path in Fig. 6.12B. Large amplitude deflections are seen to occur along this path so it is important to investigate the possibility of having plastic excursions in steel. This has been done for several values of $p_0$, and the results are shown in Fig. 6.13.

For $p_0 = 35$ kPa a wavy pattern may be seen in Fig. 6.13A, for which the material remains elastic. Although the shell has elastic behavior at this load level, elastic buckling is of concern because of the possibility of blocking the vertical displacements of an interior floating roof, such as what happened during the Guam accident in 2002. And in any case, elastic buckling is a design criterion in metal shells.

Localized plasticity occurs at $p_0 = 70$ kPa (Fig. 6.13B), and this affects a very small area in the shell. A larger area of the shell under plastic behavior occurs at 105 kPa (see Fig. 6.13C), in which the part of the shell where pressures are present is under material plasticity. Because just a few snapshots are shown in Fig. 6.13, one cannot state with precision at what load level there is an initiation of extended plasticity, but it surely occurs in the interval close to $p_0 = 105$ kPa. A refinement of this estimate requires computing the solution for more points within this neighborhood. This extended plasticity is of great concern to a designer because it weakens the shell stiffness. Plasticity has extended to be the dominant behavior on the part of the shell facing the explosion at $p_0 = 140$ kPa (Fig. 6.13D).

In summary, two concurrent aspects of behavior are present in tanks for increasing pressures due to an explosion. One is elastic buckling, leading to a wavy pattern around the circumference. The second one is extended plasticity that develops on top of the buckled shape, which occurs at higher load levels. They can be taken as lower and upper bounds of damage, and this would be of great use in studies of the fragility of tanks under explosions.

### 6.5.4 Effects of explosions in very large tanks

The previous analysis was carried out for a case in terms of geometry and a range of blast pressure intensity. Rather than exploring other geometries and configurations, it is interesting to review the work of other researchers in this field, for a much larger tank.

A very large open-top tank with a wind girder and intermediate ring stiffeners was recently reported by Lu et al. (2019) and the results

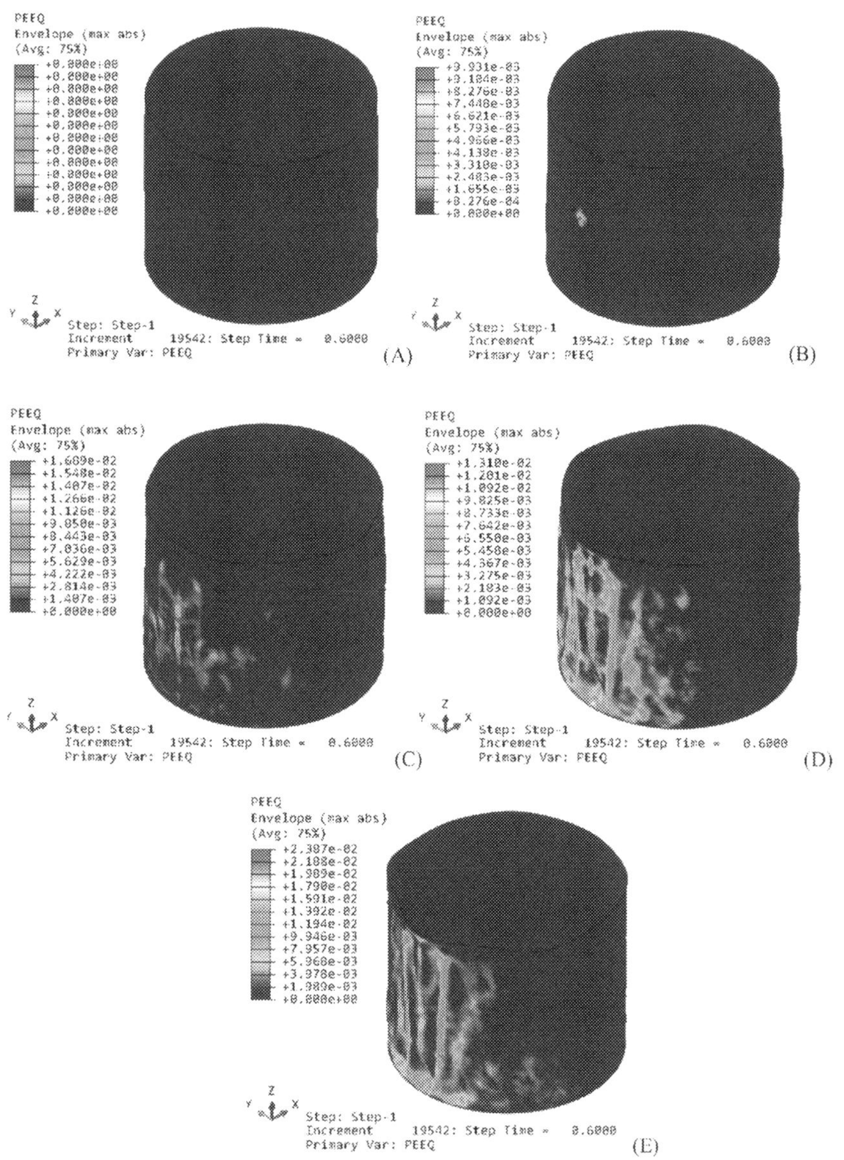


**Figure 6.13** Dynamic behavior of tank under blast pressure, computed at $t = 20t_0$. GMNIA results for $t_0 = 30$ ms, imperfection amplitude $\xi = h$, for various values of $p_0$. (A) $p_0 = 35$ kPa, (B) $p_0 = 70$ kPa, (C) $p_0 = 105$ kPa, (D) $p_0 = 140$ kPa, (E) $p_0 = 175$ kPa.

presented there are summarized in this section. The dimensions of the tank were $D = 100$ m, $H = 25$ m, which is twice the height and almost seven times the diameter of the tank studied in the last section. The tank was fabricated with eight courses of 2.7 m in height each, except for the first one which was 3 m in height.

To stabilize the shell, these large tanks need additional ring stiffeners. Two wind girders were considered at the top of the cylindrical shell followed by two ring stiffeners at a lower level; details of these rings are not given in the paper, nor their sectional modulus, but it is indicated that they were fabricated using cantilever thin-walled rings.

The influence of filling the tank with water at 25%, 50%, and 75% in elevation (liquid heights of 5.4, 11, and 16 m, respectively) was investigated.

Blast loading was assumed by a TNT equivalent mass of 380 kg placed at a run-off distance $\overline{R} = 25$ m and at $z = 3$ m from ground level (for which $p_0 \approx 87$ kPa). A second case was investigated with similar characteristics but having a 1500 kg TNT equivalent mass ($p_0 \approx 232$ kPa).

Shell deflections due to the blast load were reported as follows: Under a 25% of water filling the target tank, the 380 kg explosion induced failure of the ring stiffeners which allowed the development of a plastic buckling mode affecting half of the tank in elevation and developing three half-waves around 20% of the circumference. Effective plastic strains were in the order of $5.1 \times 10^{-2}$. The displacements (combining vertical and radial components) were in the order of 5 m at the bottom course increasing to some 14 m at the top course for an empty tank; these numbers reduce with a liquid filling of 50% to 3 m and 12 m, respectively, and 1.5 and 4 m for 75% liquid level.

For the 1500 kg of equivalent TNT explosion under 25% of water filling, the consequences were more noticeable, with significant displacements covering a larger zone in elevation (reaching the bottom of the shell). Large deformations were observed at the bottom of the shell. Effective plastic strains in the order of $8.4 \times 10^{-2}$ were reported. The displacements reported were 18 m at the bottom and 24 m at the top in the empty tank condition, which reduces to 10 and 24 m for 50% fluid, and 3 and 12 m for 75% of fluid stored. Such large displacements are associated with the fact that buckling of the wind girder had already occurred in the tank.

In both explosion intensities, the effects were less severe as the fluid level increased so that the most stringent situation was that of an empty tank. The authors state that liquid absorbs part of the energy of the blast wave caused by the explosion.

The authors found that a wind girder designed to resist wind is not sufficient to resist blast pressures, so they buckle, leaving the top of the shell under a flexible condition (Lu et al., 2019). The conditions at the bottom (the junction of the cylinder with the bottom plate) were found to influence the results: In cases of free constraint, that is, the bottom junction can deform, part of the energy due to the explosion is absorbed by deformation in the bottom plate and a less severe consequence occurs in the tank shell. On the other hand, for a fixed connection at the bottom of the cylinder, this absorption mechanism does not occur and all energy is transferred to the cylindrical shell, thus inducing larger deformations.

### 6.5.5 Domino effects under blast loads

In cases in which an accident in a tank farm is due to an explosion (possibly in one of the tanks), and the induced shock wave reaches a target tank, a second explosion or fire may occur in this second tank. In turn, this second accident may reach the third tank. Damage to this third tank may occur due to the effects of the first explosion or from the second one. In cases in which the effects of the second event dominate, then it is said that a domino effect occurs. In domino effects, the consequences of the second explosion may be even worse than those due to the original explosion. and it is said that an accident escalation occurs.

Salzano et al. (2013) state that domino effects due to explosions are not frequently considered because the original explosion is usually of a large magnitude and dominates the damage to the infrastructure; however, there are cases of accident escalations. Interest in this topic may also be associated with the need that such a complex class of accidents should be taken into account at the design stage of a refinery or tank farm. Although domino effects are usually studied within one tank farm or refinery, they can also occur between different facilities in cases where there are industrial clusters, as shown by Khan and Abbasi (2001).

A procedure to investigate domino effects under blast loads is to simulate the original explosion and its damage consequences on nearby tanks. Conditions to have a second event should be established for each target tank reached by the explosion, and cases that exceed a threshold should be investigated as a new event, leading to a second simulation.

A discussion of threshold values for domino effects has been presented by Cozzani and Salzano (2004). These authors identified states of structural damage and classified them into two qualitative categories. They also identified loss of fuel due to the shock wave and infrastructure damage. This second loss was based on empirical evidence from previous accidents and led to a qualitative classification in three categories. To characterize damage and spills simultaneously, they proposed mixed categories involving both of these effects. These are the basis of threshold values for domino effects.

Accident escalation in storage tanks is highly dependent on the levels of both structural damage and loss of contents caused by an explosion. Significant oil spills aggravate secondary scenarios in tank farms and also reduce the time to reach domino effects. A number of factors influence the probability of escalation of an accident, including plant design, type of infrastructure, and fuel stored. Salzano and Cozzani (2005) investigated domino effects using probabilistic models that link peak overpressures with structural damage in terms of run-off distance and intensity of the original explosion. This work was followed by other authors (Mingguang & Juncheng, 2008), with a more detailed classification of damage and loss of fuel.

## 6.6 Effects due to fire

### 6.6.1 Introduction to fire effects in tanks

Evidence from real events in tank farms indicates that fire usually originates in one of the tanks (known as the source tank) in the form of a flame that transfers temperatures to the neighborhood. Heat is received by another tank, called the target tank, so that the temperatures on the steel walls increase and cause deformations in the target tank.

The process of understanding the consequences of fire burning tanks in the neighborhood requires knowledge from several engineering areas: First, one needs to understand the main features of the flame at the source of fire, with characteristics that depend on the type of burning fuel, the presence of wind and other factors (Drysdale, 2011). Second, there is a heat transfer process from the flame to nearby structures, such as a target tank. Third, there is a heat exchange at the target tank, in terms of the incoming radiation, the fluid stored in the target tank, and the environment. All these effects are present in the definition of temperatures that the target tank receives. Fourth, there is a structural response to the thermal field, which in general will induce deformations and buckling of the target tank. Within this complex scenario, Chemical Engineers are interested in the propagation of fire from one tank to another and the escalation of fire in a domino effect (Landucci et al., 2009). Structural Engineers, on the other hand, are more involved in the consequences that the thermal field due to fire has on a target tank to evaluate the possibilities of reaching buckling of the structure and material failure. This chapter focuses on the second perspective, but an overview of the first approach is also given here for the sake of completeness.

Fire has received more attention than explosions in design recommendations and codes, and some guidelines are available at present (Beyler, 2002; NFPA 30, 2012; Considine, 1984; Eurocode, 2003). Thermal effects in tanks have also been investigated regarding stress levels caused by high temperatures resulting from fluid storage technology (Krol & Joswik, 2021). These authors took into account details of a tank having a shallow dome roof and its supporting structure by finite elements and modeled the thermo-mechanical behavior of various components of the supporting structure, including girders, purlins, bracings, center ring, and roof-cylinder junction. The study does not take into account temperatures due to fire, but it highlights limitations in the current design provisions.

## 6.6.2 Summary of results from small-scale tests

Only a small number of results from small-scale testing have been published in the literature. Mansour (2012) performed a series of laboratory and field tests to measure temperatures in a fluid stored in a target tank and to evaluate the flow of the radiant heat received. The aim of the study was to predict temperatures in the shell and the fuel stored, and to estimate the increment in vapor pressure accumulated in the zone between the fluid surface and the tank roof. This information is useful to evaluate the probability of fire propagation and estimate the time available to operate the vapor release valve in fixed roof tanks.

Field tests were carried out at Loughborough University in the United Kingdom to measure the radiant heat flow at various places around a tank having $D = 2.4$ m, under pool fire conditions. Another smaller tank with a conical roof, with dimensions $D = 0.42$ m and $H = 0.57$ m and with water and a 50 mm layer of gasoline, was tested by adding sensors at various points inside, to evaluate time and space changes in gasoline.

The results obtained by Mansour led to the conclusion that the solid flame model was a better representation than the point-source model to estimate the heat flow radiation around the flame. This was found to be the case for gasoline, kerosene, crude oil, and other fuels.

The radiant flow for gasoline was 2.5 times larger than for ethanol, due to the lower combustion heat and combustion speed of ethanol, which causes a shorter flame length. The surface emissive power and the view factor were lower in ethanol than in gasoline. For an 85% ethanol and 15% gasoline mixture, the values recorded were similar to those of 100% gasoline; the conclusion is that in the early stages of the flame the radiant flow is mainly due to gasoline, which is the light component of the mixture.

Finally, the wind was shown to have a significant influence on the radiant heat flow received by objects placed close to the flame, because the position of the flame changes and its distance from a target structure is modified. Data from this work was used by Espinosa et al. (2019a) to validate numerical models of flames.

## 6.7 Modeling fire effects reaching a target tank

### 6.7.1 Simplified models of temperature distribution around tanks due to a nearby fire

In its simplest form, the consequences of thermal effects reaching a target tank may be modeled by the circumferential and vertical variation of the temperature on the shell surface. Liu (2011) proposed a simple model of temperature distribution around a tank to account for the circumferential variation, in the form:

$$T(\theta) = (T_{0m} - T_{0a}) \quad \cos^{2}\left(\frac{\theta}{\theta_0}\;\frac{\pi}{2}\right) \quad \textit{if} \quad |\theta| \leq \theta_0$$

$$T(\theta) = 0 \quad \textit{if} \quad |\theta| > \theta_0 \qquad (6.6)$$

where $\theta$ is the central angle measured with respect to the meridian closest to the heat source; $\theta_0$ is the angle at which temperatures are assumed to vanish around the circumference; $T_{0a}$ is the ambient temperature, and $T_{0m}$ is the maximum temperature at the hottest meridian in the steady-state. This equation accounts for the circumferential variation of $T$, and a separate factor would be required if the vertical variation was taken into account; however, such variation does not seem to affect significantly the structural response of the tank. There are also modified expressions in Liu (2011) to account for fuel stored in the target tank.

Eq. (6.6) is convenient to carry out computations and does not involve knowledge of the chemistry of the flame or the heat transfer from the source to the target tank. This expression may be conveniently used in buckling studies, in which a given temperature distribution is used to scale the response and thus find a critical value, or even postcritical states. Most researchers in this field employed this equation in modeling the structural behavior of tanks under a nearby fire.

However, in using such a simplified approach one does not consider factors affecting the flame itself, including the type of fuel-burning at the source, effects of flame inclination due to wind, or the distance between the flame and the target.

### 6.7.2 Advanced modeling of temperature distribution around tanks due to a nearby fire

As stated before, most fire accidents in a tank farm start at one of the tanks with a flame at the source. Understanding the mechanics and chemistry of flames is a subject of Chemical Engineering, and an excellent introduction to this field is provided by Casal (2018). A few topics should be mentioned here: Pool fire models; Point-source versus Solid flame models; Single-layer versus two-layer models; Flame location aboveground; and Wind effects on flame.

- Pool fire models, in which burning occurs on top of a pool, have been widely used in the literature to describe a flame with turbulent diffusion (see, e.g., McGrattan et al., 2000, Pritchard & Binding, 1992, Rew et al., 1997). A pool fire is typically described by the geometry of the flame (length, diameter, shape, inclination) and the average emissive power. A review of pool fire was reported by Steinhaus et al. (2007), in which burning of different fuels, radiation emission, fire distribution, and soot, were considered. Wu et al. (2020) studied the time-to-failure in tanks under pool fire.
- Further assumptions in this family of models refer to the source of the flame: On the one hand, there are "point-source models," with radiation from the center of a flame, and "solid flame models," with radiation from the surface of a solid cylinder. Most finite element research in this field has been performed using solid flame models.

- The type of fuel that burns has direct consequences on the flame that develops, as shown in the literature (Beyler, 2002, Considine, 1984, Mansour, 2012, McGrattan et al., 2000). The burning of hydrocarbon fuels (as considered in this chapter) has motivated the use of two-layer flame models. The combustion of liquified gases causes clear flames without smoke, but smoke occurs in the combustion of hydrocarbon fluids. This latter situation yields a complex flame structure, named a two-layer model, which is characterized by a lower zone with a clear flame and an upper zone, in which flames are combined with dense smoke and soot particles. Each zone is part of the cylindrical solid flame but may have different lengths and emissive power. The lower zone has the maximum emissive power. Details of the implementation of a two-layer model are given by Espinosa et al. (2019a, 2019b)
- The location of the flame at the source is an important factor. Consider an example of a burning tank taken from an accident in Gibraltar, on May 31, 2011. Fig. 6.14 shows two burning tanks from this accident in which flames may be identified: in the tank shown in Fig. 6.14A fire is located at the top of a tank and the flame develops from this level. In the case shown in Fig. 6.14B, on the other hand, the flame develops at ground level. The position of the flame strongly influences its effects on the neighborhood.
- Wind modifies the inclination of a flame; this effect was considered by Sengupta et al. (2011) in a point-source model to estimate safe distances between tanks. Other authors considered wind effects on flame inclination, including Da Silva Santos and Landesmann (2014), Pantousa (2018), and Espinosa et al. (2019a, 2019b).

(A) (B)

**Figure 6.14** Examples of flame location in the source tank: (A) Flame emerging from roof level, (B) Flame reaching ground level. *Reproduced from Espinosa, S. N., Jaca, R. C., Godoy, L. A. (2019a). Thermal effects of fire on a nearby fuel storage tank.* Journal of Loss Prevention in the Process Industry, 62 *(1–12), 103990.*

There are two main possibilities to represent the radiation from a source to a target tank: Either by using heat transfer analysis or via CFD (Espinosa et al., 2017, Shokrzadeh & Sohrabi, 2016). Most researchers in this field have modeled this process utilizing heat transfer analysis and they include different levels of detail in the analysis. Regarding the heat transfer process, one should include an energy balance to control that no energy is lost in the analysis.

The incident radiation in the target tank can be estimated as a relation between the heat received by the target tank, $E_t$, and the emissive power of fire, $E_f$, in the form:

$$E_t = \alpha_t \tau \quad \varepsilon_f \quad F \quad E_f \tag{6.7}$$

where $\alpha_t$ is the absorption of heat radiation on the metal surface of the target tank; $\tau$ is the atmospheric transmissivity; $\varepsilon_f$ is the fire emission, and $F$ is the view factor. The value of $E_f$ is frequently adopted as an average value on account of the emissive power of flame and the thermal radiation of smoke. A conservative value of $\alpha_t = 1$ is usually adopted to carry out computations, and $\varepsilon_f = 1$ is taken for a flame considered as a black body.

The first work in this field known to the authors was due by Liu (2011) and Liu et al. (2012), who considered a pool fire model with one zone of burning fuel having an average emissive power, and the flame temperature was assumed to be constant (900°C). Wind effects were not considered in this analysis, and the flame developed in the vertical direction from the top of the source tank. Parametric studies were carried out to investigate the influence of flame height, flame position, the diameter of the target tank, and the level of fuel stored in the target tank. In this and other works that followed, the geometric parameters of the flame were evaluated using empirical information available in the literature.

An improved flame model was reported by Da Silva Santos and Landesmann (2014), who estimated the flame geometry using semiempirical methods and included wind effects. A single emissive power was used for the flame and temperatures in the target tank were estimated by a transient heat transfer process.

Pantousa (2018) employed a solid flame model with one zone of combustion in addressing problems of multiples sources of fire affecting a target tank, to obtain circumferential and vertical variations of temperatures on the target tank. For a single burning tank, the transient analysis showed that the temperature distribution in the target tank does not change with time. Further, Pantousa found that the distance from the source to a target tank and the diameter of the tanks involved in the accident, do not modify the circumferential distribution of temperatures, but the vertical distribution strongly depends on both variables. The target tank had a higher resistance to fire in case the burning fuel is gasoline and lower resistance is found for ethanol; this depends on the features of the flame induced by each fuel. Pantousa found that wind had a negative effect on the resistance of the target tank. The critical temperature increases for smaller diameters of the burning tank and increases the separation between flame and target tank.

A two-zone model within the flame (with clear flame and dark flame zones) was taken into account by Espinosa et al. (2019a) to represent the burning of fuels. This approach can take into account differences in the emissive power of each zone. The flame in this case can start at the top of the source tank (pool fire) or ground level (full surface fire). The incidence of wind on the flame has been also considered by Zhou (2019).

In most cases reported in the literature, the flame starts at a tank and the target tank has the same dimensions as the source tank. For the external surface of the shell, Liu (2011) took into account radiation from the flame together with convection to air; other effects taken into account were conduction in the target tank and convection from the inner surface to air and fuel stored. The temperature distribution on the surface of the target tank (Espinosa et al., 2019a) was the result of an energy balance considering all forms of heat transfer (incident radiation, energy exchange between the metal surface and the stored fuel and air in the target tank, and energy exchange with air). Improved estimates of convection coefficients as a function of temperature and stored fuel properties, were made in this work.

### 6.7.3 Main differences between simplified and advanced models

The importance of simplifications adopted by using Eq. (6.6) can be visualized made by considering results obtained using more advanced models. Attention is given in this section to results reported by Espinosa et al. (2019a) for a source tank and a target tank of the same dimensions.

The influence of the position of the flame is shown in Fig. 6.15 for a specific tank and flame. This case was computed for zero wind velocity and wall-to-wall separation between the source and target tank equal to one diameter. The results illustrate that very different temperature distributions may be obtained if the flame starts at the top of the source tank or if it starts at ground level.

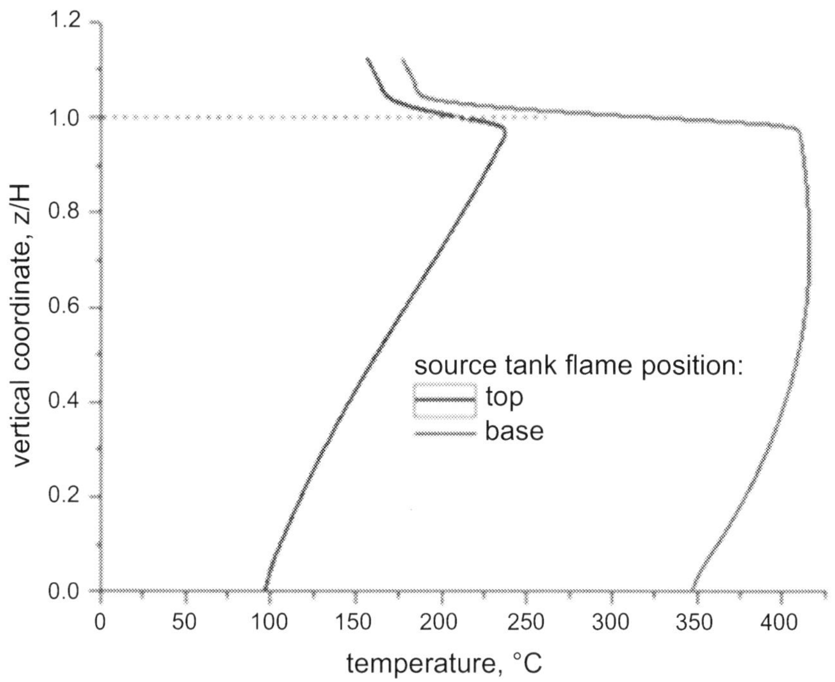


**Figure 6.15** Influence of flame positions on the temperature profile along a meridian in the target tank. *Reproduced from Espinosa, S. N., Jaca, R. C., Godoy, L. A. (2019a). Thermal effects of fire on a nearby fuel storage tank.* Journal of Loss Prevention in the Process Industry, 62 *(1–12), 103990.*

The fuel level stored in the target tank plays a significant role in the temperature distribution at the most heated meridian. This effect is shown in Fig. 6.16 for wall-to-wall separation between the source and target tank of one diameter, for zero wind speed.

In an empty tank, the temperatures increase up to a maximum at the top of the tank; if the tank is filled with fuel up to 50% in elevation, then the temperatures are significantly reduced at the lower part of the shell and at higher locations they reach the same values as in the empty tank. The same effect is observed for a tank with 94% of fuel. The fuel produces a cooling effect on the shell wall because the convection coefficient of fuel is two orders of magnitude lower than in air, and fuel is opaque to radiation. This effect was originally shown by Liu (2011).

The influence of wind may be observed in Fig. 6.17, again for two tanks having the same values of $D$ and $H$. Wind causes an inclination of the flame and if this occurs toward the target tank, then there is an

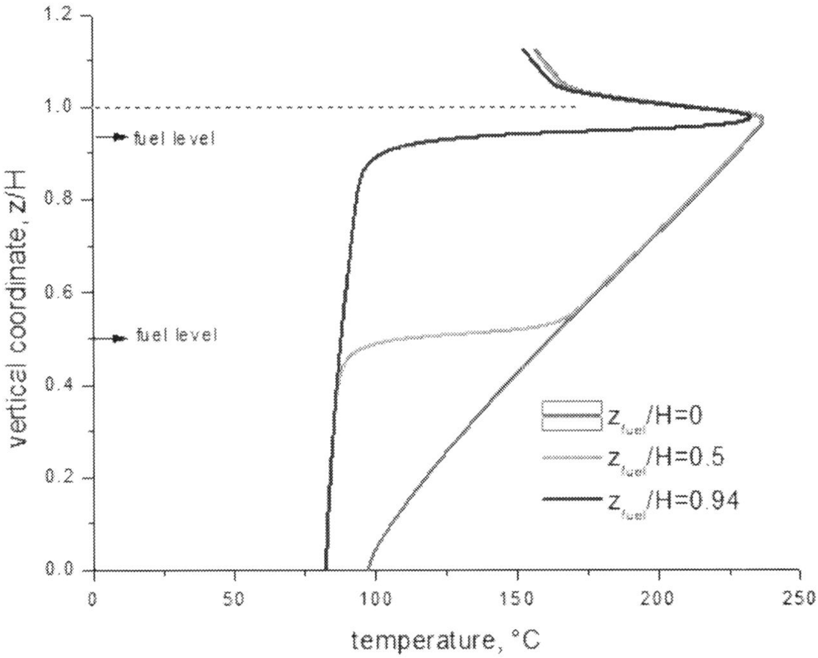


**Figure 6.16** Influence of fluid level stored in the target tank on the temperature profile along a meridian in the target tank. Flame starts at roof level in the source tank, zero wind speed *Reproduced from Espinosa, S. N., Jaca, R. C., Godoy, L. A. (2019a). Thermal effects of fire on a nearby fuel storage tank.* Journal of Loss Prevention in the Process Industry, 62 *(1–12), 103990.*

increase in incident radiation reaching the metal shell. Wind speeds between 0 and 45 km/h cause flame inclinations of 0 and 68 degrees, respectively. The temperature variation in elevation is shown in the plot. The maximum temperatures vary from about 250°C at zero wind speed to almost 500°C for a wind speed of 45 km/h. The slope in the temperature versus elevation plot increases with increasing wind speed.

Other effects, such as the separation between the source and the target tank, have been investigated by Liu (2011) and Espinosa et al. (2019a).

The results presented in this section indicate that factors taken into account by advanced models of heat transfer from the source to the target tank capture far more information than if the analysis is based on a simplified formula with assumed temperature distribution. However, advanced models require a two-step analysis, one for the heat transfer and another one for the thermo-mechanical behavior of the shell, whereas the use of a simplified model only requires the latter analysis. In practical engineering

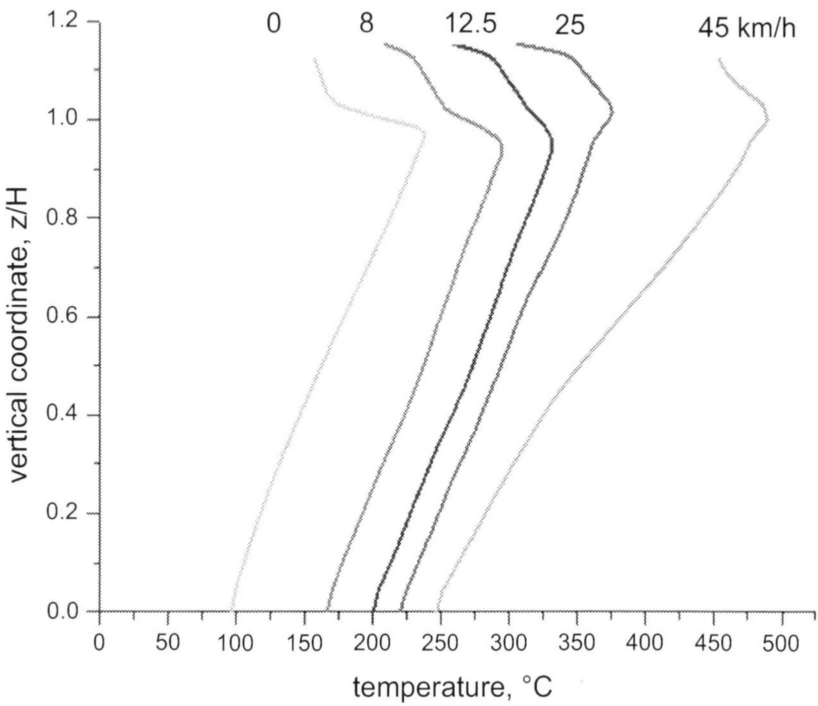


**Figure 6.17** Influence of wind speed on temperature profile, in the empty target tank for flame from roof level in the source tank. *Reproduced from Espinosa, S. N., Jaca, R. C., Godoy, L. A. (2019a). Thermal effects of fire on a nearby fuel storage tank.* Journal of Loss Prevention in the Process Industry, 62 *(1–12), 103990.*

situations in which only the critical temperature is required, a simplified model is preferred; however, in forensic studies in which more refined values are required to understand responsibilities in an accident, an advanced model should be employed.

## 6.8 Structural response and buckling under thermal loads

### 6.8.1 Types of analysis

The structural analysis of a target tank under heat has been modeled in the literature as a static problem with an assumed temperature distribution on the shell surface. The possible computational strategies identified by

Rotter and Schmidt (2008) include LBA, GNIA, and GMNIA, all of them available in general-purpose finite element programs. Special-purpose programs have also been developed to investigate thermal buckling in cylindrical shells, such as that reported by Alijani et al. (2015).

The temperature distribution used in the structural analysis was previously computed using a heat transfer analysis between the flame and the target tank, and the temperatures at the end of the heat transfer process are identified as the steady-state. This allows taking into account effects due to distance from flame to target tank, wind, fuel level stored in the target tank, and others.

An LBA approach based on the evaluation of linearized eigenvalues is the simplest approach and allows the identification of a critical temperature at which the shell reaches thermal buckling. This is the first step that any study should take into account. In most cases, the temperature in the target tank is taken utilizing the simplified Eq. (6.7), and in only a couple of cases were advanced estimates of temperatures used in the structural analysis.

To model the postbuckling behavior, one needs to employ a geometrically nonlinear analysis, in which geometric imperfections are included (GNIA). The usual strategy used to follow a nonlinear equilibrium path under mechanical loads has been the algorithm due to Riks (1979), but in cases, under thermal load, it seems that this procedure does not adequately model equilibrium states along the postbuckling path. In the context of the analysis of tanks under a thermal field, this problem was first identified by Liu (2011).

Liu shifted the analysis to an algorithm known as the Artificial Damping Method (ADM) (Kanarachos & Spentzas, 1988). The ADM is currently available in general-purpose finite element packages, such as ABAQUS (2021), ANSYS (2021), and MARC (2021). In the ADM algorithm, it is necessary to control and eventually adjust artificial damping at each time step. This methodology has been followed by several authors in the context of shell buckling, including Liu (2011), Liu et al. (2012), Kobayashi et al. (2014), Pantousa (2018), and Jaca et al. (2021).

### 6.8.2 Thermal buckling of tanks

The identification of critical states due to a thermal load in oil storage tanks has been tackled in this field by LBA or by GNIA studies. An LBA provides the solution in terms of critical temperatures and the associated buckling modes, both of which are crucial in understanding this problem and also in providing an estimate of the early stages of buckling. Most designers would tackle this problem by means of the LBA.

To illustrate the dependence of the critical temperature in tanks with a roof on their dimensions and other features, LBA results are summarized in Appendix 6.1 from various sources in the literature, all of them published during the last 10 years. The geometry of each tank is represented by its diameter $D$, the aspect ratio $H/D$ and the slenderness ratio $R/h$. All tanks reported have a fixed conical roof, although there may be differences in roof slope.

Conical roofs are usually supported by rafters and rings which are in turn supported by columns, but details of such secondary structures have only been taken into account in Godoy and Batista-Abreu (2012); in all other cases, an equivalent thickness is assumed for the roof. As explained by Burgos et al. (2015), values of roof thickness $h_r = 3$ h are considered to be adequate in terms of preserving the roof inertia. In real situations, the cylinder thickness is variable in elevation, but most studies assume a uniform thickness.

Based on the data in Appendix 6.1, the results in Fig. 6.18 were normalized using the Batdorf parameter $Z$, defined as (Batdorf, 1947)

$$Z^{Batdorf} = \sqrt{1-\nu^2}\left(\frac{H^2}{Rh}\right) \tag{6.8}$$

This parameter combines the effects of the aspect ratio $H/D$ and the shell slenderness $R/h$. Poisson's ratio is here taken as $\nu = 0.3$ for steel. In

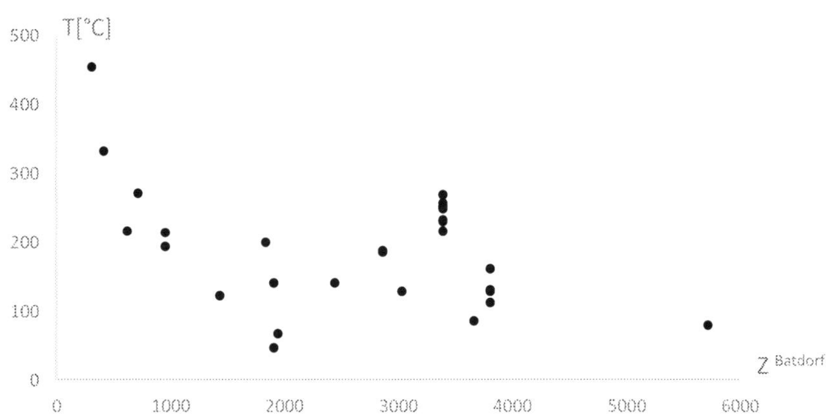


**Figure 6.18** Critical temperature of tanks with a fixed roof vs Batdorf parameter. Dots represent results listed in Appendix 6.1.

terms of the usual relations between variables adopted in this work, the $Z^{Batdorf}$ increases with the square of $H/D$ and with the $R/h$ ratio:

$$Z^{Batdorf} = 4\sqrt{1-\nu^2}\left(\frac{H}{D}\right)^2\left(\frac{R}{h}\right) \tag{6.9}$$

The results plotted in Fig. 6.18 do not include effects due to fuel stored in the target tank, and they do not account for the distance from the flame to the target tank, roof thickness, or wind effects.

The general trend seen in Fig. 6.18 is that $T^c$ decreases with increasing values of the Batdorf parameter in the same way as in cases of pressure-loaded tanks. The bounds cover a range of critical temperatures between $50°C < T^c < 450°C$, which are well within the range of temperatures that may be reached during a fire in a tank farm. This indicates that most tanks with a fixed roof directly exposed to fire at a distance of approximately one diameter from a flame will show thermal buckling. Most likely, such buckling will initially occur under elastic material behavior.

### 6.8.3 Postbuckling behavior

The previous studies of critical temperature are important for the design and safety verification of a tank, but they do not provide information regarding the postbuckling behavior of the shell, including the stability of the postbuckling path, the imperfection sensitivity, and mode changes that should be expected before failure of the shell. In forensic investigations, an analysis of the postbuckling behavior is a better tool to compare simulations with evidence obtained from postfire inspections.

A tank investigated is next considered to evaluate the buckling and postbuckling behavior. The diameter of the tank is $D = 11.44$ m, with $H/D = 1$; a uniform thickness of $h = 6.4$ mm was considered following API 650 (2021) specifications. The conical roof has a maximum elevation on top of the cylinder equal to 0.12 H, and, following previous work by Burgos et al. (2015), the roof was represented with an equivalent thickness $h_r = 3$ h. Other data of this case was reported by Jaca et al. (2021).

The flame was modeled with a solid model of two layers and assuming that gasoline burns inside the source tank. The target tank was located at a distance equal to one diameter, neglecting wind effects, and the target tank was empty in the results presented here. A heat transfer analysis was carried out to obtain the temperature distribution on the walls of the target tank. Other conditions were treated by Jaca et al. (2021).

The clear flame at the source reaches 933°C, and the dark zone reaches a maximum of 662°C; but the maximum temperature that can occur at the target tank, based on an energy balance, is $T^f = 420°C$. Thus, there is a limit to the values of temperatures that reach the target tank; such a limit can only be obtained based on an advanced model at the source.

The equilibrium path was computed in this case using the ADM approach. This is an iterative procedure under the control of the ratio between the energy of stabilization and the strain energy; this ratio should be kept at approximately 0.15. The artificial damping factor is variable as temperatures are increased, in such a way that the energy jumps are due to the activation of the artificial damping that stabilizes the response. This jump represents instability in some parts of the shell.

The initial equilibrium path is linear up to the critical temperature. The lowest instability occurs at a temperature $T^c = 248°C$, at which the buckling mode shape shown in Fig. 6.20A is obtained.

Other jumps may be seen in Fig. 6.19: These occur at 283°C, and there are a few jumps before $T = 383°C$. Jumps are characterized by

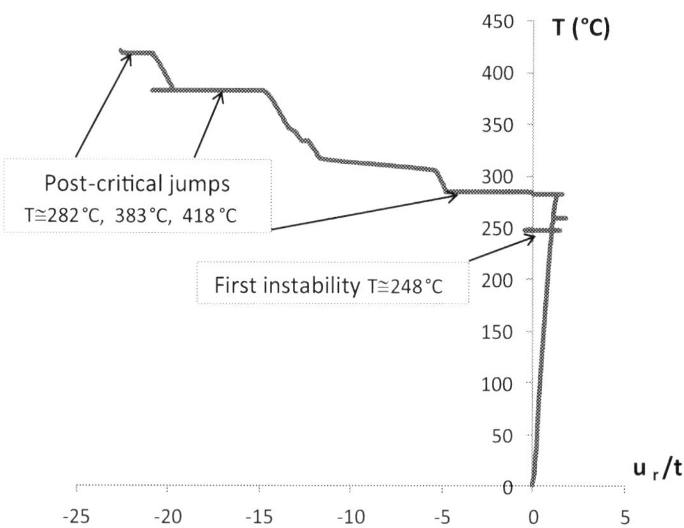


**Figure 6.19** Equilibrium path computed with ABAQUS, ADM algorithm, of an empty tank under a temperature field for flame at 1D separation from the target tank. *Reproduced from Jaca, R. C., Godoy, L. A., Calabro, H. D., & Espinosa, S. N. (2021). Thermal post-buckling behavior of oil storage tanks under a nearby fire.* International Journal of Pressure Vessels and Piping, 189, *104289.*

horizontal segments in the equilibrium path, with a significant increase in displacements for a small increase in temperature. The final jump was computed at $T = 418°C$. The computation of the path, in this case, was terminated at $T^f = 420°C$ because the fuel considered (gasoline) cannot provide temperatures higher than that. Notice that establishing a limit to the temperatures that can be reached cannot be obtained if a simplified model is used.

The mode shape changes at each jump in the equilibrium path: These shapes are shown in Fig. 6.20. In the first case, $T^c = 248°C$, the buckling mode affects the lower part of the shell, and the bulges increase in size and extend upwards into two rows of bulges. The deflected mode of the shell in the last instability, $T = 418°C$, affects less than half of the shell in elevation.

A measure of the postbuckling capacity of the tank may be found as the ratio $T^f/T^c$ and in this case, the ratio is 1.69. This shows that following the buckling temperature $T^c$, the shell can still take another 70% of the increase in temperature, but this occurs at the cost of large radial displacements, exceeding 20 times the shell thickness in the final configuration. This level of displacements will most likely be unacceptable for the normal functioning of the tank and in the most favorable scenario, it should be taken out of production.

Finally, it is well-known that shell buckling is sensitive to geometric imperfections in cases of the pressure of wind effects. However, the work of (Godoy and Pantousa, 2021) showed that the thermal buckling of tanks is not sensitive to deviations in the geometry. This can also be seen in the

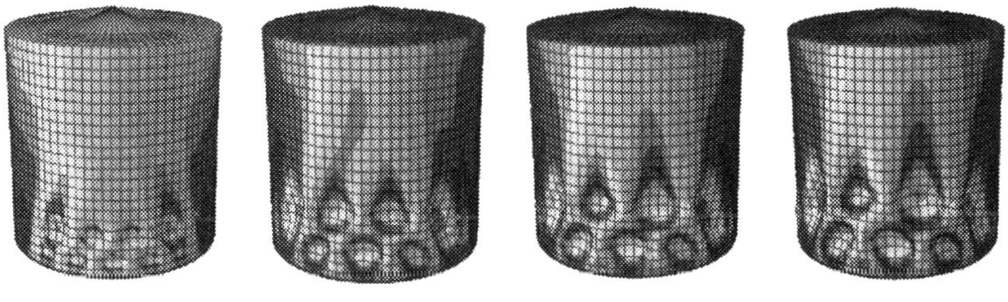

(A) (B) (C) (D)

**Figure 6.20** Modes of instability in the tank considered in Fig. 6.19. Radial displacements are plotted for (A) State at $T = 248°C$, (B) State at 282°C, (C) State at 383°C, (D) State at 418°C. *Reproduced from Jaca, R. C., Godoy, L. A., Calabro, H. D., & Espinosa, S. N. (2021). Thermal post-buckling behavior of oil storage tanks under a nearby fire.* International Journal of Pressure Vessels and Piping, 189, *104289.*

equilibrium path in Fig. 6.19, which is always ascending and no decrease in temperature capacity occurs at any stage. For this reason, imperfections in geometry are not reported here.

### 6.8.4 Other tank features that modify the structural response

Fixed conical roofs in large tanks are not self-supported and require the use of an additional supporting structure with radial beams and ring rafters, which are in turn supported by columns. This is a complex substructure because it should not obstruct the presence of an internal floating roof, in case there is one.

Modeling of the complete substructure as part of the tank has been done in a couple of cases to account for fire effects (Burgos et al., 2015, Godoy & Batista-Abreu, 2012). The structural equivalence between a tank modeled with its full substructure and an equivalent roof thickness was investigated by Burgos et al. (2015) based on the equivalence of the inertia of both cases. It was thus shown that a roof with an equivalent thickness was sufficient to represent the effects of the supporting structure on the thermal and mechanical properties of the tank. Finite element studies were reported to show that both, the simplified and the "real" structure, had the same critical temperatures and mode shapes at buckling. For an equivalent roof having a thickness equal to 2.25 times the thickness of the shell at the top, the critical temperature is 7% lower than in the more accurate model including the substructure. Larger differences are found if the equivalent thickness is increased to values of four or higher times the shell thickness. The recommendation has been the use of three times the value of the thickness in the cylinder shell of the real structure. This has been adopted by most researchers interested in the behavior of tanks under thermal effects or wind pressures.

Another common simplification found in the literature refers to open-top tanks with a wind girder. In this case, instead of representing the details of the girder, it is tempting to substitute it with a boundary condition at the top (with constraints on the radial displacements) or with an equivalent thickness at the top course. However, the results of Burgos et al. (2015) showed that the errors caused by the use of boundary constraints are not acceptable. On the other hand, an equivalent thickness of five times the thickness of the shell at the top in the zone of the wind girder was found to lead to good results in terms of critical temperatures and mode shapes.

Another simplification adopted by most researchers is to neglect the influence of a helicoidal ladder attached to the external part of the shell to access the roof (as specified, e.g., by API 650, 2021). It has been shown that such a ladder plays a role in tanks by strengthening the shell under wind; however, this strengthening effect is not present under uniform pressure (Shokrzadeh & Sohrabi, 2016). The influence of neglecting the effects due to stairways in tanks under a thermal field due to a nearby fire was investigated by Calabró et al. (2014) using LBA finite element modeling of the simplified and complete structure. The stairway was represented at various positions with respect to the most heated meridian. For a tank with a fixed roof, a 6% increase in critical temperature was found due to the stairway, with some changes in the eigenmode at buckling (inclined lobes of deformation were found in the zone under the stairway). Less noticeable effects were detected in open-top tanks with a wind girder.

### 6.8.5 Effect of multiple sources of fire

Pantousa (2018) investigated fire scenarios in which more than one burning tank and interest focuses on the response of a single target tank. Several fire scenarios of four tanks surrounding a target tank with a conical roof were considered, in which one or more source tanks were under pool fire, to investigate what factors influence the buckling of the target tank.

In this case, it was not possible to use a simplified thermal model and each flame was modeled as a solid flame with a heat transfer process to the target tank. This analysis includes the transient response to evaluate the thermal behavior in the time and temperature domains. Several scenarios were investigated, to explain the temperatures reaching the target tank based on the number of surrounding tanks that were burning. Parametric studies addressed the type of fuel (ethanol or gasoline), wind, the separation between tanks, and tank size.

The overall conclusion is that for several flames present in the scenario, the time at which the target tank fails increases. This effect may be related to having a wider zone of temperatures acting around the circumference; a related effect was studied by Godoy and Batista-Abreu (2012) for a single tank in which fire was assumed to cause temperatures on a larger central angle of the target tank. Upper bounds to critical temperatures are found for two and three burning tanks, and the critical temperature reduces for four burning tanks.

### 6.8.6 Domino effects under fire

The review of cases reported in the literature in Section 6.2 of this chapter shows that there are simple cases involving one target tank and complex cases involving many tanks burning from the same source, in what is known as a domino effect. A state of the art in this field has been published by Reniers and Cozzani (2013), and a review of historical cases may be found in Swustea et al. (2019).

The term "domino effect" refers to accidents that start at an initial event but propagate and cause an increase in the scale of the accident, with more destructive consequences. The main feature of this class of accidents is the propagation of fire in space to other tanks and buildings, and in time, by increasing the temperature as the accident develops. There is a primary scenario with an event, a propagation, and a second scenario in which several tanks may be burning. There are cases in which tanks that started burning in the second scenario propagate their effects to other tanks, leading to a tertiary scenario. Thus, a multilevel chain may occur in domino effects.

The unfolding of a domino effect requires knowledge of the causes of propagation from the primary event to secondary or multilevel events, and a scale vector is often used to model the process. Most primary events are due to a pool fire, with a scale vector given by heat transfer by radiation, followed by events involving a mechanical explosion or a vapor cloud explosion, in which scaling was due to overpressure. The induced damage should be sufficiently important to be considered as a propagation, and this effect depends on the separation between tanks and the number of tanks involved in the process. In a direct scaling, the damage is due to radiation, shock waves, and debris impact. An analysis of risk in domino effects was recently reported by Ji et al. (2018).

Several representative scenarios in tank farms were considered by Espinosa et al. (2019b), with different tank arrangements in plan and at various distances between them. As a primary event, a pool fire was assumed and the escalated vector was net incident radiation with a threshold value of 15 $kW/m^2$. Once the threshold is reached, it is assumed that the target tank starts a new fire. Heat transfer results (solved via finite element analysis) were used to obtain the radiation in each case. Wind increases the probability of having a domino effect due to the inclination of the flame.

## 6.9 Areas for further research

Two of the most dangerous loading conditions that one may have in oil storage tanks, from which oil spills can occur, are not taken into account by the current American Petroleum Institute regulations. This is in contrast with evidence from real accidents, in which it has been shown that fire and explosions are the most frequent causes of accidents in such tanks. The main features of the state of the art in both load cases were reviewed in this chapter, in an attempt to show what is known at present, and in this last section we will attempt to mention existing voids in the literature and what lines of research and development should perhaps be pursued in the coming years.

### 6.9.1 Tests on small-scale tanks under thermal loads

It was shown in this chapter that proper structural analysis of tanks requires using improved definitions of loads. Evidence from small-scale testing is very limited at present, and this is an area where it would be greatly advantageous to have more complete empirical information.

In only a couple of cases, there were research efforts to assess the effects due to fire in tank models, and in some cases, these tests had other variables in mind. In the case of fire, the equivalent data is the temperature distribution in the target tank from a single source of fire or several sources acting simultaneously.

Further, the tank response to temperatures leading to buckling has only been achieved via computational modeling, but empirical evidence is still lacking.

### 6.9.2 Tests on small-scale tanks under blast loads

In the case of explosions, testing of rigid shells should provide pressure intensities together with their space and time variation. This would help in refining the loads and should serve as a way to calibrate numerical models. Progress along these lines has been made during the last decade, and they are reported in this chapter.

Next, it is believed that the structural response of small-scale models of flexible shells under blast loads should also be investigated in laboratory conditions, to understand the sequence of buckling and plastic failure that is expected to occur in full-size, real situations. The available empirical results for flexible tanks (see, e.g., Duong et al., 2012b) were performed for specific values of blast pressures, with the consequence that it is not possible to evaluate the onset of buckling or failure. Thus, it would be important to have an experimental/computational program in which the intensity of the explosion was designed at values close to those at which there are expected changes in the structural response (either material or geometric critical states).

### 6.9.3 Modeling tanks under fire

Most results for advanced fire models at present have been obtained using sequential analysis, in which the temperature distribution is obtained in the first stage and the structural response is subsequently computed from them. An alternative approach would be to perform a coupled (multiphysics) analysis and this will surely be attempted with the improvement of computational resources.

The topics covered in this chapter address only part of what is known in this field. For example, the pressurization of tanks due to fire has been treated by Fouillen and Duplantier (2009). The coupling of thermal effects, as discussed in this chapter, with the impact of debris is the subject of Li et al. (2019). The presence of smoke coupled with thermal effects on the buckling of tanks was discussed by Pourkeramat et al. (2021). The influence of equipment attached to a tank on the thermal response has been investigated by Calabró et al. (2018). All these lines of research are very important and will surely develop further in the following years.

### 6.9.4 Modeling tanks under blast loads

Computational models have been used to represent shock waves due to an explosion, but their effects on tanks have not been investigated at present. This is an area of research that will be developed during the next few years. At present, there are several commercially available codes (known as hydro-codes) that have the capabilities of coupling an explosion and the structural effects. Comparison between such models with others or with simplified equations would be of great importance.

### 6.9.5 Design recommendations

The establishment of design recommendations is usually carried out by national committees or by professional organizations, but this field may advance using developments from the research area to improve our design possibilities.

The European Commission has led this area regarding buckling of shells (see Rotter & Schmidt, 2008), in which loading conditions due to wind and pressures were addressed for the design of shells. These efforts could be extended to deal with the loading situations discussed in this chapter, namely methodologies to consider the effects of thermal loads due to a nearby fire and blast pressures caused by explosions.

As part of this effort, a practical methodology of analysis based on what is known as the LBA/MNA has been reported (Rotter, 2011). This methodology is an intermediate approach between the GMNIA and the simplest form of buckling as given by LBA. Two conditions are investigated in the LBA/MNA: An elastic bifurcation, as given by a linearized eigenvalue analysis, and a condition of plasticity, as defined by a material nonlinear analysis in which the geometric behavior remains linear. Notice that imperfections in the geometry of a shell, which have been found to cause a reduction in the buckling capacity of the shell, are not part of the LBA/MNA approach.

An excellent discussion of this methodology, together with ways to make an efficient implementation and the challenges found in doing that, may be found in Rotter (2011). The present authors believe that the application of this criterion to the design and understanding of bounds for the thermal buckling of tanks is a promising avenue that would provide a less onerous and more practical approach than a fully nonlinear analysis, as computed via GMNIA.

### 6.9.6 Fragility and risk assessment

There has been a shift in emphasis since 1990 from evaluating the safety of isolated structures to considering the vulnerability of classes of structures and structural types in a geographical area. This is currently done by using fragility curves, which evaluate the probability of reaching a certain damage level and loss as a consequence of a given natural hazard. The use of fragility curves has been greatly increased thanks to their inclusion in US Federal Emergency Management Agency (FEMA) documents. Recent versions of Hazard software HAZUS (FEMA) allow estimating fragility and risk of structures (including tanks) under seismic loads, hurricanes, and flooding. Work along these lines has been published in recent years on the fragility of tanks (Kameshwar & Padgett, 2018a, 2018b). Another common application of this methodology is in forensic studies Bernier and Padgett (2018). However, such studies are not available at present for explosions or fire.

The type of analysis that is required in fragility studies may involve solving a large number of cases under different loads: For example, Kameshwar and Padgett (2018a) reported studies based on 1800 cases that were solved using finite element simulations. Because of the computational demand of such studies, simplifications are frequently used, either in the structural model or in the features of the behavior that are considered in the analysis. This way one may reduce the computational resources needed by formulating a simplified model of the structure (often by what is known as a surrogate model) and obtaining the solution under many different conditions. Hence, the importance of having simple equivalent models of tanks for both explosions and fire, which are not available at present.

This justifies our next current need: the development of a simple mechanical model with a behavior showing selected features of their equivalent finite element solutions. Such surrogate models may take the form of simple representations of tanks, such as by considering arches or ring elements; or by carefully chosen mechanical models with two or three degrees of freedom. Some efforts along this line have been made recently by Ameijeiras (2020) using simplified arch models to represent the shell response to blast loadings, and by Godoy and Pantousa (2022) for thermal loads.

## Acknowledgments

The authors thank the support of grants from the Science and Technology Research Council of Argentina (CONICET), and Universidad Nacional del Comahue in Argentina. Thanks are due to a number of researchers who contributed to various aspects reported in this chapter, including Jean C. Batista-Abreu, Carlos A. Burgos, Horacio *D.* Calabro, Susana N. Espinosa, Fernando G. Flores, Daphne Pantousa, Juan C. Virella, David C. Weggel, and Matthew J. Whelan. The authors greatly benefited from communications with Profs. James G. A. Croll, Jean L. Hanus, Alexandre Landesmann, Bibiana Luccioni, Chrysanthos Maraveas, Jamie *E.* Padgett, J. Michael Rotter, Luis *E.* Suárez, José L. Torero, and Alphose Zingoni.

The authors thank several institutions for granting permission to reproduce figures in this chapter: (1) Elsevier for Figs. 6.2B, 6.3, 6.14–6.17, 6.19, 6.20; (2) American Society for Civil Engineers for Fig. 6.2A; (3) US Chemical Safety Board for Fig. 6.1.

## Nomenclature

| | |
|---|---|
| $\mathbf{c^P}_n$ | Fourier coefficients of pressures due to an explosion |
| $D$ | diameter of a tank |
| $E$ | modulus of elasticity of steel |

| | |
|---|---|
| $E_f$ | emissive power of fire |
| $E_t$ | incident radiation in the target tank |
| $F$ | view factor |
| $H$ | height of the cylindrical shell of a tank |
| $h$ | thickness of the cylindrical shell |
| $\mathbf{h_r}$ | thickness of the roof |
| $\mathbf{k_1}$ | pressure decay coefficient around the circumference |
| $\mathbf{k_2}$ | pressure decay coefficient with time |
| **k-ε** | turbulence model |
| $p_0$ | peak reflected pressure due to an explosion |
| $\mathbf{p_r}$ | reflected pressure due to an explosion |
| **R** | radius of the cylindrical shell |
| $\overline{R}$ | stand-off distance, from explosion to tank wall |
| **T** | temperature |
| $\mathbf{T_{0a}}$ | ambient temperature |
| $T_{0m}$ | maximum temperature at the most heated meridian |
| $T^c$ | critical temperature |
| $T^f$ | maximum temperature that can be reached with a given fuel |
| $t$ | time |
| $t_0$ | positive duration of blast pressure |
| $t_a$ | arrival time of a shock wave |
| $t_f$ | finishing time of a shock wave |
| $t_n$ | fundamental period of a tank |
| $u_r$ | radial displacement |
| **W** | mass of equivalent TNT charge of an explosion |
| $Z$ | modulus of the wind girder cross-section |
| $\overline{Z}$ | normalized distance from explosion to tank |
| $\mathbf{Z^{Batdorf}}$ | Batdorf parameter |
| **z** | vertical coordinate measured with respect to ground level |
| $\alpha_t$ | heat radiation absorption on a metal surface |
| $\varepsilon_f$ | fire emission |
| $\nu$ | Poisson's ratio |
| $\theta$ | circumferential angular coordinate |
| $\theta_0$ | angular coordinates at which temperatures vanish around the circumference |
| ρ | density of steel |
| $\sigma_y$ | yield stress |
| $\tau$ | atmospheric transmissivity |
| ω | parameter in the Fourier analysis of blast pressures |

## Acronyms

| | |
|---|---|
| **ADM** | Artificial Damping Method |
| **CFD** | Computational Fluid Dynamics |
| **FEMA** | Federal Emergency Management Agency (United States) |

**GMNIA** Geometrically and Material Nonlinear Analysis with Imperfections
**GNIA** Geometrically Nonlinear Analysis with Imperfections
**HAZUS** Hazard FEMA software
**LBA** Linear Bifurcation Analysis

## References

ABAQUS (2021), *Simulia: Unified FEA. Dassault systems*. Warwick, RI.

Agboola, O. O., Akinnuli, B. O., Kareem, B., & Akintunde, M. A. (2021). Optimum detailed design of 13,000m$^3$ oil storage tanks using 0.8 height-diameter ratio. *Materials Today: Proceedings, 44*, 2837–2842.

Alijani, A., Darvizeh, M., Darvizeh, A., & Ansari, R. (2015). On nonlinear thermal buckling analysis of cylindrical shells. *Thin-Walled Structures, 95*, 170–182.

Ameijeiras, M. P. (2020). *Análisis de tanques de almacenamiento de petróleo bajo la acción de cargas debidas a ondas explosivas* (PhD tesis). Argentina: Universidad Nacional de Córdoba (In Spanish).

Ameijeiras, M. P., & Godoy, L. A. (2021a). Quasi-bifurcation and imperfection-sensitivity of cylindrical shells under pressures due to an explosion. *Journal of Applied and Computational Mechanics*, 7(2), 984–992.

Ameijeiras, M. P., & Godoy, L. A. (2021b). Quasi-bifurcation of discrete systems with unstable post-critical behavior under impulsive loads. In N. Challamel, J. Kaplunov, & I. Takewaki (Eds.), *Modern trends in structural and solid mechanics* (1, pp. 159–175). ISTE-Wiley.

Ameijeiras, M. P., Godoy, L. A., Weggel, D., & Whelan, M. J. (2014). Incidencia del tiempo de arribo de onda en la respuesta de tanques sometidos a explosiones externas. *Mecánica Computacional, 33*, 931–942. (In Spanish).

ANSYS. (2021). Structural analysis software. Canonsburg, PA: Ansys Inc.

API 650. (2021). *Welded steel tanks for oil storage*. Washington, DC: American Petroleum Institute.

Assael, M. J., & Kakosimos, K. E. (2010). *Fires, explosions and toxic gas dispersions: Effects calculation and risk analysis*. Boca Raton, FL: CRC Press.

Atkinson, G. (2011). Blast damage to storage tanks and steel-clad buildings. *Process Safety and Environmental Protection, 89*, 382–390.

Baker, W. E., Kulesz, J. J., Ricker, R. E., Bessey, R. L., Westine, P. S., Parr, V. B., & Oldham, G. A. (1977). *Workbook for predicting pressure wave and fragment effects of exploding propellant tanks and gas storage vessels*. Washington, DC: NASA CR-134906.

Batdorf, S.B. (1947). A simplified method of elastic-stability analysis for thin cylindrical shells. In *NACA technical note 874*. Washington, DC: National Advisory Committee for Aeronautics.

Batista-Abreu, J. C., & Godoy, L. A. (2013). Thermal buckling behavior of open cylindrical oil storage tanks under fire. *ASCE J. Performance of Constructed Facilities, 27*(1), 89–97.

Batista-Abreu, J., & Godoy, L. A. (2011). Investigación de causas de explosiones en una planta de almacenamiento de combustible en Puerto Rico, Revista Int. de Desastres Naturales. *Accidentes e Infraestructura Civil, 11*(2), 109–122. (in Spanish).

Bernier, C., & Padgett, J. E. (2018). Forensic investigation of aboveground storage tank failures during Hurricane Harvey using fragility models. In *Proceedings of the eighth congress on forensic engineering*. Austin, TX: American Society of Civil Engineers.

Beyler, C. L. (2002). *Fire hazard calculations for large, open hydrocarbon fires. SFPE handbook of fire protection engineering* (3rd ed., pp. 3.268–3.314). Quincy, MA: National Fire Protection Association.

Bhagat, V., & Jeyaraj, P. (2018). Experimental investigation on buckling strength of cylindrical panel: Effect of non-uniform temperature field. *Int. Journal of Non-Linear Mechanics, 99*, 247–257.

Budiansky, B. (1967). Dynamic buckling of elastic structures: criteria and estimates. In G. Herrmann (Ed.), *Dynamic stability of structures* (pp. 83–106). Pergamon Press.

Budiansky, B., & Roth, R. S. (1962). Axisymmetric dynamic buckling of clamped shallow spherical shells. In *Collected papers on instability of shell structures* (pp. 597–606). Washington, DC: NASA TN D-1510.

Buncefield Major Incident Investigation Board. (2008). *The Buncefield incident 11 December 2005. The final report of the major incident investigation board.* (Vol. 2). London: Health and Safety Executive.

Buratti, N., & Tavano, M. (2014). Dynamic buckling and seismic fragility of anchored steel tanks by the added mass method. *Earthquake Engineering & Structural Dynamics, 43*(1), 1–21.

Burgan, B. A. (1995). Containment of explosions and fires. In B. Simpson (Ed.), Containment structures: Risk, safety and reliability (pp. 155–170). London: E & FN Spon.

Burgos, C. A., Batista-Abreu, J. C., Calabró, H. D., Jaca, R. C., & Godoy, L. A. (2015). Buckling estimates for oil storage tanks: Effect of simplified modeling of the roof and wind girder. *Thin-Walled Structures, 91*, 29–37.

Burmeister, A., & Ramm, E. (1990). Dynamic stability analysis of shell structures. In W. B. Kratzig, & E. Oñate (Eds.), *Computational mechanics of nonlinear response of shells* (pp. 152–163). Berlin: Springer.

Calabró, H. D., Godoy, L. A., & Jaca, R. C. (2014). Inestabilidad de tanques de almacenamiento de petróleo con techo cónico debido a fuego. *Rev. Sudamericana de Ingeniería Estructural, 11*(1), 7–28. (in Spanish).

Calabró, H. D., Godoy, L. A., & Jaca, R. C. (2018). Efecto de Escaleras Perimetrales en tanques de almacenamiento de hidrocarburo frente a acciones térmicas. *Mecánica Computacional, 36*, 327. (in Spanish).

Casal, J. (2018). *Evaluation of the effects and consequences of major accidents in industrial plants* (2nd ed.). Elsevier.

Chang, J. I., & Lin, C. C. (2006). A study of storage tank accidents. *Journal of Loss Prevention in the Process Industries, 19*, 51–59.

Considine, M. (1984). Thermal radiation hazard ranges from large hydrocarbon pool fires. In *Safety and Reliability Directorate*. Oxfordshire: UK Atomic Energy Authority.

Cozzani, V., & Salzano, E. (2004). Threshold values for domino effects caused by blast wave interaction with process equipment. *Journal of Loss Prevention in the Process Industries, 17*, 437–447.

CSB (2015). *Caribbean petroleum tank terminal explosion and multiple tank fires, final investigation report.* Washington, DC: U.S. Chemical Safety and Hazard Investigation Board.

Da Silva Santos, F., & Landesmann, A. (2014). Thermal performance-based analysis of minimum safe distances between fuel storage tanks exposed to fire. *Fire Safety Journal, 69*, 57–68.

Drysdale, D. (2011). *An introduction to fire dynamics* (3rd ed.). Chichester: Wiley.

Duong, D. H., Hanus, J. L., Bouazaoui, L., Pennetier, O., Moriceau, J., Prod'homme, G., & Reimeringer, M. (2012a). Response of a tank under blast loading—part I: experimental characterization. *European Journal of Environmental and Civil Engineering, 16*(9), 1023–1041.

Duong, D. H., Hanus, J. L., Bouazaoui, L., Regal, X., Prod'homme, G., Noret, E., Yalamas, T., Reimeringer, M., Bailly, P., & Pennetier, O. (2012b). Response of a tank under blast loading – Part II: Experimental structural response and simplified analytical approach. *European Journal of Environmental and Civil Engineering, 16*(9), 1042–1057.

Espinosa, S. N., Jaca, R. C., & Godoy, L. A. (2019a). Thermal effects of fire on a nearby fuel storage tank. *Journal of Loss Prevention in the Process Industry, 62*(1–12), 103990.

Espinosa, S. N., Jaca, R. C., & Godoy, L. A. (2019b). Propagación de fuego en una batería de tanques de almacenamiento de combustible. *Mecánica Computacional, 37*, 169–178. (in Spanish).

Espinosa, S. N., Jaca, R. C., & Godoy, L. A. (2017). Thermal and structural analysis of a fuel storage tank under an adjacent pool fire. *Fire Research, 2*(1), 31–36.

Fingas, M. (2013). *The basics of oil spill cleanup*. Boca Raton, FL: CRC Press.

Eurocode 3 (2003). Design of steel structures-part 1-2: general rules-structural fire design. In *European Committee for Standardization EN 1993-1-2*. Bruxelles, Belgium.

Fouillen, F., & Duplantier, S. (2009). Pressurization of fixed roof storage tanks due to external fires. *AIChE Spring national meeting*. Tampa, FL.

Gelfand, B. E., Silnikov, M. V., Medvedev, S. P., & Khomik, S. V. (2012). *Thermo-gas dynamics of hydrogen combustion and explosion*. Springer.

Godoy, L. A. (2016). Buckling of vertical oil storage steel tanks: review of static buckling studies. *Thin-Walled Structures, 103*, 1–21.

Godoy, L. A., & Batista-Abreu, J. C. (2012). Buckling of fixed roof aboveground oil storage tanks under heat induced by an external fire. *Thin-Walled Structures, 52*, 90–101.

Godoy, L. A., & Pantousa, D. (2022). A simplified model for the thermal buckling of oil storage tanks. *International Journal of Structural Engineering, 12*(2), 125–143.

Hu, K., & Zhao, Y. (2016). Numerical simulation of internal gaseous explosion loading in large-scale cylindrical tanks with fixed roof. *Thin-Walled Structures, 105*, 16–28.

Islam, M. N., Kormi, K., & Al-Hassani, S. T. S. (1992). Dynamic response of a thin-walled cylinder to side pressure pulse. *Engineering Structures, 14*(6), 395–412.

Jaca, R. C., Godoy, L. A., Calabro, H. D., & Espinosa, S. N. (2021). Thermal post-buckling behavior of oil storage tanks under a nearby fire. *International Journal of Pressure Vessels and Piping, 189*, 104289.

Ji, J., Tong, Q., Khan, F., Dadashzadeh, M., & Abbassi, R. (2018). Risk-based domino effect analysis for fire and explosion accidents considering uncertainty in processing facilities. *Industrial & Engineering Chemistry Research., 57*(11), 3990–4006.

Kameshwar, S., & Padgett, J. E. (2018a). Fragility and resilience indicators for portfolio of oil storage tanks subjected to hurricanes. *ASCE Journal of Infrastructure Systems, 24*(2), 04018003.

Kameshwar, S., & Padgett, J. E. (2018b). Storm surge fragility assessment of above ground storage tanks. *Structural Safety, 70*, 48–58.

Kanarachos, A. E., & Spentzas, C. N. (1988). An artificial damping method for the determination of the steady state of harmonically excited non-linear systems. *Journal of Sound and Vibration, 120*(3), 597–608.

Khan, F. I., & Abbasi, S. A. (2001). An assessment of the likelihood of occurrence, and the damage potential of domino effect (chain of accidents) in a typical cluster of industries. *Journal of Loss Prevention in the Process Industries, 14*, 283–306.

Kleiber, M., Kotula, W., & Saran, M. (1987). Numerical analysis of dynamic quasi-bifurcation. *Engineering Computations, 4*, 48–52.

Kobayashi, T., Mihara, Y., & Fujii, F. (2014). Artificial damping method for local instability problems in shells. In G. Pietraszkiewicz (Ed.), *Shell structures: Theory and applications*. London: Taylor and Francis.

Krol, P. A., & Joswik, R. (2021). Design and modeling of storage tanks exposed to thermal actions. In P. Akimov, & N. Vatim (Eds.), *Lecture notes in civil engineering* (189, pp. 53–62). Springer.

Kroplin, B., & Dinkler, D. (1986). Dynamic vs static buckling analysis of thin-walled shell structures. In T. Hughes, & E. Hinton (Eds.), *Finite element methods for plate and shell structures* (2, pp. 239–251). Swansea: Pineridge Press.

Kubiak, T. (2013). *Static and dynamic buckling of thin-walled plate structures*. Springer Int.

Landucci, G., Gubinellia, G., Antonioni, G., & Cozzani, V. (2009). The assessment of the damage probability of storage tanks in domino events triggered by fire. *Accident Analysis and Prevention, 41*, 1206–1215.

Lee, L. H. N. (1981). On dynamic stability and quasi-bifurcation. *International Journal of Nonlinear Mechanics, 16*(1), 79–87.

Li, Y., Jiang, J., Bian, H., Yu, Y., Zhang, Q., & Wang, Z. (2019). Coupling effects of the fragment impact and adjacent pool-fire on the thermal buckling of a fixed-roof tank. *Thin-Walled Structures, 144*, 106309.

Lindberg, H. E. (1964). Buckling analysis of very thin cylindrical shell due to an impulsive pressure. *Journal of Applied Mechanics*, 267–272.

Lindberg, H. E. (1987). *Dynamic pulse buckling. Theory and experiment.* Dordrech, The Netherlands: Martinus Nijhoff Publishers.

Liu, Y. (2011), *Thermal buckling of metal oil tanks subject to an adjacent fire* (PhD thesis). UK: The University of Edinburgh.

Liu, Y., Chen, J. F., Rotter, J. M., Torero, J. T., & Ai, J. (2012). Thermal buckling of metal oil tanks subject to an adjacent fire. In Proceedings of the first international conference on performance-based and life-cycle structural engineering, (pp. 156–165). Hong Kong: The Hong: Kong Polytechnic University.

Lu, S., Wang, W., Chen, W., Ma, J., Shi, Y., & Xu, C. (2019). Behavior of thin-walled cylindrical shell storage tank under blast impacts. *Shock and Vibration*, 6515462.

MacMillan, D. (2021). Despite spills and air pollution, fossil fuel companies award CEOs for environmental records. *The Washington Post*, October 10th.

Mansour, K. A. (2012). *Fires in large atmospheric storage tanks and their effect on adjacent tanks* (PhD thesis). Leicestershire, UK: Loughborough University.

Maraveas, C. (2014). Thermal buckling analysis of thin-walled steel oil tanks exposed to an adjacent fire. In S. T. Smith (Ed.), Proc. 23rd Australasian conference on the mechanics of structures and materials (pp. 9–12). Bay, Australia: Byron.

MARC (2021). *MSC Software*. Irvine, CA.

McGrattan, K. B., Baum, H. R., & Hamins, A. (2000). *Thermal radiation from large pool fires. NISTIR 6546*. Gaithersburg, MD: National Institute of Standards and Technology.

Mingguang, Z., & Juncheng, J. (2008). An improved probit method for assessment of domino effect to chemical process equipment caused by overpressure. *Journal of Hazardous Materials, 158*, 280–286.

Mishra, K. B., Wehrstedt, K. D., & Krebs, H. (2014). Amuay refinery disaster: The aftermaths and challenges ahead. *Fuel Processing Technology, 119*, 198–203.

Myers, P. (1997). *Aboveground storage tanks*. McGraw Hill.

NFPA 30. (2012). *Flammable and combustible liquids code*. Quincy, MA: National Fire Protection Association.

Ngo, T., Mendis, P., Gupta, A., & Ramsay, J. (2007). Blast loading and blast effects on structures—an overview, electronic journal of structural engineering. *Special Issue: Loading on Structures*, 76–91.

Nolan, D. (2014). *Handbook of fire and explosion protection engineering principles*. Elsevier.

Noret, E., Prod'homme, G., Yalamas, T., Reimeringer, M., Hanus, J. L., D.-H., & Duongb, D. H. (2012). Safety of atmospheric storage tanks during accidental explosions. *European Journal of Environmental and Civil Engineering, 16*(9), 998–1022.

Pantousa, D. (2018). Numerical study on thermal buckling of empty thin-walled steel tanks under multiple pool-fire scenarios. *Thin-Walled Structures, 131*, 577–594.

Pantousa, D., & Godoy, L. A. (2019). On the mechanics of thermal buckling of oil storage tanks. *Thin-Walled Structures, 145*, 106432.

Pantousa, D., Tzaros, K., & Kefaki, M. A. (2018). Thermal buckling behavior of unstiffened and stiffened fixed-roof tanks under non-uniform heating. *Journal of Constructional Steel Research, 143*, 162–179.

Persson, H., & Lönnermark, A. (2004). *Tank fires: Review of fire incidents 1951–2003*. Swedish National Testing and Research Institute.

Planas-Cuchi, E., Vílchez, J. A., & Casal, J. (1999). Fire and explosion hazards during filling/emptying of tanks. *Journal of Loss Prevention in the Process Industries*, *12*(6), 479–483.

Pourkeramat, A., Daneshmehr, A., & Aminfar, K. (2020). Investigations on the influence of aspect ratio on thermal buckling response of large-scale tanks. In *Proceedings of the third international conference on civil engineering, structures and earthquakes*. Tehran, Iran.

Pourkeramat, A., Daneshmehr, A., Jalili, S., & Aminfar, K. (2021). Investigation of wind and smoke concentration effects on thermal instability of cylindrical tanks with fixed roof subjected to an adjacent fire. *Thin-Walled Structures*, *160*, 107384.

Pritchard, M. J., & Binding, T. M. (1992). FIRE2: A new approach for predicting thermal radiation levels from hydrocarbon pool fires. *IChemE Symposium*, *130*, 491–505.

Putelat, T., & Triantafyllidis, N. (2014). Dynamic stability of externally pressurized elastic rings subjected to high rates of loading. *International Journal of Solids and Structures*, *51*, 1–12.

Reniers, G., & Cozzani, V. (2013). *Domino effects in the process industries: Modelling, prevention and managing*. Oxford: Elsevier.

Rew, P. J., Hulbert, W. G., & Deaves, D. M. (1997). Modelling of thermal radiation from external hydrocarbon pool fires. *Trans IChemE*, *75*, 81–89.

Riazi, M. R. (2021). *Oil spill occurrence, simulation, and behavior*. Boca Raton, FL: CRC Press.

Riks, E. (1979). An incremental approach to the solution of snapping and buckling problems. *International Journal of Solids and Structures*, *15*, 529–551.

Rotter, J. M. (2011). Shell buckling design and assessment and the LBA-MNA methodology. *Stahlbau*, *80*, 791–803.

Buckling of steel shells: European design recommendations In J. M. Rotter, & H. Schmidt (Eds.), *European convention for constructional steelwork* (5th ed.). Lisbon, Portugal: Mem Martins.

Ruis, C., Salvatorelli, F., & Thompson, V. K. (1989). Elastic response of thin-walled cylindrical vessels to blast loading. *Computers and Structures*, *32*(5), 1061–1072.

Salzano, E., & Cozzani, V. (2005). The analysis of domino accidents triggered by vapor cloud explosions. *Reliability Engineering and System Safety*, *90*, 271–284.

Salzano, E., Hoorelbeke, P., Khan, F., & Amyotte, P. (2013). Overpressure effects. In G. Reniers, & V. Cozzani (Eds.), *Domino effects in the process industries* (pp. 43–69). Elsevier.

Sengupta, A., Gupta, A. K., & Mishra, I. M. (2011). Engineering layout of fuel tanks in a tank farm. *Journal of Loss Prevention in the Process Industries*, *24*, 568–574.

Shokrzadeh, A. R., & Sohrabi, M. R. (2016). Strengthening effects of spiral stairway on the buckling behavior of metal tanks under wind and vacuum pressures. *Thin-Walled Structures*, *106*, 437–447.

Simitses, G. J. (1990). *Dynamic stability of suddenly loaded structures*. New York, NY: Springer-Verlag.

Steinhaus, T., Welch, S., Carvel, R. O., & Torero, J. L. (2007). Large-scale pool fires. *Thermal Science*, *11*, 101–118.

Swustea, P., van Nunen, K., Reniers, G., & Khakzad, N. (2019). Domino effects in chemical factories and clusters: An historical perspective and discussion. *Process Safety and Environmental Protection*, *124*, 18–30.

Taveau, J. (2012a). Explosion of fixed roof atmospheric storage tanks, Part 1: Background and review of case histories. *Process Safety Progress*, *30*(4), 381–392.

Taveau, J. (2012b). Explosion of fixed roof atmospheric storage tanks, Part 2: Analytical methods to assess safety distances. *Process Safety Progress*, *31*(1), 72–82.

Taveau, J. (2012c). The Buncefield explosion: Were the resulting overpressures really unforeseeable? *Process Safety Progress*, *31*(1), 55–71.

US Army Corps of Engineers (2008). *Unified facilities criteria, structures to resist the effects of accidental explosions*. UFC 3-340-02.

Vilchez, J. A., Sevilla, S., Montiel, H., & Casal, J. (1995). Historical analysis of accidents in chemical plants and in the transportation of hazardous materials. *Journal of Loss Prevention in the Process Industries*, *8*(2), 87–96.

Virella, J. C., Godoy, L. A., & Suárez, L. E. (2006). Dynamic buckling of anchored steel tanks subjected to horizontal earthquake excitation. *Journal of Constructional Steel Research*, *62*(6), 521–531.

Weggel, D., & Whelan, M. J. (2013). Rigid tank testing summary and procedures for estimating blast overpressure distribution on a cylindrical tank surface. *ISSERT Report*. Charlotte, NC: University of North Carolina.

Wu, Z., Hou, L., Wu, S., Wu, X., & Liu, F. (2020). The time-to-failure assessment of large crude oil storage tank exposed to pool fire. *Fire Safety Journal*, *117*, 103192.

Zhou, F. (2019). Numerical simulation of large crude oil storage tank fire under various wind speeds. In *Proceedings of the third international conference fluid mechanics and industrial applications* (pp. 1–13). IOP Journal of Physics: Conf. Series 1300, 012003.

Zingoni, A. (2015). Liquid-containment shells of revolution: A review of recent studies on strength, stability and dynamics. *Thin-Walled Structures*, *87*, 102–114.

## Appendix 6.1: Summary of critical temperatures for tanks with a conical roof

A summary of critical temperatures computed via LBA or GNIA is presented in Table 6.A1. The results were collected from various sources

**Table 6.A1** Tanks with a fixed roof: critical temperatures obtained from the literature.

| *D* [m] | *H/D* | *R/h* | Fuel level *z/H* | Tank separation *s/D* | Wind speed [km/h] | LBA $T^c$ [C] $h_r$ <10 h | LBA $T^c$ [C] $h_r$>10 h | References |
|---|---|---|---|---|---|---|---|---|
| 20 | 1.0 | 1000 | 0 | N/A | N/A | 128 | | Liu (2011) |
| 20 | 1.0 | 1000 | 0.25 | N/A | N/A | 160 | | |
| 20 | 1.0 | 1000 | 0.40 | N/A | N/A | 410 | | |
| 20 | 1.0 | 1000 | 0.50 | N/A | N/A | 560 | | |
| 20 | 1.0 | 1000 | 0.60 | N/A | N/A | 675 | | |
| 14.64 | 0.83 | 1153 | 0 | N/A | N/A | 128 | | Godoy and Pantousa (2021) |
| 14.64 | 0.83 | 1153 | 0.125 | N/A | N/A | 151 | | |
| 14.64 | 0.83 | 1153 | 0.25 | N/A | N/A | 188 | | |
| 14.64 | 0.83 | 1153 | 0.375 | N/A | N/A | 261 | | |

(*Continued*)

**Table 6.A1** (Continued)

| *D* [m] | *H/D* | *R/h* | Fuel level *z/H* | Tank separation *s/D* | Wind speed [km/h] | LBA $T^c$ [°C] $h_r$ <10 h | LBA $T^c$ [°C] $h_r$>10 h | References |
|---|---|---|---|---|---|---|---|---|
| 14.64 | 0.83 | 1153 | 0.5 | N/A | N/A | 491 | | |
| 30.48 | 0.4 | 1200 | 0 | N/A | N/A | 230 | | |
| 25.92 | 0.5 | 2040 | 0 | N/A | N/A | 67.7 | | Calabró et al. (2014) |
| 40 | 0.5 | | 0 | N/A | N/A | 46.1 | | Pantousa et al. (2018) |
| 20 | 0.5 | | 0 | N/A | N/A | 141.2 | | |
| 20 | 1 | | 0 | N/A | N/A | 112.2 | | |
| 20 | 2 | | 0 | N/A | N/A | 99.8 | | |
| 15 | 1.0 | 750 | 0 | N/A | N/A | 188.3 | | Maraveas (2014) |
| 15 | 1.0 | 750 | 0.50 | N/A | N/A | 446.1 | | |
| 20 | 1.0 | 1500 | 0 | N/A | N/A | 80.1 | 79.6 | Pantousa and Godoy (2019) |
| 20 | 1.0 | 1000 | 0 | N/A | N/A | 130.0 | 135.8 | |
| 20 | 1.0 | 750 | 0 | N/A | N/A | 186.6 | 200.9 | |
| 15 | 0.8 | 1500 | 0 | N/A | N/A | 86.2 | 81.7 | |
| 15 | 0.8 | 1000 | 0 | N/A | N/A | 140.7 | 139.7 | |
| 15 | 0.8 | 750 | 0 | N/A | N/A | 199.8 | 207.6 | |
| 20 | 0.5 | 1500 | 0 | N/A | N/A | 122.8 | 102.4 | |
| 20 | 0.5 | 1000 | 0 | N/A | N/A | 194.5 | 172.5 | |
| 20 | 0.5 | 750 | 0 | N/A | N/A | 271.3 | 254.5 | |
| 30 | 0.33 | 1500 | 0 | N/A | N/A | 217.3 | 145.7 | |
| 30 | 0.33 | 1000 | 0 | N/A | N/A | 332.9 | 244.8 | |
| 30 | 0.33 | 750 | 0 | N/A | N/A | 455.8 | 360.8 | |
| 11.44 | 1.0 | 890 | 0 | 1.0 | N/A | 248 | | Jaca et al. (2021) |
| 11.44 | 1.0 | 890 | 0 | 1.0 | 8.3 | 257 | | |
| 11.44 | 1.0 | 890 | 0 | 1.0 | 12.6 | 251 | | |
| 11.44 | 1.0 | 890 | 0 | 1.0 | 25.2 | 231 | | |
| 11.44 | 1.0 | 890 | 0 | 1.0 | 45.0 | 253 | | |
| 11.44 | 1.0 | 890 | 0 | 0.33 | N/A | 217 | | |
| 11.44 | 1.0 | 890 | 0 | 1.0 | N/A | 248 | | |
| 11.44 | 1.0 | 890 | 0 | 1.5 | N/A | 269 | | |
| 11.44 | 1.0 | 890 | 0 | 2.0 | N/A | 232 | | |
| 11.44 | 1.0 | 890 | 0.5 | 1.0 | N/A | 603 | | |
| 11.44 | 1.0 | 890 | 0.94 | 1.0 | N/A | 912 | | |
| 20 | 0.5 | | 0 | N/A | N/A | 215 | | Pourkeramat et al. (2020) |
| 20 | 1.0 | | 0 | N/A | N/A | 161 | | |
| 10 | 2.0 | | 0 | N/A | N/A | 283 | | |

and were computed for tanks with a fixed conical roof. Because of the variety of sources and methods used by each author, some inconsistencies between results may be found. This information is the basis of the plot shown in Fig. 6.18.

The first three columns contain basic information about the geometry of the shell investigated; column 4 contains the fuel level; column 5 refers to the wall-to-wall separation between the source tank and the target tank, measured as a fraction of the diameter of the target tank. Wind speed is specified in column 6. Notice that values in columns 5 and 6 can only be obtained if an analysis of the flame and a heat transfer are carried out. The assumed thickness of the roof is denoted by $h_r$.

PART 2

# Case histories

CHAPTER SEVEN

# The Ashland oil spill

**John Joeckel**
Seaconsult Inc., VA, United States

## 7.1 Incident summary

- This is one of the worst inland oil spills in the nation. According to Coast Guard statistics, as reported by the New York Times, only a 14-million-gallon spill into the Delaware River in 1975 and a 2-million-gallon spill after an explosion in Brooklyn the next year involved larger quantities.[1]
- By Monday evening, January 4th, 1988, the Pittsburgh oil spill was among the top stories on the NBC Evening News with Tom Brokaw. The network's Cassandra Clayton reported from the Monongahela River showing emergency crews at work with attempted cleanup efforts and noting that Pennsylvania Governor Robert P. Casey had declared the region a disaster area. Some residents appeared on camera expressing concern about drinking water safety, and Pittsburgh's public safety director, Glenn Cannon, commented that cleanup costs would be substantial. Lt. Gov. Mark S. Singel visited evacuees the day in a shelter set up at a local high school.[2]
- The US Coast Guard closed the Monongahela River to vessel traffic between the Ashland facility in West Elizabeth and Pittsburgh. The smell of diesel fuel could be detected as far north as Pittsburgh. Rail and motor vehicle traffic was halted along some routes near the river due to concerns about human health and fire hazards.[3]
- The known effects of the introduction of diesel fuel into the environment include the death of at least 11,000 fish and 2000 birds, and the contamination of dozens of miles of shoreline.[4]

[1] Pittsburgh Post-Gazette, The 1988 Monongahela Oil Spill, April 2014

[2] Jack Doyle, "Disaster at Pittsburgh—1988 Oil Tank Collapse," PopHistoryDig.com, April 6, 2015.

[3] Jack Doyle, "Disaster at Pittsburgh—1988 Oil Tank Collapse," PopHistoryDig.com, April 6, 2015

[4] Jack Doyle, "Disaster at Pittsburgh—1988 Oil Tank Collapse," PopHistoryDig.com, April 6, 2015

*Above Ground Storage Tank Oil Spills.*
DOI: https://doi.org/10.1016/B978-0-323-85728-4.00003-6

- There was no established protocol for wildlife response for this event. Nonetheless, the Pennsylvania Game Commission responded quickly, calling Tri-State Bird Rescue & Research, Inc., less than 48 h after the spill, and providing all possible support. Funds were made available to outfit an excellent facility. Ashland Oil accepted responsibility for and promptly reimbursed all expenses. Of the birds delivered to the facility, 94% were released. This rate would be commendable in any spill; given the harsh weather, lack of an established facility, and inexperience of the volunteers, the success of this effort was exceptional.[5]
- Studies conducted after the crisis did not account for the emotional anguish of citizens, personal inconveniences, unknown health risks, or the economic losses on commercial activities.
- 1 million people were impacted, primarily drinking water system disruptions since the drinking water supply from the river system had become contaminated by the spill.
- Water shortages occurred for areas as far as 200 miles downriver. By the time the spill passed Cincinnati, oil levels in the Ohio River dropped to the point where immediate concern regarding drinking water had subsided.
- Approximately 20% of spilled oil that entered the waterway was recovered.
- Over 20,000 feet of boom, more than 200 on-site responders, and 600 miles of river impacted.
- Ground Water Contamination at the facility site, recovery wells drilled and activated.
- On the first night following the spill, some 242 families—about 1200 people—were evacuated from their homes near the Ashland facility. Firefighters from the Floreffe Volunteer Fire Co. went door-to-door telling people they had to leave their homes as a precaution for safety reasons due to fear of an explosion in the area from a gasoline leak. During the tank collapse, some debris from the spill had punctured a gasoline line at another big million-gallon tank, which had then leaked some 20,000 gallons of gasoline. A mixture of diesel and gasoline fumes created a danger of an explosion, which was the primary reason for the evacuation.
- $32.5 million spill costs: $14 million to settle more than 5000 third-party claims as well as damages, expenses; $11 million for cleanup;

[5] Establishing a Wildlife Response After the Ashland Oil Spill, International Oil Spill Conference, Frink & Dalton, Tri-State Bird & Recue, 1989

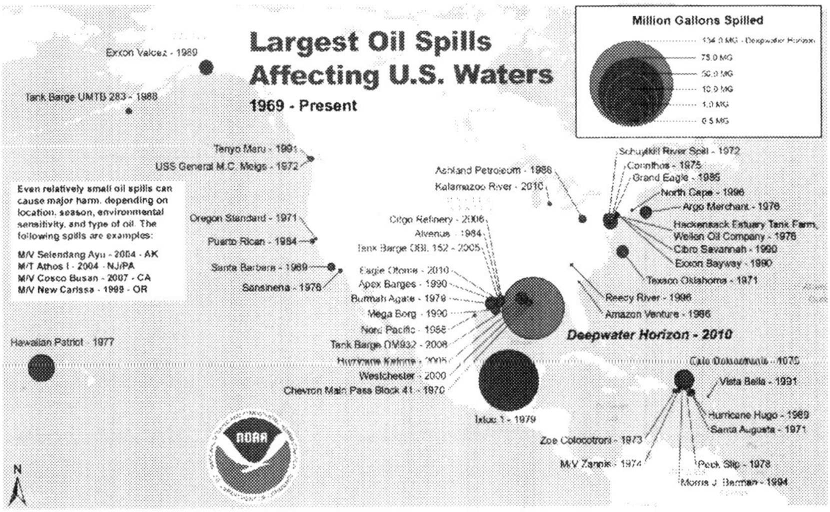


**Figure 7.1** Largest oil spills affecting US waters since 1969–2017 (NOAA/Office of Response and Restoration). *Office of Response and Restoration.* National ocean service, national oceanic and atmospheric administration, *Incident Photo Gallery.*

$5.25 million in legal and administrative fees to handle class-action suits and $2.25 million in criminal fines paid for violations of the Federal Clean Water Act.

- The aftermath of the Ashland Oil spill resulted in state and federal hearings, legislation, and regulations, including the Oil Pollution Act of 1990 (OPA'90) with regulatory activities and Environmental Protection Agency (EPA)'s Spill Prevention, Control, and Countermeasure (SPCC) regulations (Fig. 7.1).

Note: This graphic depicts the Ashland spill at only 2 million gallons, however, the total spill both to water and on land was closer to 4 million gallons, with approximately 1 million gallons entering the waterway.

## 7.2 Background

January 2, 1988, was a cold windy snowy Sunday morning in Ashland, Kentucky, when I awoke early and turned on the TV news. The leading headline was a story of an overnight oil spill in the

Pittsburgh, PA area from a bulk oil storage terminal owned by Ashland Oil.

At the time of this incident, I was Director of Fleet Operations for Ashland Petroleum, a wholly-owned subsidiary of Ashland Oil, Inc. My areas of responsibilities included marine transportation for Inland Waterways, Great Lakes, and Ocean Tankers, including an inland river repair yard/barge cleaning facility, for the Marine Transportation Department of Ashland Petroleum. I was also President of Cleveland Tankers, the Great Lakes tanker company of Ashland Petroleum. Marine Transportation was a division in the Ashland Petroleum Supply and Transportation Group, and part of the Facilities and Surface Transportation Group along with the bulk oil storage terminals. The bulk oil storage facility at Floreffe, PA suffering the spill, was one of the terminals in the Facilities and Surface Transportation Group and not part of the responsibility of the Marine Transportation group.

Ashland Oil Incorporated, at the time of this incident, was the largest independent oil refiner in the country. The Ashland Petroleum Floreffe, Pennsylvania, terminal sits on the left descending bank of the Monongahela River, approximately twenty (20) miles upstream from Pittsburgh, and consists of asphalt and clean product bulk storage tanks (Fig. 7.2).

On the evening of January 2nd, 1988, at approximately 1700, a facility employee gauged the tank level and verified the tank was almost full. Shortly thereafter, when the employee was walking away from the tank, a loud cracking noise was heard. The tank suffered an instantaneous catastrophic structural failure and collapsed (Fig. 7.3).

As stated in the EPA FOSC IOSC 1989 paper, "On January 2, 1988, a storage tank at the Ashland Oil Terminal, Floreffe, Pennsylvania, near the Monongahela River 24 miles upstream of Pittsburgh, suffered an instantaneous and complete failure, releasing 90,000 barrels (about 3.9 million gallons) of diesel oil into the environment." The speed and volume of the release made "first aid" almost impossible. Within hours, an estimated 18,000 barrels (750,000 gallons) of diesel had entered the river. Responders were faced with a power and communication lines shutdown, contaminated potable water intakes, oil dispersion from river currents and turbulence resulting from numerous river dams, and life-threatening weather conditions, with temperatures often below 0°F. With the US Coast Guard as the first federal official on the scene and the EPA acting as

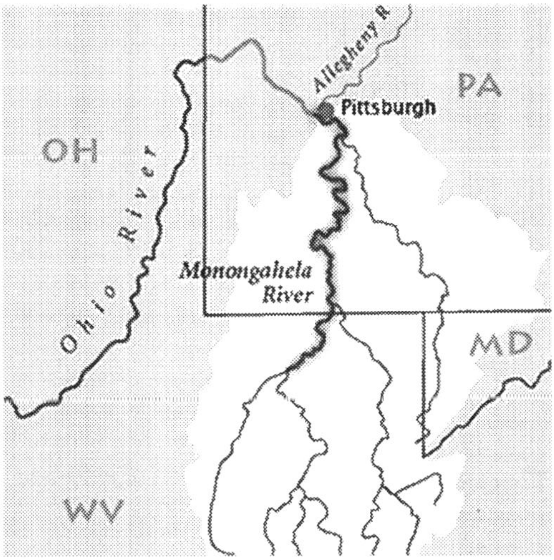


**Figure 7.2** 1988 oil tank collapse (Doyle, 2015). *Doyle, J. (2015).* Disaster at Pittsburgh—1988 oil tank collapse. *Available from PopHistoryDig.com.*

**Figure 7.3** Ashland terminal in relation to the Monongahela River at right, following the tank collapse. *Doyle, J. (2015).* Disaster at Pittsburgh—1988 oil tank collapse. *Available from PopHistoryDig.com.*

on-scene coordinator, a thoroughly integrated response organization of federal, state, local, and interstate agencies evolved.

Once the situation was stabilized, aggressive restoration of the Ashland site by traditional removal techniques began. However, significant factors challenged the scientific community throughout river recovery areas. First, as contamination became suspended throughout the water column, water supply intakes at all depths were affected, resulting in water shortages for most of the population downstream as far as 200 miles. Supplying potable water and technical assistance to water authorities prompted questions about the use of 311 (k)[6] for activities not clearly defined as cleanup[7].

Additionally, cold weather caused extensive freezing, and ice cover limited full use of absorbent materials and equipment. Finally, river access points also hampered boom deployment and surface oil recovery.[8]

## 7.3 Initial incident and response actions

Ashland personnel immediately called the National Response Center (NRC) as required by the Clean Water Act. By late evening, local fire officials determined that the spilled oil had crossed the Ashland property lines on the nearby highway and surrounding wetlands, as well as flowed onto the adjoining properties.

Clean-up commenced that evening on land but was terminated on water due to swift currents and subzero temperatures. The Monongahela River was largely ice-covered and in flood stage.

Ashland Oil's initial response team arrived at the facility on Sunday morning. Initially, I was not part of that team. As a marine transportation employee, I was not part of the terminal group. At the time, Ashland's expectations were that the department experiencing the spill event would manage the incident within that department, thus the terminal group was initially expected to manage the event alone.

Since I had substantial oil spill response experience, due to responding to multiple marine oil spill incidents over the years, I went to the company headquarters early that morning to offer my help to the terminal group. A couple of days later, due to the magnitude of the event, I was requested to become involved in the response and became the de facto

[6] CWA Section 311 and later OPA created a range of response tools to deal with oil and hazmat spills on the waters of the US, establishing a public/private solution to spill response.

[7] The Spill Prevention, Control, and Countermeasure (SPCC) regulations are far-reaching requirements promulgated by the Environmental Protection Agency under Section 311 of the Clean Water Act (CWA).

[8] The Ashland Oil Spill, Floreffe, Pa—Case HTORY and Response Evaluation, Cdr. E. A. Miklaucic, US Coast Guard. J. Saseen, US Environmental Protection Agency, International Oil Spill Conference Proceedings (1989)

Responsible Party (RP) Incident Commander for Ashland, with a primary focus on shoreline cleanup and the on-water response activities, providing management oversight to the operations, planning, logistics, and some financial activities for the response. I reported to Ashland Petroleum senior management, specifically to the Group Vice President, Supply and Distribution who was responsible for the Marine Transportation and the Facilities Groups. Media relations and legal services continued to report to Ashland Oil, Inc., senior management. The terminal group focused primarily on the on-land response in and around the terminal and getting the terminal back in operation.

At the time of this incident, Ashland had minimal oil spill response planning in place. There was no structured oil spill response organization that utilized all the response resources corporate-wide and no experience working within an Incident Command/Unified Command System which at this time, was not yet an accepted industry-wide response structure, nor was ICS/UCS widely used and accepted by local, state, and federal agencies. As a result, the response management to this incident, amongst the RP, local, state, and federal response agencies was often chaotic and confused as to who was "in charge", especially before the arrival of the Federal on Scene Coordinator (FOSC), Jerry Saseen of the EPA (Fig. 7.4).

With the arrival of the EPA FOSC, there remained organizational confusion since there was no predetermined response organizational structure that defined operational, planning, and logistical responsibilities and utilized as a formalized combined response management organization the resources of the private sector and the various governmental agencies as we have all been trained to do today with utilizing ICS/UCS.

River cleanup oil recovery operations spanned 30 river miles and were conducted at times in extremely cold conditions. However, questions and concerns from officials about drinking water quality safety issues plagued the response for up to nearly 200 river miles dominated response actions.

After the arrival of the initial Ashland response management team, government officials and the press arrived. Later that afternoon, the first of several joint news conferences took place. The oil slick was now nearly 33 miles long, already passing by Pittsburgh and moving downriver 10 to 20 miles per hour. Both the Monongahela and Ohio Rivers were in a high-water stage with high-velocity currents and with 50%–90% of the surface ice-covered. Another factor impacting the response, the oil had become emulsified and entrained in the water column due to flowing

Figure 7.4 The confluence of three rivers downtown Pittsburgh, the Monongahela to the right showing ice coverage, the Allegheny to the left, and the Ohio River at the bottom of the photo. *Office of Response and Restoration.* National ocean service, national oceanic and atmospheric administration, *Incident Photo Gallery.*

through several dams on the Monongahela and Ohio. These factors negatively impacted containment operations.

Cold January temperatures impeded the cleanup process by not only creating mechanical issues with the equipment but also causing hypothermia and increasing the probability of contamination because oil emulsified faster in the cold.[9]

The very extreme weather conditions prevented extensive use of recovery methods such as the application of sorbent materials because ice cover prevented the sorbent materials from contacting the oil. In addition, the ice cover and high risk of injury for the work crews prevented the placing of booms in some otherwise strategic locations.[10]

Flowing ice also reduced the effectiveness of the containment boom with ice lifting the boom skirt allowing the oil to entrain below and escape past the boom (Fig. 7.5).

Conventional containment booming was utilized whenever feasible considering the high-velocity flow of the river and the variable surface ice conditions. As the oil moved downstream and flowed through the various

[9] Pittsburgh Post-Gazette, The 1988 Monongahela Oil Spill, April 2014

[10] US Environmental Protection Agency, Region III, The Ashland Oil Spill of January 1988, An EPA Perspective, International Oil Spill Conference, Laskowski and Voltaggio, 1989

**Figure 7.5** Ice Conditions on the Monongahela River near the Floreffe terminal. *Office of Response and Restoration.* National ocean service, national oceanic and atmospheric administration, *Incident Photo Gallery.*

dams, increasing the emulsification, continuing to reduce the surface oil, the booming strategy had to deal with a high degree of sheening versus thick oil, so increased use of absorbent boom was used when feasible, for the sheen in conjunction with the hard containment boom (Fig. 7.6).

## 7.4 Findings and lessons learned concerning the response

- The prompt notification by Ashland to the local response authorities and the NRC was fundamental to the establishment of the command post on the evening of the spill.
- A lack of immediately available resources, such as containment and monitoring equipment, hindered the initial response. There were no existing inventories of private and public equipment and resources. Response delays were caused by the need to locate and then transport needed resources to the site.
- The lack of a predeveloped emergency response plan showed the importance of having a predeveloped response plan that includes a

**Figure 7.6** Combination sorbent and containment boom configuration. *Office of Response and Restoration.* National ocean service, national oceanic and atmospheric administration, *Incident Photo Gallery.*

wide range of local, regional, and national response resources of equipment and technical expertise.

- The initial 18 h of response to the emergency were characterized as a loosely organized, but relatively effective attempt to minimize the dangers and damage from the spill. As local, county, state, and federal agencies responded, each exercised its responsibilities independently as a liaison was established at the scene. At no time in the initial hours of the response, did any agency deem it necessary to "take charge" of the entire response effort. A response organization then developed, which, while not without some temporary shortcomings, served as an effective structure throughout the remainder of the response period.
- The lack of a trained emergency response organization throughout the incident response showed the importance of having a corporate-wide response organizational structure that could mobilize all resources of a company across all departmental boundaries, train in all the needed competencies of the Incident Command System, and be able to work within a Unified Command System structure with responding local, state, and federal agencies. The lack of RP, local, state, and federal agency knowledge and training in Incident Command/Unified Command hindered the overall efficiency and effectiveness of the response.

- I have been the responsible individual managing emergency responses before the implementation of the Incident Command/Unified Command System on major responses, for example, the Ashland Oil Floreffe incident in 1988 and the Tanker Jupiter fire, explosion with the spill in 1990. I must emphatically state, the difference in the ability of an Incident Commander, and the RP to manage a minor and major incident via IC/UC, significantly enhances the effectiveness and efficiency of the entire response organization ensuring all participants, the RP, federal, state, and local agencies work cooperatively instead of at odds with each other.
- The importance of having a preplanned/prelocated adequately sized and equipped space to act as an Incident Command Post (ICP). Off-site may be preferable to alleviate the response traffic and congestion that will ensue at the impacted facility. Essential that the designated ICP be fully equipped with communication systems, for example, landline telephones, cell phones, computers with external access to the internet, and an intranet for internal response team communication.
- Due to the significant river miles of contamination and spread-out locations of work crews along those river miles, twice daily helicopter overflights were critical to maintaining an operational grasp of the response activities. Radio communications were difficult due to the distance and elevated terrain between field operations and the ICP, adding to the importance of the helicopter overflights. Today, improved radio systems with repeaters and cell phones would minimize many of the communication difficulties.
- The FOSC can take steps to determine the availability of resources and to standardize analysis among the agencies.[11]
- Although the response to the Ashland Oil spill was effective in protecting public health, it is evident that more timely health effects data on spilled hazardous substances were needed along with assistance in interpreting their significance.[12]
- This was the first major oil pollution accident in Ashland's 64-year history. No matter how well one manages and operates their day-to-day business, emergencies can happen anytime to anybody and without any warning! Preparedness is the key to a successful safe response!

[11] The Ashland Oil Spill, Floreffe, PA—Case History and Response Evaluation, Cdr. E. A. Miklaucic, US Coast Guard. J. Saseen, US Environmental Protection Agency, International Oil Spill Conference Proceedings (1989).

[12] Economic and Policy Implications of the January 1988 Ashland Oil Tank Collapse in Allegheny County, *Pennsylvania Final Report*, Allegheny County Planning Department, Trauth et al. (1989).

Preparedness = Personnel Training (Competence not just Compliance) + Real Life Exercises + Pre-Determined Availability of Sufficient Response Resources.

- One prime oil spill response contractor was utilized who managed multiple subcontractors. This alleviated the multiple contractor oversight workloads on the RP response team.
- This incident highlighted the need for rapid response, adherence to protocols, interagency communication, and hands-on experience in establishing a successful response to wildlife contaminated by oil.
- Importance of continuously tracking response resources and costs. Compiling costs daily and cumulatively throughout the response to provide corporate management and USCG/EPA with the ongoing costs/burn rate. A risk management company to manage claims and track resources were utilized to assist in 3rd party claims and tracking response resources. This service was extremely helpful in tracking resources on standby, assigned, and/or available and assisted in the justification of operational decisions and associated expenditures when requesting reimbursements from the insurance underwriters.
- Know your insurance deductible and if spill costs may exceed those deductible levels, consider having an insurance underwriter representative sitting with the Finance Section, sooner rather than later. This will significantly enhance your reimbursement process.
- The RP should consider purchasing large quantities of the boom, absorbents, and other resources directly from the manufacturer/vendors to avoid surcharges added on by a response contractor.
- The USEPA Regional Response Team concluded that "even under the most perfect conditions, it would have been difficult to obtain the personnel and equipment necessary to block off the drains and the outfall to the river in such a short time" and thereby mitigate the effects of the spill (Stanley & Thomas, 1988, 1). There were several factors beyond the control of emergency responders which contributed to the difficulty including darkness, river conditions, and extremely cold temperatures. Because of darkness and the lack of power at the site (which was turned off for precautionary reasons), the severity of the spill was not realized until 12–18 h after the incident occurred (Trauth, 1989).[13]

[13] Trauth, J. M. (ed.), (July 1989). Economic and Policy Implications of the January 1988 Ashland Oil Tank Collapse in Allegheny County, Pennsylvania, University Center for Social and Urban Research, University of Pittsburgh, Pittsburgh, PA.

- The willing cooperation between all regional elements was the outstanding factor that resulted in the successful protection of the public health and the environment during the Ashland major oil spill. All responding agencies, groups, and individuals are to be commended for their performance throughout the emergency. Despite the magnitude of the spill, the rapid entry of oil into the river, and adverse weather conditions, all public water supplies were protected.[14]

## 7.5 Drinking-water response actions

Due to the river being in flood stage, the diesel oil going over the Elizabeth lock and dam, approximately two (2) miles downstream from the oil entry to the river, emulsified and became entrained in the water column, resulting in most of the oil not being seen on the surface and thus, not available to be removed and or be contained by the conventional booming. This became a primary factor in the response due to the negative impact on drinking water quality for residents and businesses along the Monongahela and Ohio Rivers from Floreffe to Cincinnati and beyond to Louisville. A total of approximately 625 miles from the initial spill point (Fig. 7.7).

As the Associated Press reported at the time, "*approximately 23,000 suburban Pittsburgh residents lived for a week without tap water while the river carried the pollution past their water intakes.*" Since many municipal and industrial water systems had supply intakes in the river, potable and industrial water supplies were disrupted for many miles downriver from the spill site.

At the time, Ashland had the largest inland river petroleum tank barge fleet in the nation, well over 300 tank barges ranging in size from 10,000 barrels to 30,000 barrels. Ashland also operated a boat and barge repair facility coupled with a tank barge cleaning facility in Catlettsburg, KY. Ashland selected several petroleum tank barges, processed them through the tank cleaning facility, and prepared them for temporary water storage/ supply service.

Two tank barge groups were formed, with towboats, fitted with US Coast Guard Strike Team pumps. The barge groups were stationed ahead

[14] US Environmental Protection Agency, Region III, The Ashland Oil Spill of January 1988, An EPA Perspective, International Oil Spill Conference, Laskowski and Voltaggio, 1989.

**Figure 7.7** Ashland Oil tank failure, Floreffe, Pennsylvania, January 1988. Dam on the Monongahela River. *Office of Response and Restoration.* National ocean service, national oceanic and atmospheric administration, *Incident Photo Gallery.*

of the spill with a towboat to maneuver the tow[15] up freshwater creeks and filled with clean water. When the tow was full, the barges were maneuvered to various downriver water intakes, hooked up to the intake, and just before when the oil "plume" reached that water intake, the water intake would shut down and the barges would discharge their water to that municipal system until the plume passed by providing clean water to that municipal water system.

When the oil plume moved past, as advised by the Ohio River Valley Sanitation Commission (ORSANCO), the barge tow would move downriver to another clean creek and begin taking on a new load of water and when full head downstream, leapfrogging the second barge tow to get to another downstream intake ahead of the oil plume. Thus, the two tows would continue to leapfrog each other, filling themselves with water, hooking up to a water intake, supplying water as the oil plume passed, then disconnect and travel downstream to refill with clean water.

Most of the diesel oil was not visible on the surface since most of the oil was emulsified in the water column, making it difficult to keep accurate track of the oil plume as it transited down the Ohio River.

[15] Towboats push barges lashed together to form a "tow". A tow may consist of a small number of barges on smaller waterways and up to over 40 barges on the Mississippi River below its confluence with the Ohio River.

Three types of monitoring activities took place during this spill event:[16]

- monitoring of the river to define the spill mass and track its movements,
- monitoring the effects on fish and wildlife, and
- monitoring at intakes to protect water supplies.

River flow and velocity forecasts by the National Weather Service were initially utilized to predict the progress of the oil plume.

Ashland teamed with the eight-state Ohio River Valley Water Sanitation Commission (ORSANCO), US EPA, the National Oceanic and Atmospheric Administration (NOAA), Pennsylvania Department of Environmental Resources (PADER), and the Army Corps of Engineers to set up and implement a program of sampling river water to monitor the spill plume as it moved downriver. This was necessary in order to provide an accurate definition of the extent of oil pollution and for the purpose of implementing risk management actions, for example, the water systems shutting down their river intakes, activating alternate supply sources, and ensuring the water barge operation was always ahead of the spill plume arrival providing water to the municipal water systems if needed until the plume had passed by that location.

Each evening on the national news, position locations of the oil plume were reported. People would go down to the river to see the spill but there was little on the surface to see. The plume position reports were the result of the ORSANCO sampling/monitoring, there was little to observe on the surface since the emulsified oil was entrained throughout the water column impacting the water intakes.

## 7.6 Findings and lessons learned water supplies

- The Ashland oil spill could have been far more devastating had public water supplies been contaminated or water shortages more severe. Municipal and industrial water systems should preplan for the availability of alternate contingency water supplies and equipment, as well as

[16] US Environmental Protection Agency, Region III, The Ashland Oil Spill of January 1988, An EPA Perspective, International Oil Spill Conference, Laskowski and Voltaggio, 1989

**Figure 7.8** Oil saturated Debris Upstream Lock and Dam. *Office of Response and Restoration.* National ocean service, national oceanic and atmospheric administration, *Incident Photo Gallery.*

enhance water quality monitoring capabilities pursuant to hydrocarbon contaminants.

- This incident raised several technical, legislative, and administrative issues-such as assessing long- and short-term environmental damage, evaluating regulations regarding oil tanks, and examining spill response procedures. An important research need is the development of computer models that can better predict the travel time and the concentration of contaminants should a spill occur, especially on inland freshwater waterways.
- Water monitoring coordination and communication suffered initially because no lead agency was assigned. The Ohio River Valley Water Sanitation Commission (ORSANCO) ultimately accepted the lead. The early establishment of a single water monitoring data coordinating agency can serve to improve the focus on communications so that efficiency and data relevance are optimal (Fig. 7.8).

## 7.6.1 Contaminated marine debris

There are 16 locks and dams between the spill site and Louisville, two on the Monongahela River, and fourteen on Ohio. The two on the Monongahela and those locks and dams on upper Ohio were locations where there was considerable oil-soaked debris on the upriver sides of the structures that required removal. Opening and closing the dam gates to walk the debris from the farthest location away from the shore toward the nearshore where the debris was then clam shelled into oil-tight barges and

**Figure 7.9** Oil Contaminated Debris Upriver Side of Dam. *Office of Response and Restoration*. National ocean service, national oceanic and atmospheric administration, *Incident Photo Gallery*.

trucks for transport to approved landfills. This was a major undertaking in the response operations (Fig. 7.9).

## 7.7 Crisis management response actions

The crisis management skills of the Chairman of Ashland Oil Inc, John R. Hall, garnered national attention during this incident and became a primary aspect characteristic of the Ashland response and has been noted as a positive example of senior management responses to emergency situations.

When he was notified that an Ashland Oil bulk storage oil tank had leaked diesel oil into the Monongahela River in Pittsburgh and the severity of the situation became clear, the Chairman went to Pittsburgh to investigate. He inspected the site of the collapse and discovered that the tank was not up to company standards. He gave a news conference apologizing for the incident, taking responsibility, and promising that Ashland would pay for all cleanup costs and reimburse communities for expenses incurred because of water shortages. "We're doing everything we know how to clean up the danger as quickly as possible and restore water supplies," he said. "Ashland is a responsible corporate citizen that is also responsible for community concerns in the areas where our terminals are located." Hall drew praise for being so forthright and direct in confronting the crisis.

"The further I dug, the madder I got," he recalled. Over a million gallons had spilled, the worst inland oil spill in US history at that time, leaving the city's suburbs without drinking water. The tank in question had been moved from Cleveland and rebuilt in pieces in Pittsburgh. "Everywhere I looked, we had done something wrong," Hall said. Harry Wiley, former director of advertising and communications for Ashland Oil, recalled that Ashland's communications department was waiting for its attorneys to tell them what the company was going to do. "And John Hall went up there and said it's our mistake. We did it. We'll fix it. We'll take care of it. And the lawyers went Aw,—Aw,—." Hall held a press conference and told the crowd the spill was Ashland Oil's fault. At the end of the conference, he explained that "you always want to do what is right." "That was not a talking point," said Dan Lacy, former vice president of corporate affairs for Ashland. "That was John, under pressure, at the moment, speaking from his heart." (West, 2016).[17]

When Mr. Hall learned that the Western Pennsylvania Water Co. shut down one of its facilities whose water intake was downriver from the spill he began to inquire as to available options to address this water supply issue. This supply disruption meant water shortages for residences and businesses since it is common that municipal and industrial water intakes to take their water supply from the Ohio River and its tributaries.

Mr. Hall directed Ashland to pay for a temporary pipe to be laid across the affected area to secure freshwater from the Allegheny River, which merges with the Monongahela River downriver. He also authorized flying in the Coast Guard Strike Force on Ashland company planes.

When it became clear that the collapsed storage tank was not new but had been reconstructed from a used, 40-year-old steel tank, that an alternative testing method had been used to confirm its structural integrity once reconstructed, and that the construction permit had been applied for but had received only verbal permission before reconstruction and initial use, John Hall publicly admitted these facts.

The company Chairman's admission that Ashland made mistakes in erecting the diesel fuel tank catastrophically collapsed was problematic for it made the company's legal team concerned about liability. However, his

[17] West, J. (2016). John Hall: The Kentucky Commodore. Kentucky Educational Television <https://www.ket.org/promos/kentucky/john-hall-the-kentucky-commodore/>.

honesty served the company well with the public. Concern, contrition, openness, providing the chief executive to the public, being willing to pay for the damages within limits, and supporting an outside investigation, are all issues that should be considered in crisis management and crisis communication.

Ashland opened a Pittsburgh office to handle claims and staffed it with a senior executive. John Hall went there regularly along with other senior executives to keep communication channels open. The company provided affected communities with drinking water for weeks, and commissioned the Battelle Memorial Institute of Columbus, Ohio, and an environmental consulting firm, to examine the long-term environmental impact of the spill. Battelle also was engaged to conduct an independent investigation into the collapse.

Hall showed that he genuinely cared about the spill, the water shortage, and the people in the community. Also, the fact that he did not know about the permit or testing issues, showed his honesty in dealing with the public and the regulatory agencies.

John Hall's handling of this crisis won praise from financial analysts just days after the spill and from other observers as time went on. He would be named Crisis Manager of the Year by Carnegie Mellon University for his handling of the spill's aftermath. Hall's approach to the crisis became the subject of a Harvard Business School case study and today remains a textbook example of effective crisis management.[18]

The example of how John Hall managed this crisis has stayed with me all the subsequent years of managing emergencies and I have used those lessons learned during my management of subsequent emergencies and have attempted to pass on these crisis management strategies and tactics to those individuals I have been privileged to train in emergency management.

## 7.8 Crisis management findings and lessons learned

- Concern, contrition, openness, serving up the chief executive to the public and media, being willing to pay for the damages within limits, and supporting an outside investigation were all components of the

[18] Harvard Business School, Ashland Oil Inc.: Trouble at Floreffe, 9–390–017, January 19, 1990

crisis management approach taken by Ashland Oil and its Chairman John Hall.

- In a significant event, senior management must get involved. The public, media, and elected officials will want to see and hear from the senior executive of the company, not just a media relations representative. Ensure senior management is briefed totally before facing the media, public, and elected officials.
- It the importance of having a crisis communication strategy with individuals trained in the implementation of that strategy. Get any bad news out before you are outed by the media or agencies. The truth will eventually come out, best if you bring it out; the coverup is usually worse than the incident. Talk about the good aspects of what you are doing to mitigate incident risks to the public and environment. Do not ever promise more than you can deliver! You will be held accountable if you do not deliver as promised.
- Be honest, forthright, and maintain credibility! Lose credibility with regulatory agencies, the media, and the public, you will lose both the battle and the war. Keep the high ground, keep public support and if warranted, accept responsibility. The lawyers may not necessarily agree with that of course!
- Studies conducted after the crisis did not account for the emotional anguish of citizens, personal inconveniences, unknown health risks, or the economic losses on commercial activities. Today, these are important factors for the RP to consider in their overall spill management.
- Have empathy for what the public may be going through according to potential stress, economic hardship, life disruptions, etc. Keep the affected population informed. Express your concern and find solutions to alleviate the disruptions and hardship imposed on the public by the incident.
- Educate and inform the public and other stakeholders directly about real or perceived risks from the event. Use town hall meetings, and social media (Facebook/Twitter), do not just rely on the printed or televised press to accurately tell your story. Face-to-face communication between the RP and affected stakeholders, is more effective than issuing press releases and establishing sufficient resources to accomplish.
- The investigation commences at the same time as the emergency response. The RP must be prepared to handle both the response and the investigation simultaneously. Ensure legal representation is available onsite immediately, especially for employees involved in the incident!

## 7.9 The tank that failed

The tank that failed, designated No.1338 in the Floreffe facility, was being filled with diesel fuel oil for the first time since its re-erection at the Floreffe site the previous August. At approximately 1700, when the oil level in the tank was almost at the operating maximum, the tank shell fractured vertically without warning, separating from the bottom plate at the connecting welds from the force of the escaping oil. The tank instantaneously catastrophically collapsed with the oil flowing out in a tidal wave, overflowing the earthen dike containment, structurally damaging adjacent tankage, and flowed into storm sewers, ultimately into the Monongahela River and then from there downstream to the Ohio River. The total volume spilled was estimated at 4 million gallons (956,000 barrels), causing harm to the environment, and affecting the drinking water supply of residents and businesses along the Monongahela and Ohio Rivers. (Fig. 7.10).

The tank that catastrophically failed had been built originally at Whiskey Island in Cleveland, Ohio in the early 1940s timeframe. During the late 1970s and 1980s, the domestic oil industry underwent significant consolidation with nationwide shutdowns of refineries and terminals. It was a relatively routine practice in those days to dismantle good condition bulk oil storage tanks in the facilities being shut down and transfer the dismantled tankage to a terminal that was slated to remain open and then re-erect the tank for reuse.

This tank was a cylindrical tank, approximately 118 feet in diameter, with a flat bottom and supported a conical roof. The shell was approximately 47 feet high with the sides consisting of the welded plate.

I had commenced my career as a First-Class Pilot onboard Cleveland Tankers Great Lakes tanker fleet in the early 1970s, and as a deck officer had discharged cargo into this tank at Whiskey Island. Thus, it was ironic that this very tank would be the cause of the Floreffe spill that I had to respond to about 15 years later.

The total tank capacity was roughly 4.23 million gallons (100,000 barrels) and until 1986 it had been used to hold distillate oils and heavier distillates. In 1986 the tank was taken down by oxyacetylene cutting adjacent to the original welds and then reassembled by welding in Floreffe, re-erecting the plates in the same order/positions they were dismantled.

**Figure 7.10** Ashland terminal in relation to the Monongahela River at right, following the tank collapse. *Doyle, J. (2015).* Disaster at Pittsburgh—1988 oil tank collapse. *Available from PopHistoryDig.com.*

This 4-million-gallon tank was being used for the first time after being dismantled and re-erected. The former Whiskey Island bulk oil storage tank, holding approximately 4 million gallons of diesel oil, failed, and catastrophically collapsed, dumping nearly 1 million gallons of the oil into a storm sewer that led to the Monongahela River. The spill contaminated drinking water sources for over a million people in Pennsylvania, West Virginia, and Ohio.

The remaining 3 million gallons either remained in the dike containment or overflowed outside of the containment area inundating the area of the terminal as well as escaping to ground areas outside of the terminal boundaries. Groundwater contamination occurred, eventually requiring wells to be drilled to remove contaminated groundwater/oil over time.

The American Petroleum Industry (API) has a Standard 650 for proper testing of tanks. Ashland was required to use a hydrostatic method of testing—filling the tank with water to settle the foundation and to test the strength of the tank's welds. Instead of this mandated test, Ashland personnel filled the tank with only three feet of water to settle the foundation. They had sprayed oil on the welds inside the tank and then vacuum suctioned the outside to determine whether any oil could be pulled through possible leaks in the weld. This is the testing model presented by 650 for desolate locations with scarce water supplies (Fig. 7.11).

**Figure 7.11** Ashland collapsed tank, Floreffe, PA. *Office of Response and Restoration.* National ocean service, national oceanic and atmospheric administration, *Incident Photo Gallery.*

Examination of the fracture faces in the tank shell showed them to be flat and perpendicular to the plate surfaces, with the characteristic chevron markings of brittle fracture. The flaw was 'dime-size" and its orientation was in the vertical direction of the tank. Metallographic studies of the flaw revealed it to be due to flame cutting, rather than welding but not the flame cutting of the dismantling procedure. The flaw had been present in the steel plate before being welded when the tank was originally built.[19]

It was finally concluded that the failure was due to the material immediately surrounding the flaw being of particularly low toughness, with crack initiation occurring under the combined effect of hydrostatic and residual stresses. As the tank was operating below the nil ductility temperature of the shell plate, the crack emerging from the locally embrittled area could not be arrested.[20]

The original welds and the welds used during the reassembly made the tank's steel walls more brittle. Other factors in the collapse were the relative weakness of the World War II-era steel, pressure from the diesel fuel,

[19] Catastrophic failures of steel structures, Ashland Storage Tank, Hayes, B., & Phaal R., TWI, Classic brittle failures in large-welded structures, *Engineering Failure Analysis*, *3*(2), 1996.

[20] Catastrophic failures of steel structures, Ashland Storage Tank, Hayes, B., & Phaal R., TWI, Classic brittle failures in large-welded structures, *Engineering Failure Analysis*, *3*(2), 1996.

and cold weather. The flaw, a small, rusting cavity about one-eighth-inch deep, was located near a weld about 8 feet from the bottom of the tank's 48-foot-high wall.[21]

An engineering firm (Battelle) was hired by Ashland Oil to determine the cause of the tank failure. The company report stated that all those factors contributed to the collapse; however, the one factor that had largely been discounted was a flaw in the plate itself, probably caused by an errant blowtorch.

The report stated the flaw that triggered the spill was merely one of a set of circumstances:

- Rewelding. When the tank was rebuilt, the welding near the flaw caused "hoop stress" weakening—a strain caused by the steel plate contracting after the area had cooled.
- Old, brittle steel. The metal of the 1940s had a low-carbon, high-phosphorus composition that made it inherently less tough than modern-day steel.
- Cool-weather. The temperature of the steel in the cold January air was 38 degrees, which made it more susceptible to brittle fracture like "a piece of taffy breaking off."
- Load stress. The 4-million-gallon tank was filled to near the top when it collapsed. The 3.8 million gallons was more than it had held before. Had it not been for the combination of all these factors, it was speculated that the oil might have drained out like a slow leak, not a sudden, catastrophic failure. Specifically, company representatives took responsibility for several problems:
- Not building a new tank. New steel could have withstood temperature and load stresses better.
- The tank's ability to hold its capacity in oil was not tested properly by filling it with water.
- Proper construction permits had not been obtained from county officials.[22]

https://www.youtube.com/watch?v=Bl8Unowv5XA&feature=youtu.be: Animation Video Ashland Oil Company Diesel Fuel Spill 1988 Allegheny County, Pennsylvania[23]

[21] Dime-Size Flaw Blamed for Tank Collapse, Catherine Dressler, AP, May 1988, quoting Ray Mesloh, Battelle Memorial Institute researcher

[22] US Environmental Protection Agency, Region III, The Ashland Oil Spill of January 1988, An EPA Perspective, International Oil Spill Conference, Laskowski and Voltaggio, 1989

[23] United States Environmental Protection Agency (EPA) 1994 video Spill Prevention Control and Countermeasure Training Series, Highlights: Protecting the Environment from Oil Spills.

## 7.10 Causes of tank failure findings and lessons learned

- The National Institute of Standards and Technology conducted an independent investigation into the physical cause of the Ashland tank collapse. It was determined that the failure was caused by a brittle fracture of the tank shell, which was initiated by a defect that existed before the reconstruction of the tank. Complete rupture of the tank shell occurred because the steel did not possess adequate toughness at the operating temperature to prevent brittle fracture propagation. The collapse shows the importance of using steel with sufficient fracture toughness to prevent the propagation of a brittle fracture in tanks when sudden failure would mean unacceptable human, environmental, or economic losses. Highlighted the potentially serious problem of locally intensified strain-aging embrittlement associated with rewelding and weld repairs of older steels.
- The importance of compliance with the procedures contained in the American Petroleum Institute (API) Standard 650, Welded Tanks for Oil Storage. This standard provides the industry with guidance to ensure tanks are of adequate safety and a reasonable economy for use in the storage of petroleum, petroleum products, and other liquid products.
- The secondary containment volume design capacity may have been sufficient but failed to contain the bulk of the spilled oil due to the catastrophic instantaneous collapse of the tank. This caused an eruption of oil in a tidal wave that overflowed the berm damaging nearby tanks and spilling on the ground outside of the containment area causing groundwater contamination as well as escaping via a storm drain into the Monongahela River.

## 7.11 Followup activities and the aftermath of the Ashland oil spill incident

In the wake of the Ashland release, several followup investigations, assessments, and legal actions were initiated that required a significant amount of effort and time by various regulatory agencies as well as Ashland Oil.

A brief list of some of these activities includes[24]:

- Governor's Task Force in Pennsylvania to investigate the causes of the spill.
- State senate hearings
- US Congressional hearings and briefings
- A long-term environmental assessment by ORSANCO
- Regional Response Team evaluation of emergency response actions
- Multi-media facility compliance inspection by EPA and PADER
- SPCC inspection by EPA and DER
- National SPCC Task Force to review the SPCC program and regulations.
- Enforcement negotiations with Ashland for long-term soil and groundwater cleanup and start-up of the facility
- Citizen lawsuits for damages

Although the 1988 Floreffe oil spill 4-million-gallon spill was one of the largest inland oil spills to that date in the United States, it was only one of several catastrophic disasters involving the oil companies in the 1980s and 1990s, the worst being the spill of nearly 11 million gallons of crude oil into Alaska's Prince William Sound in 1989 from the tanker Exxon Valdez.

The Ashland and Exxon incidents were followed by:

- 1989 World Prodigy, Narragansett Bay oil spill, Rhode Island, approximately 300,000 gallons of fuel oil
- 1989 Presidente Rivera, Marcus Hook, Pennsylvania, approximately 306,000 gallons of crude oil
- 1990 Arthur Kill pipeline, Sewaren, New Jersey, approximately 567,000 gallons of #2 fuel oil.
- 1990 American Trader, Bolsa Chica State Beach, California, approximately 417,000 gallons of crude oil
- 1990 Mega Borg oil spill, off Galveston, Texas in the Gulf of Mexico, approximately 4.6 million gallons of light crude oil
- 1990 Apex barges 3417 & 3503, Galveston Bay, Texas, approximately 714,000 gallons of # 5 oil

[24] US Environmental Protection Agency, Region III, The Ashland Oil Spill of January 1988, An EPA Perspective, International Oil Spill Conference, Laskowski and Voltaggio, 1989

This series of 1988–1990 oil spill incidents resulted in the US Congress passing the Oil Pollution Act of 1990 (OPA'90), which was signed into law in August 1990 by President George H. Bush.

The Tanker Jupiter fire/spill incident in latter 1990 at Bay City, Michigan, occurred just a few months after OPA '90 came into force, of which at the time, I was the President of the shipping company, Cleveland Tankers as well as the RP Incident Commander. The Tanker Jupiter was the first major US oil spill incident covered by OPA'90.

Following the Floreffe, Pennsylvania oil spill in 1988, US EPA formed the SPCC Task Force to examine federal regulations governing oil spills from aboveground storage tanks. The SPCC Task Force recommended that EPA:

- clarify certain provisions in the Oil Pollution Prevention Regulation,
- establish additional technical requirements for regulated facilities, and
- require the preparation of facility-specific response plans.

In response to the Task Force's recommendation, US EPA proposed revisions to the Oil Pollution Prevention Regulation in the 1990s. The goal of the SPCC Regulation are to prevent oil from reaching navigable waters and adjoining shorelines and to contain discharges of oil. The regulation requires facilities to develop and implement SPCC Plans and establishes procedures, methods, and equipment requirements (Subparts A, B, and C).

The OPA'90 landmark legislation was followed by the US Coast Guard establishing the Negotiated Rulemaking (Reg-Neg) Committee to assist in promulgating regulations to the OPA'90 legislation. I was privileged to be one of the American Petroleum Institute's representatives on this Committee.

In 1990, the Oil Pollution Act amended the Clean Water Act to require some oil storage facilities to prepare Facility Response Plans. On July 1, 1994, US EPA finalized the revisions that direct facility owners or operators to prepare and submit plans for responding to a worst-case discharge of oil.

OPA'90 and its subsequent rulemakings and the SPCC regulations, forever changed how the oil industry (production, transportation, storage, and distribution), fundamentally prepares for and responds to oil spills in the United States.

## References

Doyle, J. (2015). *Disaster at Pittsburgh—1988 oil tank collapse.* Available from PopHistoryDig.com.

Office of Response and Restoration. *National ocean service, national oceanic and atmospheric administration*, Incident Photo Gallery.

Trauth, et al. (1989). Economic and policy implications of the January 1988 Ashland oil tank collapse in Allegheny County. In *Pennsylvania final reports* (pp. 57–58). Allegheny County Planning Department.

Stanley, L., Thomas, C. V. (1988). The Ashland oil spill of January 1988, An EPA perspective. *International oil spill conference proceedings*, 39–43.

West, J. (2016). John Hall: The Kentucky Commodore. In *Kentucky Educational Television.* Available from <https://www.ket.org/promos/kentucky/john-hall-the-kentucky-commodore/>.

## Further reading

Gross, J.L., Smith, J.H., & Wright, R.N. (1989). Ashland tank-collapse investigation. *Journal of Performance of Constructed Facilities, 3* (3)

Hayes, B., & Phaal, R. (1996). TWI, Classic brittle failures in large-welded structures. *Engineering Failure Analysis, 3*(2).

L.A. TIMES ARCHIVES. (1988). *Crisis control: Ashland chairman's conduct on oil spill wins public support.* Associated Press.

Catherine Dressler. (1988). Dime-size flaw blamed for tank collapse. In Ray Mesloh, Battelle Memorial Institute.

Establishing a Wildlife Response After the Ashland Oil Spill, (1989). Tri-State Bird Rescue and Research. In *International oil spill conference*, Frink & Dalton.

Saseen, J. (1988). *On-scene coordinator's report ashland oil terminal major oil spill Floreffe, Allegheny County, Pennsylvania.*

Harvard Business School, (1990). Ashland Oil Inc.: Trouble at Floreffe, *9*, 390–017.

The Ashland Oil Spill, (1989). The Ashland oil spill, Floreffe, PA—Case history and response evaluation. *International Oil Spill Conference Proceedings, 1989* (1), 45–51.

Clark, R.M., Vicory, A.H., & Goodrich, J.T. (1990). The Ohio river spill. *Journal American Water Works Association (AWWA).*

United States Environmental Protection Agency (EPA). (1994). *Spill prevention, control, and countermeasure (spcc) for the upstream (oil exploration and production) sector.*

Pittsburgh Post-Gazette. The 1988 Monongahela oil spill.

PART 3

# Legislation

CHAPTER EIGHT

# An overview of typical legislation governing the design, construction, and operation of storage tanks

**Merv Fingas**
Spill Science, Edmonton, Alberta, Canada

## 8.1 Introduction

Storage tanks can be a significant source of spills into the environment. The extensive regulations that have been created for storage tanks, reduce this spill risk considerably while also considering the risk of fires. Storage tank regulations have been developed over many years and are now similar in many regions of the world. This chapter will provide a generalized outline of what is typically in these regulations with some examples of content.

This information has been generalized from a number of sources. Specific values are given as examples (e.g.,) to provide the reader with values used in certain jurisdictions. The generalized information is derived from many sources (EPA, 2010; CCME, 1994; ERCB, 2001). Those wishing to find more specific information are advised to consult their local legislation.

## 8.2 Basics of regulation

The legislation varies with specific situations. Often, there are variations in legislation that depend on:

- Stipulations on the type of material to be stored
- The length of time the material is stored,—for example, There may be differences for temporary storage,
- the environmental sensitivity of the site where the material is being stored,
- the nature and integrity of the primary containment tank,

*Above Ground Storage Tank Oil Spills.*
DOI: https://doi.org/10.1016/B978-0-323-85728-4.00008-5

- Other risks such as fire and explosion,
- Site conditions and consideration of other risks such as seismic, storms, and
- the volume of material being stored.

## 8.3 Siting

The location where the storage tanks are to be located typically has several specific specifications:

- A site that minimizes the potential for environmental concerns,
- Readily accessible for fire-fighting and other emergency procedures,
- Not located in a floodplain, unless appropriate alternative secondary containment, measures are incorporated into the design and installation,
- Chosen to minimize any threats to the integrity of the storage facility, the quality of soils, surface water, and groundwater, and the health of humans, animals, and plants during the construction, operation, and closure of the storage area/facility. and
- Not located within a certain distance (e.g., 100 m) of the normal high-water mark of a body of water, permanent stream, or water well used for domestic purposes.

## 8.4 Separations

The tanks should be separated from each other by a certain distance dependent on size (e.g., 100 m) and also from other features such as roads (e.g., 100 m), pumping stations (e.g., 60 m), heaters, or other process equipment (e.g., 25 m). The containment dykes have separation requirements from roads (e.g., 60 m), from each other (e.g., 10 m) and from streams (e.g., 100 m).

## 8.5 Identification of storage facilities

All storage facilities should have signs at the entrance to the facility indicating the approval holder or owner's name, emergency phone number, and legal description. Within a storage facility and at storage areas that form part of an operating upstream petroleum site, signs should indicate the materials that are stored, warnings, and any general housekeeping practices that should be followed in the storage area (e.g., segregation).

## 8.6 Construction

General construction and installation factors for aboveground storage tanks typically include:

- Tanks should be designed, fabricated, tested, and installed to applicable engineering, manufacturing, and regulatory standards. Some applicable standards are listed in Appendix B of this paper,
- Tanks should either be made from or externally coated with weather-resistant material,
- Steel tanks should be externally coated (e.g., painted, galvanized), and if storing corrosive liquids, they should be internally coated or lined to minimize corrosion. In corrosive environments, it may be appropriate to apply cathodic protection to aboveground steel tanks,
- Transfer lines and hoses should be compatible with the material being transferred and should have leak-proof connections,
- Spill control devices should be used around hose connections at fluid transfer points to help prevent the contamination of soil, surface runoff water, and groundwater. Spill control devices should include methods to keep precipitation or other materials out of the spill control device, prevent rusting and allow for easy inspection of their integrity (e.g., elevation above ground level), and recover any spilled or leaked fluids from the device,
- Tank loading and unloading areas should be designed to contain any spills or leaks,
- Sites should be appropriately contoured to prevent the collection of surface water on the ground immediately surrounding the secondary containment system (i.e., tank farm area),
- Tanks that have been withdrawn from service are permitted to be reused if they comply with regulations, and
- Measures should be incorporated to prevent the overfilling of tanks. Examples include automatic sensing devices for interconnection with

shutoff equipment at the supply point, automatic overfill shutoff devices of a float valve or other mechanical type, vent restriction devices, and overfill alarm devices of audible or visual type.

## 8.7 Dike construction

The area around a single-walled aboveground storage tank should have a secondary containment system designed to contain leakage and prevent it from impacting the surrounding environment (ERCB, 2001). Secondary containment systems should consist of an impervious liner and a dike. The area within the secondary containment system should be graded to a sump or low-lying area (within the diked area) to allow for the collection of rainwater, snow-melt water, and any possible leakage from the tanks. No uncontrolled discharge of collected fluids or discharge of untested fluids is typically allowed,

A dike should

- be constructed of soil, steel, concrete, solid masonry, or synthetic material and designed to contain liquids within the diked area, to be able to withstand the hydrostatic head associated with it being full of liquid, and so that it will not deteriorate or develop leaks during the projected life of the structure,
- be sized to have a volumetric capacity of not less than 110% of the capacity of the tank when the diked area contains one tank or when the diked area contains more than one tank of not less than the sum of (1) the capacity of the largest tank located in the diked area, and (2) 10% of the greater of the capacity of the largest tank, or the aggregate capacity of all other tanks located in the diked area. The volume of support structures, etc. in the dyke area must be subtracted from the volumetric capacity,
- have no openings in it (e.g., dike drains to the surrounding area); and
- be maintained in good condition. The area encompassed by the dike should be kept free from weeds, debris, and extraneous combustible material.

### 8.7.1 Liners

A liner should meet the following criteria (ERCB, 2001):

- consist of a material that is inert to or compatible (chemically resistant) with the material being stored in the tank,

- be impervious (i.e., a compacted clay liner meeting the criteria required), a natural liner meeting the criteria specified in the regulation, or a synthetic liner meeting the requirements,
- be durable and appropriate for the operating and ambient conditions, and
- cover the area within the dike, including the area beneath the tanks, and be keyed into the dike walls.

Natural and synthetic liners are used in storage and treatment areas to impede the movement of materials that could adversely impact soil or groundwater. The initial decision in the installation of any liner is whether the liner type is appropriate for the given application. It is important to remember that the purpose of a secondary containment liner is not to function as a tank or other primary containment. When the leak detection system is activated, the tank or storage system should be shutdown, inspected, and repaired.

There are two modes of mass transfer through a barrier such as a liner, liquid transfer (through hydraulic conductivity), and vapor transfer (through molecular diffusion). Hydraulic conductivity applies to natural materials, where mass transfer depends on the movement of liquid through the pore structure of soil and the driving force is hydraulic pressure or head. Vapor transfer applies to polymer barriers, where the driving force is the concentration gradient of the permeating substance across the barrier.

Although no material is completely impermeable, the type, design, and installation of a liner are extremely important in achieving the desired level of impermeability for both primary and secondary containment liners.

#### 8.7.1.1 Compacted clay liners

Clay soils may be suitable material from which to construct a compacted clay liner (ERCB, 2001). For secondary containment, the clay soil should be compacted to achieve a hydraulic conductivity of $1 \times 10^{-6}$ cm/s or less determined in-situ or $1 \times 10^{-7}$ cm/s or less as determined in a laboratory from a representative disturbed sample (the material should meet hydraulic conductivity requirements under the full hydrostatic head). For use as primary containment, compacted clay liners should have a hydraulic conductivity that is at least one order of magnitude less than that required for secondary containment.

Hydraulic conductivity of $\leq 1 \times 10^{-7}$ cm/s is achievable if suitable starting material (clayey soil) is excavated, reworked, or homogenized and laid down in lifts following appropriate construction protocol on a properly prepared sub base. Key properties in achieving low hydraulic conductivity are the Plasticity Index and the clay content of the soil. There are reasons to prefer a soil of low plasticity over highly plastic clay, as soils with low plasticity are often easier to mix, hydrate, and homogenize in the field and tend to be less susceptible to desiccation cracking. The ideal situation involves small, soft, weak clods of clay that are easily remolded and compaction with a heavy roller that effectively remolds and melds the clay clods together.

The literature identifies that the most important factors in achieving low hydraulic conductivity in compacted clay liners include:

1. using suitable clayey soils that meet the following specifications (examples in brackets):

- greater than 30% fines (defined as dry weight percentage passing the No. 200 sieve),
- greater than 20% clay (0.002 mm or smaller as determined by the hydrometer method),
- well-graded (no excess of particles in any size range and no intermediate sizes lacking),
- Liquid Limit (LL) equal to or greater than 30, and
- Plasticity Index (PI) equal to or greater than 10, laying the clayey soil down in lifts.

2. minimum of four lifts, with each being 15 to 20 cm thick (loose thickness) properly preparing the surface to receive a lift of soil.

- if placing the soil on subgrade, the subgrade should be adequately compacted,
- if placing the soil on a previously compacted lift, the surface should be sacrificed to a nominal depth of 2.5 cm before placing the next lift of soil using the clayey soil at the correct water content, further:
- each lift should be placed at approximately 2 to 3% wet of optimum moisture,
- compacting each lift to a minimum of 95% of the Standard Proctor maximum dry density using the proper type of compactor with an appropriate number of passes,
- the best type of compactor in most instances is a heavy, footed roller with feet that fully penetrate the loose lift of soil,
- the compactor should be heavy enough to ensure that adequate compactive energy is delivered to the soil and that the feet fully penetrate the full depth of the lift to kneed it and bond it to the previous lift,

- the number of compactor passes over a given area varies between soil and compactor type, but sufficient passes should be conducted to achieve the desired density,
- placing down sufficient lifts to achieve a final compacted thickness of 0.6 m or greater protecting each compacted lift from damage,
- a smooth-drum roller is often used to compact the surface of a completed lift, as this forms a hard skin, which helps to minimize desiccation and sheds water, and
- the smooth surface should be roughened with a disc before placing the next lift.

3. upon completion of the liner, an appropriate overlying material should be used to protect the liner from mechanical damage and weathering.

The construction of a compacted clay liner requires application by qualified personnel overseen by a professional geotechnical engineer. The specifications of the clay material used for the liner and the details of the liner construction (quality assurance/quality control [QA/QC] data) should be documented.

#### 8.7.1.2 Natural liners

Natural liners involve scarification and re-compaction of in situ clay, without excavating the underlying clay, and placing it in lifts as for a compacted clay liner (ERCB, 2001). Natural liners may only be used at sites that have a deposit of appropriate clayey soils with a minimum thickness of 0.9 m and where the seasonal high groundwater table is greater than 1 m below the expected bottom of the liner.

The potential for in-situ clayey deposits to serve as natural liners should only be investigated when sites are located in relatively low-permeability clay or till. Delineation of the in situ clayey deposit requires a site investigation by a qualified person. Attention should be focused on looking for hydraulic defects, such as sand seams, cracks, and fissures. A minimum of three boreholes, arranged in an approximate equilateral triangle, is required to establish the orientation of any significant geologic plane. The depth of sampling from the surface to characterize underlying soil materials should be at least 3 m. One sampling should be extended to establish depth to groundwater. The clayey soil should be analyzed in a lab to determine Liquid Limit, Plasticity Index, clay content, and fines content.

Literature shows that the most important factors in achieving suitable natural liners

Include:

1. preparing the site for the construction:

- remove the topsoil from the site and appropriately salvage and store it,
- remove rocks or clumps greater than 50 mm in size,
- scarifying and recompacting the in-situ clayey deposit to 95% of the Standard Proctor maximum dry density,
- only scarify the in-situ clayey deposit to a depth that can be recompacted using select equipment,
- the in-situ clay should be at 2% to 3% wet of optimum, and
- smoothing out the completed liner with a smooth barrel compactor and applying overlying material to protect the liner from mechanical damage and weathering.

The completed liner should achieve a hydraulic conductivity as specified. A variety of geochemical, geophysical, and engineering tools are available for investigating the hydraulic integrity of natural liners. Literature indicates that for natural liners, in situ hydraulic conductivity tests can be more accurate than laboratory tests. The construction of a natural liner requires application by qualified personnel overseen by a professional geotechnical engineer. The specifications of the clayey deposit, including the site delineation information, and the details of the liner construction (QA/QC data) should be documented.

#### *8.7.1.3 Synthetic liners*

A wide array of synthetic liners is available, many of which were developed for applications such as waste containment (ERCB, 2001). The long-term integrity of a synthetic liner is dependent on the physical strength of the liner, its resistance to effects of aging or environmental degradation, upkeep and maintenance of cover, and its resistance to the substance contained in the storage system. As secondary containment application does not require long-term, continuous contact with the contained substances, the requirements for liner performance in secondary containment systems may be less rigorous than those for primary containment systems.

Most synthetic liners are impermeable to liquid transfer but are permeable to vapor to a degree that depends on the solubility of the liquid in the polymer, temperature, and the thickness of the membrane. The most important physical and mechanical attributes of the liner that determine its suitability for a given application are thickness, density, mass per unit area, tensile properties, tear resistance, hydrostatic resistance, and puncture resistance. Other key physical properties include linear expansion properties, cold temperature properties, resistance to ultraviolet light, resistance to soil burial, and dimensional stability.

The physical configuration of the liner determines the seaming and construction procedures used to install it. Therefore, the installation of synthetic liners should follow the manufacturer's specifications and be conducted by qualified personnel. The specifications of the liner material (suitability for its intended use) and the details of its installation (QA/QC data) should be documented and made available to regulators if requested. To be suitable for secondary containment, the synthetics liner should be a minimum of 30 mils. Synthetic liners used for secondary containment in the petroleum industry include

- coated fabrics or laminates,
- extruded film or sheet, and
- spray-on coatings.

#### 8.7.1.3.1 Coated fabrics and laminates

These geomembranes include polymer films coated or laminated onto a textile substrate using a manufacturing process such as coating. Polymer formulations include chorosulfonated polyethylene, neoprene, ethylene, interpolymer alloy, butyl rubber, epichlorohydrin rubber, ethylene propylene diene monomer (EPDM), and various combinations. The coatings are typically elastomeric or rubbery, and the substrates are usually high-strength textiles with a broad weave (e.g., nylon, polyester).

#### 8.7.1.3.2 Extruded film or sheet

Geomembranes of this kind are manufactured in a one-step process without the use of a textile backing or substrate and are made from polyvinyl chloride (PVC), high-density polyethylene (HDPE), polyethylene of lower densities, and elastomers. Because of its chemical resistance, HDPE is widely used and is available in thicknesses ranging from 20 mils to greater than 100 mils.

#### 8.7.1.3.3 Spray-on coatings

These products are usually installed by spraying elastomers (e.g., polysulfide, polyurethane) onto a geotextile or other material for backing. The coating thickness is variable and is a function of the spray dwell time, flow rate, and operator technique. Both polysulfide and polyurethane have good resistance to petroleum products. The resulting sprayed-on coating has added durability and strength because of the geotextile backing.

## 8.8 Discharge of water from dyked area

Discharge is typically regulated and should meet environmental authority's approval (ERCB, 2001). Provided the water has not been contaminated, surface run-on/runoff waters collected on an upstream petroleum site (e.g., within a diked area of a tank farm, within the surface water collection system) should be released back into the environment. Collected waters should be tested and meet the following criteria before being released in a controlled fashion to adjacent lands (example specifications):

- chloride content 500 mg/L maximum (e.g., test strips),
- pH 6.0 to 9.0 (e.g., test strips and/or meter readings),
- no visible hydrocarbon sheen (roughly equates to less than 10 mg/L),
- no other chemical contamination (e.g., clean operating conditions such that collected waters are not impacted by spills/releases),
- the landowner or occupant consent,
- water is not allowed to flow directly into any watercourse, and
- each release recorded, including the prerelease test data and the estimated volume of water, released.

Contaminated water should not be released into the environment. It should be sent to an approved facility for treatment and/or disposal or, if possible, treated on-site and then released. The minimal parameters listed above are intended as screening parameters for sites exhibiting good housekeeping practices. On sites where spills or releases have occurred, the collected surface water should be tested for parameters that would demonstrate that the water has not been affected.

The discharge of collected surface waters into a watercourse is not permitted unless otherwise approved by an environmental authority. Approval holders or licensees wishing to use collected surface waters in a facility's process should consult with authorities regarding the need for a water diversion license.

## 8.9 Double-walled tanks

Aboveground storage tanks with double walls should have the primary tank separated from the secondary containment to provide a continuous interstitial space below and around the primary tank;

- be equipped with a method of overfill protection that incorporates an audible or visual alarm that alerts the user of a potential overfill condition or an automatic shutoff mechanism to prevent the overflow of the primary tank;
- be equipped with an effective spill control device at the fill/delivery connection;
- have a system to monitor the interstitial space between the tank walls (e.g., pressure, vacuum, electronic, or vapor monitoring or manual sampling);
- be protected against damage from vehicular traffic (e.g., controlled access to the site, bollards, guard rails, or concrete barricades);
- be equipped with a valve as close as practical to the tank to prevent draining of the tank should a leak or break occur in the piping; and
- for systems designed with delivery connections at grade level (bottom load), be equipped with provisions to allow the delivery hose to be emptied and with a drip catchment device for the hose.

Operators should check the interstitial-space monitoring device at least monthly to ensure that the tank system is not leaking and should document any abnormal circumstances, as well as any corrective actions, are taken. Monthly checks may not be required if the interstitial space is equipped with a continuous monitoring system that will indicate when the primary or secondary tank is leaking. Automatic shutdown systems should be checked monthly and maintained to ensure continuous functionality, and documentation about this and any abnormal circumstances from the monitoring/sampling of the interstitial space should be retained.

Any spills or releases should be cleaned up immediately and reported if required. As required, corrective action should be initiated and abnormal circumstances and corrective actions should be documented.

## 8.10 Piping systems

Piping systems are typically required to follow several standard specifications. Piping systems should have properties consistent with the tank design and requirements. Inspection of piping systems should be carried out regularly and should be documented. Piping systems, although integral to storage tank facilities have a separate set of standards and a separate set of practices.

### 8.10.1 Standards applicable

The standards applicable to piping systems include:

The applicable fire code standard; ASTM A 53, "Pipe, Steel, Black and Hot-Dipped, Zinc-Coated, Welded and Seamless";

Canada/Canadian Standards Association (CAN/CSA) Z245.1–98, "Steel Line Pipe"; Canada/Underwriters' Laboratories of Canada (CAN/ULC)-S633–1999, "Flexible Underground Hose Connectors";

ORD-C107.7–1993, "Glass-Fibre Reinforced Plastic Pipe and Fittings";

ORD-C107.4–1992, "Ducted Flexible Underground Piping Systems";

ORD-C107.14–1992, "Non-Metallic Pipe and Fittings"; or

ORD-C536–1998, "Flexible Metallic Hose".

The standards applicable also depend on the attachment. If for example, one end of a storage tank pipe is connected to a heater, there are standards for piping connected to heaters.

Piping must contain provisions to prevent siphoning of liquid from the tanks. Piping is also required to have seals to prevent either liquid or vapor leaks

### 8.10.2 Above-ground piping

Above-ground piping must have shields to protect it from impact should this be an issue. Joints must be visible for inspection and maintenance.

### 8.10.3 Below-ground piping

Below-ground piping follows other standards: American Petroleum Institute (API) RP 1632–96, "Cathodic Protection of Underground Storage Tank and Piping Systems" and API Std 2610–94, "Design, Construction, Operation, Maintenance and Inspection of Terminal and Tank Facilities".

Secondary containment is typically required for below-ground piping. Drainage should be in a monitored sump.

## 8.11 Leak detection

Approval holders or licensees should be able to demonstrate the integrity of their tanks and verify whether any material has escaped. Leak detection methods for aboveground storage tanks include the following (further technical information on this below):

- Incorporation of a layer of porous material, such as sand, over the liner and underneath the tanks to protect the liner and to allow any leakage to move preferentially through the porous material to a collection area within the diked area,
- Monthly visual inspections of tanks and the surface of the diked area for evidence of problems, damage, or leakage. Any spills or leaks should be cleaned up immediately and reported if required and corrective action should be initiated as required. Abnormal circumstances and corrective actions should be documented, and
- Additional leak detection provisions may also include the incorporation of subliner leakage detection devices (e.g., weeping tile system).

## 8.12 Corrosion protection

Corrosion protection is typically required in several jurisdictions and may be specified to comply with standards (see Appendix B).

## 8.13 Inspection

Authorities require companies to maintain records to demonstrate compliance. Inventory records for production materials are typically handled through the use of standard reporting forms (ERCB, 2001).

Tank owners should:

- Maintain inventory records and retain the records on-site or at the local field office for a specified time period (e.g., 2 years). Where applicable, this includes copies of dockets for material received and shipped,
- Maintain inspection and corrosion monitoring programs to provide an indication of the integrity of tanks and piping. Records of test or maintenance checks should be retained for a minimum period of time (e.g., 5 years), but preferably for the lifetime of the tank or facility,
- Document and retain for a minimum period of time (e.g., 5 years), but preferably for the lifetime of the tank(s), any abnormal circumstances identified from the monthly visual inspections of aboveground

storage tanks, the monthly interstitial space monitoring of double-walled aboveground tanks, as well as any corrective actions taken to remedy the situation and prevent it from reoccurring, and
- Document and retain for a minimum period of time (e.g., 5 years), but preferably for the lifetime of the storage device(s), the monitoring results from the leak detection system for lined earthen excavations and bulk pads, as well as any investigative work or corrective actions, are taken to remedy a breach of the storage devices.

Note that in the event that a storage device has overflowed or its leak detection system indicates that it may be leaking, the owner or operator should investigate the situation, verify the integrity of the storage device, report the release if required, and, if required, implement corrective actions. The actions should be documented and may include:
- repairing and testing the storage device,
- replacing the storage device, or
- implementing cleanup activities as required, including assessing the soil for contamination,
- Where applicable, retain groundwater monitoring records for a minimum of 5 years, but preferably for the lifetime of the upstream petroleum site,
- Maintain records from alternative leak detection systems (e.g., electromagnetic surveys)
- Keep all required approvals, licenses, and permits on-site or at the field/plant offices.
- Maintain records on excavation or nearby construction that could affect the integrity of the storage system. and survey (soil vapor surveys, weeping tile monitoring wells, inventory reconciliation, etc.) for a minimum period (e.g., 5 years), but preferably for the lifetime of the upstream petroleum site,
- Maintain the names of all people who conducted the inspection and monitoring programs.

## 8.14 Record keeping

Regulations often state that records on various aspects of the tank operations must be kept and should be available for inspection. Retention time on these varies but often are 2 or 5 years, sometimes for the life of the tank. Records required include:

- Inventory records for wastes and chemicals,
- Groundwater monitoring records,
- Alternative leak detection records (e.g., vapor surveys, EM surveys),
- Copies of all required approvals, licenses, registrations, and permits on-site or at the Field Centre,
- Names of all persons conducting inspection and monitoring programs,
- Records of surface water discharges,
- Tank inspection records/results,
- Corrosion monitoring records,
- Abnormal circumstances and corrective actions from monthly visual tank and dike inspections,
- Tank inspection records/results,
- Monthly monitoring well results and any corrective actions from single-wall tanks with secondary containment,
- Abnormal circumstances from monthly visual inspections,
- Abnormal circumstances and corrective actions from monthly interstitial monitoring of
- double wall tanks,
- Monitoring results (including field and laboratory analytical results) and any subsequent corrective actions from leak detection systems, and
- Contamination was found in the leak detection system.

## 8.15 Leak testing or integrity testing

Leak testing or integrity testing is often specified to take place at regular intervals (e.g., 3 to 5 years). This testing is designed to establish if tanks have a leak that isn't detected by observation (e.g., a gross leak). There are many tests available, each of which has advantages and disadvantages. Some of these tests at listed in Table 8.1 (adapted from ERCB, 2001).

## 8.16 Withdrawal of storage tanks from service

Sometimes tanks must be taken out of service. Special rules, as will be summarized below, often apply to such situations

Table 8.1 Summary of some leak or integrity tests.

| Name | Brief description |
|---|---|
| **Ultrasonic test** | There are different kinds of ultrasonic tests. The **A-Scan** point-to-point technique, historically used, involves taking a thickness reading at one point on a section of the tank and repeating this process several times on the same tank to generate a sample<br>The **B-Scan** technique can take ultrasonic thickness readings over every 0.04 inch of the surface, which results in almost 100% coverage of the tank surface. This eliminates the problems associated with the A-Scan technique missing problem areas. The B-Scan can also scan through coatings, unlike the A-Scan technique. |
| **Internal visual inspection** | An internal visual inspection can be performed by emptying and cleaning the inside of the tank and then visually inspect it for signs of weakness or holes. |
| **External visual inspection** | An external visual inspection can be performed on all aboveground storage tanks that are visible on all sides. |
| **Vacutect system** | All openings to the tank are sealed off and a low vacuum (less than half of 1 psi) is applied to the tank using a vacuum pump. The vacuum level is constantly monitored and maintained by the computer in the testing unit. While under vacuum, Vacutect monitors for three things: water level, noise (via a hydrophone), and the pressure in the tank |
| **Mass-based systems** | The test is designed to measure any changes in the buoyancy force acting on a probe inserted into the tank. Uses mass measurement technology to determine if the product is entering or leaving the tank. The technology is based on the fact that buoyancy force only varies as a direct result of a change in the mass of the liquid. The buoyancy force is not affected by changes in product temperature, since the change in volume due to temperature change is offset by a corresponding change in liquid density. |
| **Vigilant test** | The method is based on sensing the vacuum changes that occur in the interstitial space between an outer rigid tank and an inner wall formed by installing a |

*(Continued)*

Table 8.1 (Continued)

| Name | Brief description |
|---|---|
| | flexible membrane liner in the tank. Vacuum changes are analyzed continuously with a microprocessor to determine the rate of change. Very slow changes occur on tight tanks due to molecular permeation through the membrane into the interstitial space. This baseline behavior is determined experimentally for each tank after the installation of the membrane is complete. The vacuum behavior will vary significantly if a leak is present. Both liquid and air leaks may be detected using this method. |
| **Pressure test** | This test involves the introduction of slight pressure (nitrogen gas) to the tank. A decrease in pressure is measured over a time interval. If the pressure decreases, the tank may have a leak. This test requires the pneumatic isolation of the tank and/or lines being tested. The tank must also be empty of fluids |
| **Vacuum test** | A vacuum test involves the introduction of a slight vacuum to the tank. A decrease in a vacuum is measured over a time interval. If the vacuum decreases, the tank may leak.<br>This test requires the pneumatic isolation of the tank and/or lines being tested. The fluid level in the tank should be noted. |
| **In-fill test** | This test involves the overflowing of a tank (preferably with water) and the subsequent recording of liquid levels over time. This requires the hydraulic isolation of the tank. |
| **Hydrostatic test** | The tank is filled and stabilized. The tank pressure is raised by 5 to 7 psi by a pump or by adding a similar hydrocarbon. If pressure is maintained for 1 h, the tank is leak-free. |
| **Tracer test** | A tracer gas (or liquid) is injected into the tank. Soil gas samples are taken from probes installed into the ground around the tank. A leak is declared if a tracer is detected outside the tank. |
| **Mass integrity test** | A small flow of nitrogen is forced into the bottom portion of the tank (in the product) and the pressure required to maintain a continuous flow of bubbles is measured (measuring for head pressure). |

(*Continued*)

Table 8.1 (Continued)

| Name | Brief description |
|---|---|
| | The test procedure measures the change in the product mass during an overnight data collection. The rate of mass change is determined and described in a leak rate. |
| **Inventory** | The owner or user maintains records on all of the product that enters and leaves the tank. |
| **Reconciliation** | By examining these records, the owner or user should be able to tell if there is a leak in the tank. There are a number of software programs that use leak detection algorithms for analyzing inventory, sales, and delivery data to conduct leak-detection testing. |
| **Robotic inspection** | A visual inspection can be performed internally on an aboveground storage tank while it is in operation. The robot is lowered into the tank and performs ultrasonic testing on the floor of the tank, providing video footage of the tank bottom for analysis. The robot also has the capability of cleaning the tank. |
| **Permanent leak detection devices** | There are ways of installing a tank so that when there is a leak it will be noticed by the approval holders or licensees |
| **Pressure decline test procedure** | Involves the introduction of slight pressure (nitrogen gas) to the tank (approximately 5 psi or less). A decrease in pressure is measured over a time interval. If the pressure decreases over time, the tank may have a leak. The test requires the pneumatic isolation of the tank and lines being tested. The fluid level in the tank should be noted |
| **Vacuum decline test procedure** | Involves the introduction of a slight vacuum to the tank. A decrease in the vacuum is measured over a time interval. If the vacuum decreases, the tank may have a leak. This test requires the pneumatic isolation of the tank and lines being tested. The tank must also be empty of fluids. |

### 8.16.1 Temporary withdrawal from service (usually time specified—e.g., <180 days)

When an aboveground or underground storage tank is taken out of service for a period not exceeding the time limit, the owner or operator should

- isolate the tank,
- empty the tank or measure and record the fluid level in the tank and then repeat this procedure monthly, making records available to the

authorities upon request, maintain the impressed current cathodic corrosion protection system, if applicable, and
- maintain monthly leak detection monitoring.

### 8.16.2 Temporary withdrawal from service exceeding a certain time (e.g., >180 days)

When an aboveground or underground storage tank is taken out of service for a period exceeding the time limit specified, the operator should remove all liquids and vapors from the storage tank and its connecting piping, isolate the tank and mark it to clearly indicate that it is empty, maintain the impressed current cathodic corrosion protection system, if applicable, and verify the integrity of the tank before reactivation if the tank has been out of service for longer than 1 year and then appropriately relabel the tank.

Should the approval holder or licensee wish to change the service of a tank, before reactivation the tank should be cleaned and refurbished if necessary and then verified for compatibility with the new service.

### 8.16.3 Permanent withdrawal from service

Aboveground and underground tanks permanently taken out of service should have all fluids and sludges removed and be purged of all combustible vapors. It is expected that aboveground tanks will be removed from the active part of the site and either be relocated to an appropriate storage area on the site or sent for disposal. A sampling protocol should be in place to ensure that contamination is a lot left behind.

### 8.16.4 Replacement of an existing aboveground storage tank or addition of a new tank to an existing tank farm

Upon removal of the old tank, any contamination should be managed, and then the ground should be compacted and appropriately prepared for placement of the new tank. Should an existing multitank farm have sufficient capacity to accommodate an additional tank, it may be added. However, if the addition of a new tank results in the reconstruction of the tank farm area, it is expected that the reconstruction will meet the secondary containment and leak detection requirements.

Owners or operators replacing a tank or adding a new one to an existing tank farm should meet the construction requirements outlined above. These tanks should be integrity verified every five years unless the tank farm area has been reconstructed to meet the secondary containment and leak detection requirements.

## References

CCME (1994). Environmental Code of Practice for Aboveground and Underground Storage Tank Systems Containing Petroleum and Allied Petroleum Products (PN 1326), Canadian Council of Environmental Ministers, Publication CCME-EPC-LST-71E, August.

EPA (2010). Spill Prevention, Control, and Countermeasure (SPCC) Regulation 40 CFR part 112, EPA 540-K-09-001.

ERCB (2001). Directive 055: Storage Requirements for the Upstream Petroleum Industry, Energy Resources Conservation Board, December.

# Appendix A Glossary of storage terms (ERCB, 2001)

**Above-ground storage tank**: A tank that sits on or above the ground surface and whose top and complete external sides can be visually inspected.

**Adverse effect**: An impairment of or damage to the environment, human health or safety, or property.

**Bulk pads**: A ground surface area designated for the segregated storage of materials without the use of a container or tank.

**Cathodic protection**: A method of preventing corrosion to a metal surface by introducing another metal (anode) into the ground to create a corrosion cell in which the surface to be protected becomes a cathode. If deterioration or corrosion occurs at the anode (introduced metal), the cathodic protection may be of a sacrificial type of impressed current design.

**Condensate**: A mixture mainly of pentanes and heavier hydrocarbons that may be contaminated with sulfur compounds, that is recovered or is recoverable at a well from an underground reservoir, and that may be gaseous in its virgin reservoir state but is liquid at the conditions under which its volume is measured or estimated.

**Container**: Any portable aboveground containment device (e.g., drums, pails, bags, boxes, totes) with a capacity not exceeding 1 $m^3$.

**Crude bitumen**: A naturally occurring viscous mixture, mainly of hydrocarbons heavier than hexadecane, that may contain sulfur compounds and that in its naturally occurring viscous state will not flow to a well.

**Crude oil**: A mixture mainly of heavier hydrocarbons that may be, that is recovered or is recoverable at a well from an underground reservoir, and that is liquid under the conditions under which its volume is measured or estimated and includes all other hydrocarbon mixtures except raw gas, condensate, or crude bitumen.

**Environment**: All components of the earth including air, land, and water; all layers of the atmosphere; all organic and inorganic matter and living organisms; and interacting natural systems.

**Facility**: Any building, structure, installation, equipment, or appurtenance over which the Government in question, has jurisdiction and that is connected to or associated with the recovery, development, production, handling, processing, treatment, or disposal of hydrocarbon-based resources or any associated substances or wastes and includes, without limitation, a battery, a processing plant, a gas plant, an oilfield waste management facility, a compressor, and similar equipment

**Freeboard**: The unused upper portion of a primary containment device.

**Impervious**: A natural material that demonstrates a hydraulic conductivity of $10^{-6}$ cm/s or less as determined in situ or of $10^{-7}$cm/s or less as determined in a laboratory from a representative disturbed sample, or a synthetic membrane liner or barrier appropriately selected to control the migration of specific fluids.

**Leachate**: Interstitial fluids separated from materials or fluids generated by the percolation of liquids (e.g., water) through materials.

**Leachate collection system**: A seepage pathway and collection system constructed on the surface of the primary containment device to allow for the drainage, collection, and removal of any generated leachate.

**Leak detection system**: A system designed for the early detection of any leakage from a primary containment device; may include visual, electronic, or statistical inventory methodologies.

**Monitoring well**: A well used to detect liquid or vapor leakage from a primary or secondary containment device or to sample a groundwater aquifer or unsaturated zone to detect the presence of any contaminants.

**Petroleum product**: A single product or a mixture of at least 70% hydrocarbons refined from crude oil, with or without additives, that is or could be used as a fuel, lubricant, or power transmitter. Without restricting the foregoing, such products include gasoline, diesel fuel, aviation fuel, kerosene, naphtha, lubricating oil, fuel oil, and engine oil (including used oil) and exclude propane, paint, and solvents.

**Primary containment device**: A device used to physically contain materials produced, generated, or used in regulated processes. Primary containment devices include but are not limited to, single-walled tanks, the internal wall of double-walled tanks, containers, and the liners of lined earthen excavations and bulk pads.

**Secondary containment system**: An impervious barrier or liner used to contain and prevent any leakage from the primary containment device from impacting the environment.

**Site**: The area defined by the boundaries of a lease site for an upstream petroleum facility or well site.

**Spill control device**: A device (e.g., load box) used to physically collect and recover spills and leaks of materials from process equipment, piping valves, flanges, and other equipment, especially at material transfer points. Spill control devices should be maintained to ensure their integrity and that they are of sufficient capacity to be functional (e.g., free of precipitation).

**Storage**: The holding of materials produced, and generated, for some time until the products or wastes are transported, treated, or disposed of.

**Tank**: A device designed to contain liquid materials that have an internal capacity of more than 1 $m^3$ and is constructed of impervious materials that provide structural support and may include such materials as steel, but do not include piping.

**Watercourse**: The bed and shore of a river, stream, lake, creek, lagoon, swamp, marsh, or other natural body of water or a canal, ditch, reservoir, or other man-made surface feature, whether it contains or conveys water continuously or intermittently.

**Weather protection**: A structure, protective coating, or cover that ensures that the integrity of the primary containment device and its labeling are not compromised by the elements of nature.

**Cathodic protection or cathodically-protected** means a method of reducing or preventing corrosion of a metal surface by making that surface the cathode of an electrochemical cell.

**Combustible liquid or product means** any liquid having a closed cup flashpoint at or above 37.8°C and below 93.3°C.

**A contingency plan** means planned procedures for reporting, containing, removing, and cleaning up a spill or leak.

**Containment sump** means a dispenser, pump, transition, or turbine sump.

**Corrosion** means the deterioration of a metal resulting from a reaction to its environment.

**A corrosion expert** means a person recognized by NACE International (formerly the National Association of Corrosion Engineers) as a corrosion specialist, cathodic protection specialist, or a registered professional engineer experienced in corrosion protection.

**Corrosion protection** means a method of reducing or preventing corrosion of a storage tank system through cathodic protection, the application of protective coatings, or the use of a noncorroding material in its construction.

**Discharge** means releasing, spilling, leaking, pumping, pouring, emitting, emptying, or dumping petroleum or allied petroleum products into the environment, whether intentional or unintentional.

**A dispenser sump** means a container located underneath or near a dispenser or self-contained suction pump that collects or contains leaks.

**Empty** means to remove the contents of a storage tank system as far as is practicable by such means as draining, suction, pouring, or pumping.

**Existing** means that which was in place or commenced operation on or before the effective date of the code of interest.

**Flammable liquid** or **product** means any liquid having a closed cup flashpoint below 37.8°C and a vapor pressure not exceeding 275.8 kPa (absolute) at 37.8°C.

**Flashpoint** means the minimum temperature at which a liquid within a container gives off vapor in sufficient concentration to form an ignitable mixture with air near the surface of the liquid.

**Free oil** means the nonsoluble, nonemulsified petroleum and allied petroleum product layer that accumulates in an oil-water separator.

**Fuel oil** means kerosine or any hydrocarbon oil as classified in CAN/CGSB-3.2–99, "Fuel Oil, Heating" and CAN/CGSB-3.3–99, "Kerosine."

**Handling** means the storing, transmitting, transporting, or distributing of petroleum or allied petroleum products and includes putting petroleum products into a container or the fuel tank of a motor vehicle, vessel, or aircraft.

**Impermeable barrier** means a secondary storage tank wall, synthetic membrane liner, or other equivalent material.

**Internal coating** means a coating or lining of a noncorrodible material bonded firmly to the interior surface of a storage tank that does not chemically or physically degrade when in contact with the petroleum or allied petroleum products stored therein.

**Interstitial space** means the space between the primary storage tank or piping wall and the impermeable barrier within a secondary containment system.

**A leak** means any loss of liquid petroleum or allied petroleum products because of a defect in a storage tank system.

**Leak detection** means a device or method that is capable of detecting leaks in a storage tank system.

**Liner** means a material used as the outer barrier of a secondary containment system but does not include the outer wall of double-wall piping or storage tanks.

**Line-leak detector** means a device used in pressure piping systems to detect a leak in the piping.

**The liquid limit** is the water content where the soil starts to behave as a liquid. The liquid limit is measured by placing a clay sample in a standard cup and making a separation (groove) using a spatula. The cup is dropped till the separation vanishes. The water content of the soil is obtained from this sample.

**Motive fuel** means any fuel used to power a vehicle, aircraft, or vessel.

**Oil-water separator** means a device for collecting and separating nonsoluble, nonemulsified petroleum and allied petroleum products from water.

**Operator** means the person who is responsible for the day-to-day operation of an installation where a storage tank is located or when referring to a vehicle, the driver in charge of the vehicle.

**Out-of-service** means that a storage tank system or portion thereof is no longer serving its intended use.

**Overfill protection device** means a mechanical device, electrical device, or fill procedure system that is intended to prevent a storage tank from being overfilled.

**Owner** means the Crown, an institution, corporate entity, government department or agency, or a person who has legal ownership of the storage tank system or who has been assigned custody to control, care for, manage, or dispose of the storage tank system.

**Petroleum product** means a single product or mixture of at least 70% hydrocarbons, by volume, refined from crude oil, with or without additives, that is used, or could be used, as a fuel, lubricant, or power transmitter and without restricting the foregoing, such products include gasoline, diesel fuel, aviation fuel, kerosene, naphtha, lubricating oil, fuel oil, engine oil and used oil, and exclude propane, paint, and solvents.

**Piping** means fuel conduits, including fittings and valves that are necessary for the safe handling and storage of petroleum products and allied petroleum products and are specified by a nominal inside diameter.

**The plasticity index** is a measure of the plasticity of the soil. The plasticity index is the size of the range of water contents where the soil exhibits plastic properties. The Plasticity index is the difference between the liquid limit and the plastic limit. Soils with a high Plasticity index tend to be clay, those with a lower PI tend to be silt, and those with a Plasticity index of 0 (nonplastic) tend to have little or no silt or clay.

**Precision leak detection test** means a test capable of detecting a storage tank leak as small as 0.38 L/h with a probability of detection of 0.95 or greater and a probability of false alarm of 0.05 or less, within a period of 24 h, accounting for variables such as vapor pockets, thermal expansion of the product, temperature stratification, groundwater level, evaporation, pressure and end deflection.

**Pressure liquid media leak detection test** means a test utilizing a device to pressurize piping with a suitable test liquid to determine the presence of leaks.

**Product transfer area** means the area around the connection point between a delivery truck, railcar, or vessel and a storage tank system with a large capacity (e.g., 500 L or more).

**Protected** means having an impact, projectile, and fire resistance qualities for an aboveground storage tank system.

**Protective coating** means a coating applied to a surface to protect the substrate from corrosion.

**Secondary containment** means an impermeable barrier that prevents leaks from the primary storage tank system from reaching outside the containment area.

**Separated solid** means the particulate material that settles at the bottom of an oil-water separator.

**Site** means a lot or property where there are one or more aboveground storage tank systems within a certain distance of each other (e.g., 200 m), and all storage tanks on the property are owned by the same owner(s).

**Sludge** means the petroleum or allied petroleum product residue or material that accumulates at the bottom of a storage tank.

**A spill** means any loss of liquid petroleum or allied petroleum product from a storage tank system that is not attributable to a leak in the storage tank system.

**A spill containment device** means a container fitted to the inlet of a storage tank or to the suction coupling of a used oil storage tank that helps prevent spills from entering the environment.

**Static liquid media leak detection tes**t means a leak detection test in which a suitable test liquid is placed into the containment device and is monitored for a change in the liquid level and the rate of change.

**A storage tank** means a closed container for the storage of petroleum or allied petroleum products with a capacity of more than a specified amount that is designed to be installed in a fixed location.

**Storage tank system** means a system for the storage and dispensing of petroleum or allied petroleum product and is not limited to storage tanks, associated piping, vents, pumps, and dispensing equipment.

**An underground storage tank** means a storage tank with all of the storage tank volume below grade and the primary tank or double-wall surrounded by or in intimate contact with backfill.

**Used oil** means oil from industrial and nonindustrial sources that have been acquired for lubricating or other purposes and has become unsuitable for its original purpose due to the presence of impurities or the loss of original properties. Used oil does not include oils derived from animal or vegetable fats, crude oil or recovered fuel oils spilled onto land or water, and wastes from petroleum-refining operations.

**A vent** means an opening in a storage tank system that is specifically designed to relieve excessive internal pressure or vacuum within a storage tank system.

# Appendix B Standards applicable to above-ground storage tanks

## American Petroleum Institute Standards

API 12B, Specification for Bolted Tanks for Storage of Production Liquids

API 12D, Specification for Field Welded Tanks for Storage of Production Liquids

API 12F, Specification for Shop Welded Tanks for Storage of Production Liquids

API 12R1, Installation, Operation, Maintenance, Inspection, and Repair of Tanks in Production Service

API 334, A Guide to Leak Detection for Aboveground Storage Tanks

API 340, Liquid Release Prevention and Detection Measures for Aboveground Storage Facilities

API 570, Piping Inspection Code: Inspection, Repair, Alteration, and Re-rating of In-Service Piping Systems

API 575, Inspection Practices for Atmospheric and Low-pressure Storage Tanks

API 620, Design and Construction of Large, Welded, Low-Pressure Storage Tanks

API 650, Welded Tanks for Oil Storage

API 651, Cathodic Protection of Aboveground Petroleum Storage Tanks

API 652, Linings of Aboveground Petroleum Storage Tank Bottoms

API 653, Tank Inspection, Repair, Alteration, and Reconstruction

API 1604, Closure of Underground Petroleum Storage Tanks

API 2000, Venting Atmospheric and Low-pressure Storage Tanks

API 2015, Requirements for Safe Entry and Cleaning of Petroleum Storage Tanks

API 2021 (R2020) Management of Atmospheric Storage Tank Fires

API 2026, Safe Access/Egress Involving Floating Roofs of Storage Tanks in Petroleum Service,

API 2207, Preparing Tank Bottoms for Hot Work

API 2350, Overfill Prevention for Storage Tanks in Petroleum Facilities

API 2610, Design, Construction, Operation, Maintenance, and Inspection of Terminal and Tank Facilities

**ASCE—American Society of Civil Engineers**

ASCE 24−14, Flood Resistant Design and Construction

**American Society for Testing and Materials (ASTM)**

ASTM, D 2434, Standard Test Method for Permeability of Granular Soils (Constant Head)

ASTM, D 3385, Standard Test Method for Infiltration Rate of Soils in Field Using Double-Ring Infiltrometer

ASTM, D 4044, Standard Test Method (Field Procedure) for Instantaneous Change in Head

(Slug) Tests for Determining Hydraulic Properties of Aquifers

ASTM, D 5084, Standard Test Method for Measurement of Hydraulic Conductivity of Saturated Porous Materials Using a Flexible Wall Permeameter

ASTM, D 5126, Standard Guide for Comparison of Field Methods for Determining Hydraulic Conductivity in the Vadose Zone,

ASTM E1930/E1930M—17 Standard Practice for Examination of Liquid-Filled Atmospheric and Low-Pressure Metal Storage Tanks Using Acoustic Emission

ASTM E2256−19 Standard Guide for Hydraulic Integrity of New, Repaired or Reconstructed Aboveground Storage Tank Bottoms for Petroleum Service

ASTM F3063/F3063M-20 Standard Specification for Aircraft Fuel Storage and Delivery

**Canadian Council of Ministers of the Environment**

Canadian Council of Ministers of the Environment, 1993. Environmental Code of Practice for the Measurement and Control of Fugitive VOC Emissions from Equipment Leaks,

ISBN: 1−895925−12-6

Canadian Council of Ministers of the Environment, 2003, Environmental Code of Practice for Aboveground Storage Tank Systems Containing Petroleum Products. (Publication CCME-EPC-LST-71E, August 1994).

Canadian Council of Ministers of the Environment, 1995, Environmental Guidelines for Controlling Emissions of Volatile Organic Compounds from Aboveground Storage Tanks, Publication CCME-EPC-87E, June 1995.

**DIN German National Standards**

DIN EN ISO 28300 Berichtigung 1:2011 Petroleum, petrochemical, and natural gas industries—Venting of atmospheric and low-pressure storage tanks (ISO 28300:2008)

**ISO—International Organization for Standardization**

ISO 28300:2008, Petroleum, petrochemical, and natural gas industries—Venting of atmospheric and low-pressure storage tanks

ISO 11223:2004, Petroleum and liquid petroleum products—Direct static measurements—Measurement of the content of vertical storage tanks by hydrostatic tank gauging

ISO 16961:2015, Petroleum, petrochemical, and natural gas industries—Internal coating and lining of steel storage tanks

ISO 4512:2000, Petroleum and liquid petroleum products—Equipment for measurement of liquid levels in storage tanks—Manual methods

ISO 4266−3:2002, Petroleum and liquid petroleum products—Measurement of level and temperature in storage tanks by automatic methods

ISO 7502−3:1993, Petroleum and Liquid Petroleum Products—Calibration of Vertical Cylindrical Tanks—Part 3: Optical-Triangulation Method

ISO 4268:2000, Petroleum and liquid petroleum products—Temperature measurements—Manual methods

ISO/TR 16732−3:2013, Fire safety engineering—Fire risk assessment—Part 3: Example of an industrial property

**NACE International (Formerly National Association of Corrosion Engineers)**

NACE RP0169−2002, Control of External Corrosion on Underground or Submerged Metallic Piping Systems

NACE RP0285−2002, Corrosion Control of Underground Storage Tank Systems by Cathodic Protection

NACE RP0193−2001, External Cathodic Protection of On-Grade Carbon Steel Storage Tank Bottoms

NACE TM0101−2001, Measurement Techniques Related to Criteria for Cathodic Protection on Underground or Submerged Metallic Tank Systems

NACE No. 10/SSPC-PA6, Fiberglass-Reinforced Plastic (FRP) Linings Applied to Bottoms of Carbon Steel Aboveground Storage Tanks

**National Fire Code of Canada**

NRCC 38727, National Fire Code of Canada (NFCC)—1995

**NFPA—National Fire Protection Association**

NFPA (Fire) 30, Flammable and Combustible Liquids Code

**Petroleum Equipment Institute—PEI**

PEI RP800−20, Recommended Practices for Installation of Bulk Storage Plants

PEI RP600−18, Recommended Practices for Overfill Prevention for Shop-Fabricated Aboveground Tanks

Petroleum Equipment Institute, 1994, Recommended Practices for Installation of Underground Liquid Storage Systems. (EI RP100−94).

**Steel Tank Institute**

STI SP001−00, Standard for Inspection of In-service Shop Fabricated Aboveground Tanks for the Storage of Flammable and Combustible Liquids

STI R831−98, Optional Recommended Practice for Control of Localized Corrosion Within Underground Steel Petroleum Storage Tanks

STI R893−89, Recommended Practice for External Corrosion Protection of Shop Fabricated Aboveground Tank Floors

STI RP011−01, Recommended Practice for Anchoring of Steel Underground Storage Tanks

**Underwriters' Laboratories of Canada**

ULC-S601−2000, Aboveground Horizontal Shop Fabricated Steel Tanks

ULC-S601(A)-2001, Shop Refurbishing of Aboveground Horizontal Shop Fabricated Steel Tanks

CAN/ULC-S602−1992, Aboveground Steel Tanks for Fuel Oil and Lubricating Oil

CAN/ULC-S603−1992, Underground Steel Tanks

CAN/ULC-S603.1−1992, Galvanic Corrosion Protection Systems for Underground Steel Tanks

ULC-S603(A)-2001, Refurbishing of Underground Steel Tanks

ULC-S618−2000, Magnesium and Zinc Anodes and Zinc and Copper/Copper Sulfate Reference Electrodes

ULC-S630−2000, Aboveground Vertical Shop Fabricated Steel Tanks

ULC-S630(A)-2001, Shop refurbishing of Aboveground Vertical Shop Fabricated Steel Tanks

CAN/ULC-S633−1999, Flexible Underground Hose Connectors

CAN/ULC-S643−2000, Aboveground Shop Fabricated Steel Utility Tanks

CAN/ULC-S651−2000, Emergency Valves

ULC-S652−1993, Tank Assemblies for Collection of Used Oil

ULC-S653−1994, Contained Aboveground Steel Tank Assembles

ULC-S655−1998, Aboveground Protected Tank Assemblies

ULC-S656–2000, Oil-Water Separators

**(Other Recognized Documents)**

ORD-C58.9–1997, Secondary Containment Liners for Underground and Aboveground Tanks

ORD-C58.15–1992, Overfill Protection Devices for Flammable Liquid Storage Tanks

ORD-C80.1–2000, Aboveground Non-Metallic Tanks for Fuel Oil

ORD-C107.7–1993, Glass-Fiber Reinforced Plastic Pipe and Fittings

ORD-C107.12–1992, Line Leak Detection Devices for Flammable Liquid Piping

ORD-C107.14–1992, Non-Metallic Pipe and Fittings

ORD-C107.19–1992, Secondary Containment of Underground Piping

ORD-C107.21–1992, Under-Dispenser Sumps

ORD-C142.5–1992, Aboveground Concrete Encased Steel Tank Assemblies

ORD-C142.6–2000, Storage Vaults

ORD-C142.13–1997, Mobile Refueling Tanks

ORD-C142.15–2000, Precast Concrete Tanks

ORD-C142.17–1998, Aboveground Special Purpose Relocatable Vertical Tanks

ORD-C142.18–1995, Aboveground Rectangular Steel Tanks

ORD-C142.19–1994, Spill Containment Devices for Aboveground Tanks

ORD-C142.20–1995, Aboveground Secondary Containment Tanks

ORD-C142.21–1995, Aboveground Used Oil Systems

ORD-C142.22–1995, Contained Aboveground Vertical Steel Tank Assemblies

ORD-C142.23–1991, Aboveground Waste Oil Tanks

ORD-C536–1998, Flexible Metallic Hose

PART 4

# Risk analysis

CHAPTER NINE

# Canadian storage tank spill risk analysis

**Merv Fingas**
Spill Science, Edmonton, Alberta, Canada

## 9.1 Introduction

This chapter examines the statistics available to place the risk of spills from large above-ground storage tanks in context with other sources of spills. Spill data collection from jurisdiction to jurisdiction varies, resulting in imperfect data sets. Further, there are often interruptions in data collection, resulting in data gaps. Nevertheless, it is important to look at the data to learn what the importance of various sources is. The use of data can provide spill workers with priorities in spill prevention.

## 9.2 Total spills in Canada

Spill statistics are collected by a number of agencies. In Canada, provincial offices collect data and Environment Canada maintains a database of spills. The database was started in the early 1970s and then after about three decades, was stopped. Data collection began again in 2015. The latter data set will be used for specific storage tank spill data. The original data set will be used to determine the percentage of the total spillage from storage tanks (Fingas, 2012).

It can sometimes be misleading to compare oil spill statistics exactly, because different methods may be used to collect the data. In general, statistics on oil spills are difficult to obtain and any data set should be viewed with caution. The spill volume or amount is the most difficult to determine or estimate. For example, in the case of a storage tank accident, the exact tank volume may be known before the accident, but the remaining

*Above Ground Storage Tank Oil Spills.*
DOI: https://doi.org/10.1016/B978-0-323-85728-4.00009-7

oil may have been transferred to other tanks immediately after the accident. Some spill accident data banks do not include the amounts burned, if and when that occurs, whereas others include all the oil lost by whatever means. Sometimes the exact character or physical properties of the oil lost are not known and this leads to different estimations of the amount lost.

Reporting procedures vary in different jurisdictions and organizations, such as government or private companies. Minimum spill amounts that must be reported according to various regulations depend on the product spilled. Spill statistics compiled in the past are less reliable than more recent data because few agencies or individuals collected spill statistics before about 1970. Nowadays, techniques for collecting statistics are continually improving.

The number of spills reported also depends on the minimum size or volume of the spill. In both Canada and the United States, most oil spills reported are more than 4000 L (about 1000 gallons). Historically, In Canada, there were about 12 such oil spills every day, of which only about one is spilled into navigable waters. These 12 spills amounted to about 40 tons of oil or petroleum product. Currently, there are fewer spills, mostly because of improved technologies.

Table 9.1 shows a number of sources of spillage in Canada from an earlier data set (Fingas, 2012).

Table 9.1 shows that the sources of spills by volume contain storage tanks as a source, in three different categories. If the sources that contain

**Table 9.1** The source of spill volume.

| Source of spillage | Percent of total | Includes storage tank spills |
|---|---|---|
| Pipelines | 31.5 | |
| Wells, batteries | 21.6 | X |
| Storage, refineries | 15.3 | X |
| Tank trucks | 9 | |
| Rail | 6.3 | |
| Marine terminals | 3.2 | X |
| Other motor vehicles | 2.7 | |
| Industrial plants | 2.7 | |
| Bulk carriers | 2.6 | |
| Other watercraft | 2.3 | |
| Tankers | 1.9 | |
| Service stations | 1.8 | |

**Table 9.2** The source of spill numbers.

| Source of spillage | Percent of total | Includes storage tank spills |
|---|---|---|
| Wells, batteries | 20.8 | X |
| Storage, refineries | 20 | X |
| Pipelines | 11.2 | |
| Tank trucks | 10.4 | |
| Other watercraft | 5.8 | |
| Bulk carriers | 5.6 | |
| Service stations | 4.8 | |
| Marine terminals | 4.6 | X |
| Other motor vehicles | 4 | |
| Industrial plants | 4 | |
| Tankers | 4 | |
| Other sources | 2.4 | |
| Rail | 2.4 | |

storage tanks are considered, and half of these were from storage tanks, storage tanks would be the second-most important source of spillage. Table 9.2 shows the source of spills in Canada by numbers. This table shows similar values as Table 9.1; however, the number of spills is largest from storage tanks

These data are interesting and show that storage tanks constitute a large percentage of both the volume and number of spills in Canada. One must be aware, however, that these data are pre-2000 and thus may be different in the modern era.

## 9.3 Comparison of Canadian to US data

Similar data to the above is available for the US (Schmidt-Etkin, 2011). There are important differences. First, some of the data contain releases from emissions and deliberate discharges. These, where obvious, have been removed from this consideration. The data are summarized in Table 9.3 (summarized from Schmidt-Etkin, 2011).

The data in Table 9.3 show that storage tank spills are about a third in importance in terms of volume in the U.S.A. This is similar to the Canadian situation, however, the spills from storage tanks are somewhat higher in Canada. The period of the US data is approximately a decade later.

Table 9.3 Sources of Spill Volume in the USA.

| Source | 1998–2007 tonnes | % Total 1998–2007 |
|---|---|---|
| **Production** | 2,930 | **9.19** |
| Offshore platform spills | 182 | 0.57 |
| Offshore pipeline spills | 373 | 1.17 |
| Offshore supply vessel spills | 1 | 0.00 |
| Inland production well spills | 705 | 2.21 |
| Produced water | 1,669 | 5.23 |
| **Refining** | 1,734 | **5.44** |
| Refinery spills | 1,734 | 5.44 |
| **Transport** | 13,864 | **43.48** |
| Inland pipelines | 10,965 | 34.39 |
| Tanker trucks | 1,312 | 4.11 |
| Railroads | 278 | 0.87 |
| Tank ships | 514 | 1.61 |
| Tank barges | 776 | 2.43 |
| Tank vessel operational discharge | 19 | 0.06 |
| **Storage and consumption** | 13,357 | **41.89** |
| Non-tank vessels | 229 | 0.72 |
| Other vessels | 595 | 1.87 |
| Vessel operational discharge | 2,745 | 8.61 |
| Gas stations and truck stops | 170 | 0.53 |
| Residential | 71 | 0.22 |
| **Storage tanks** and other inland EPA-regulated facilities | 8,525 | **26.74** |
| Aircraft | 49 | 0.15 |
| Coastal facilities (nonrefining) | 604 | 1.89 |
| Inland unknown | 74 | 0.23 |
| Motor vehicles | 295 | 0.00 |
| Total | **31,885** | **100.00** |

## 9.4 Analysis of storage tank spills in Canada

Data are available from a recently-reactivated database in Canada (ECCC, 2021). This database started in 2015 and replaced an older database that was started in 1971 and which data were used in Section 9.2 above. These newer data were used to further analyze the spill situation from storage tanks. The database was culled to examine only spills of greater than 1000 L. The first item is to analyze the total number of spills by volume and then by numbers. These data are shown in Fig. 9.1.

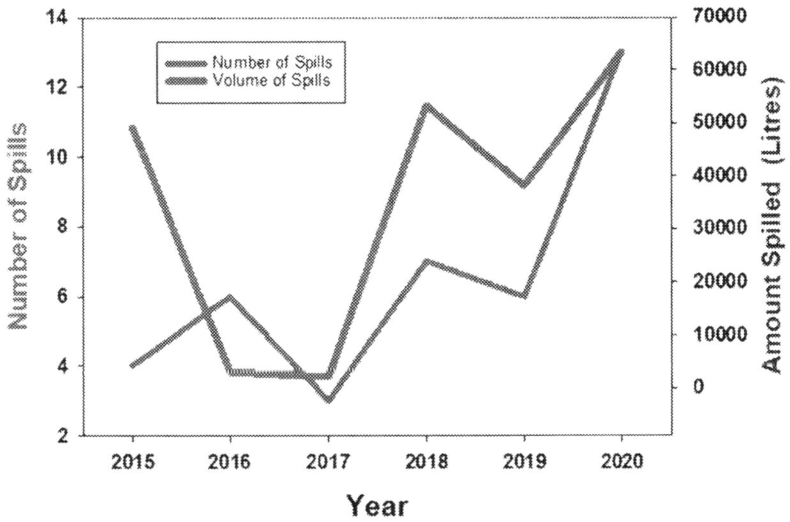


**Figure 9.1** The numbers and amounts of spills from storage tanks in Canada over recent years.

Fig. 9.1 shows that both the numbers and volume of spills rose through the period that this new database was active. The data from 2019 to 2020 are probably the most accurate. This would yield an average of 10–12 spills per year in the volume range totaling 20,000 to 60,000 L. It should be noted again that these involve spills of over 1000 L. The number of spills and the volume of spills tend to follow similar trends.

The products spilled are of particular interest. The volume of the different products spilled is shown in Fig. 9.2.

Fig. 9.2 shows that the largest spill amount is that of diesel fuel. This is no surprise as diesel fuel is one of the most commonly used products for trucks, trains, and other transportation. The presence of asphalt as a second item is not expected and is the result of one large spill. The next items of lubricant, crude oil, and gasoline are somewhat expected based on their frequent use and storage. Fig. 9.3 shows the frequency of products spilled.

Fig. 9.3 shows that diesel fuel is the most frequently-spilled product and gasoline the next. This conforms to the thinking that the fuels, which are most frequently refined and used, are also the ones most frequently-spilled. Crude oil which is the feedstock for both gasoline and diesel fuel is the next most frequently spilled product and asphalt, heating oil, and jet fuel, are the next most frequently-spilled products.

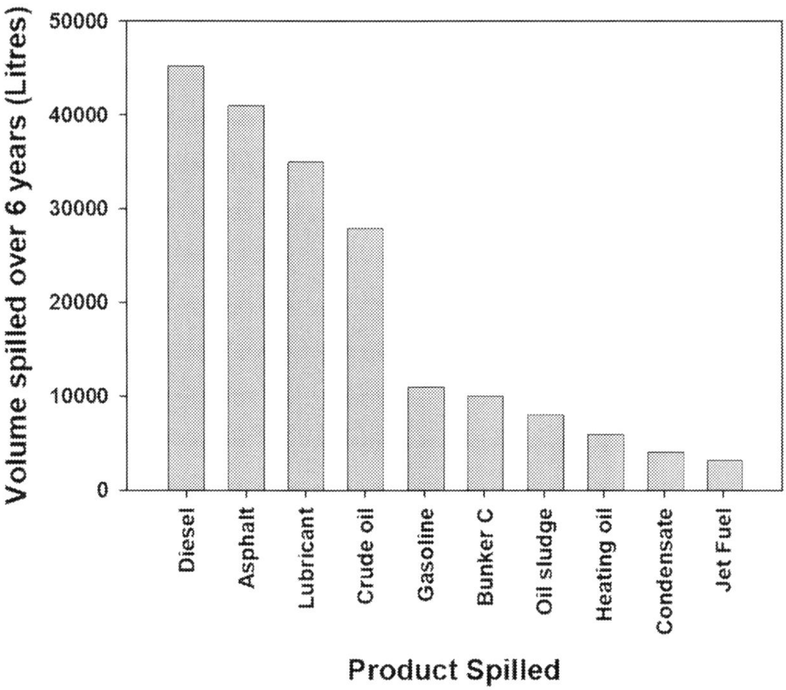


**Figure 9.2** The volume of different oil types spilled in Canada over recent years.

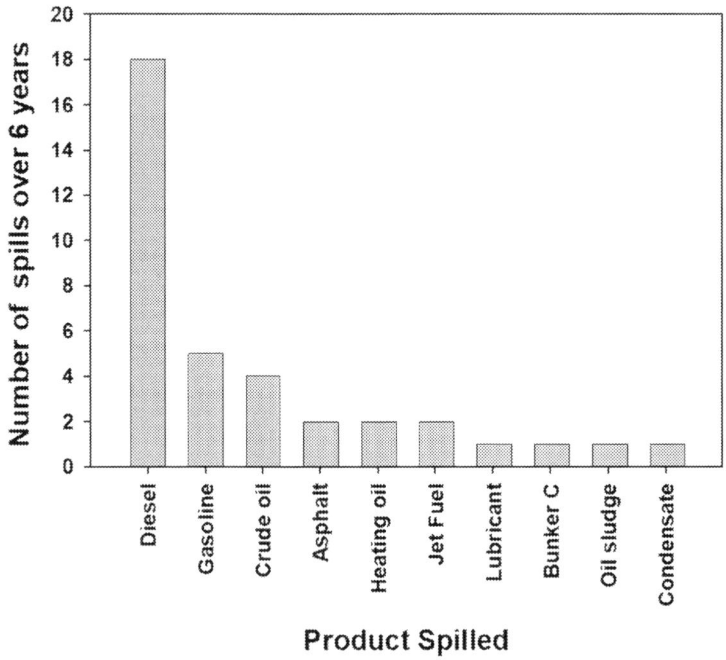


**Figure 9.3** The frequency of different oil types spilled in Canada over recent years.

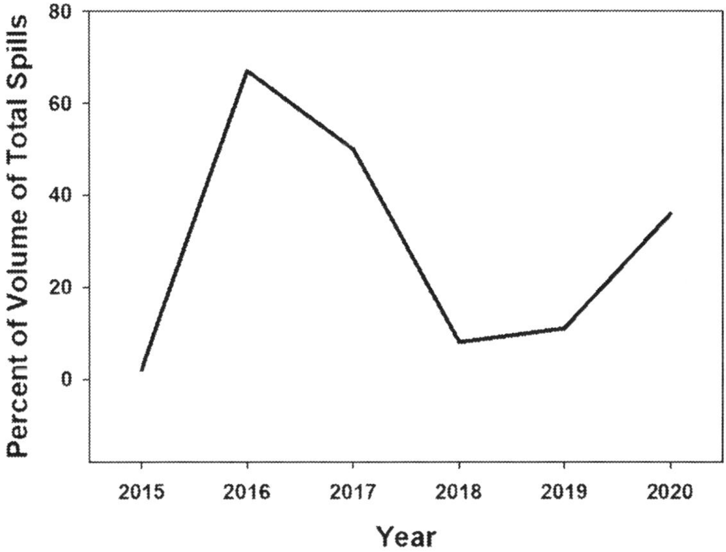


**Figure 9.4** The time trend of the percent of diesel spill volume compared to other oils.

Fig. 9.4 addresses the question of whether the most-frequently spilled product, diesel, varies much in spillage over the years as a percent of the volume of spillage.

Fig. 9.4 shows that the percent of diesel spill as a volume compared to the total volume spilled, does vary between 10 to 70% over the years under consideration. This is indicative of the fact that several other variables play a role in this statistic.

## 9.5 Summary and conclusions

Spills from storage tanks are a frequent occurrence in North America. Typically, spills from storage tanks are second or third in importance compared to all sources. In Canada, major spills (> 1000 L) occur at a rate of about 10 to 12 incidents per year, releasing 10,000 to 60,000 L of material. The most frequently-spilled materials include diesel fuel, gasoline, crude oil, heating oil, and jet fuel.

The availability of statistics is limited, as in Canada a spill database is being re-developed. To perform this analysis in this paper, some data from an older database (pre-2000) was used. The new database may not yet have a representative set of data. This is evidenced by the increase in spills from the start of 2015. However, the statistics may be adequate to provide a view of the spills from storage tanks.

## References


ECCC (2021). Data base data from national spills database, 2021. Correspondence from Kristine Brossard, January 2.

Fingas, M. F. (2012). *The basics of oil spill cleanup* (p. 245) Taylor and Francis.

Schmidt-Etkin, D. (2011). *Spill occurrences: a world overview* (pp. 7–48). Oil Spill Science and Technology.

# Index

*Note*: Page numbers followed by "*f*" and "*t*" refer to figures and tables, respectively.

## L



## M



## N



## O

## P

## T

Printed in the United States
by Baker & Taylor Publisher Services